Hollow Fibers and Nanofibers in Membrane Science

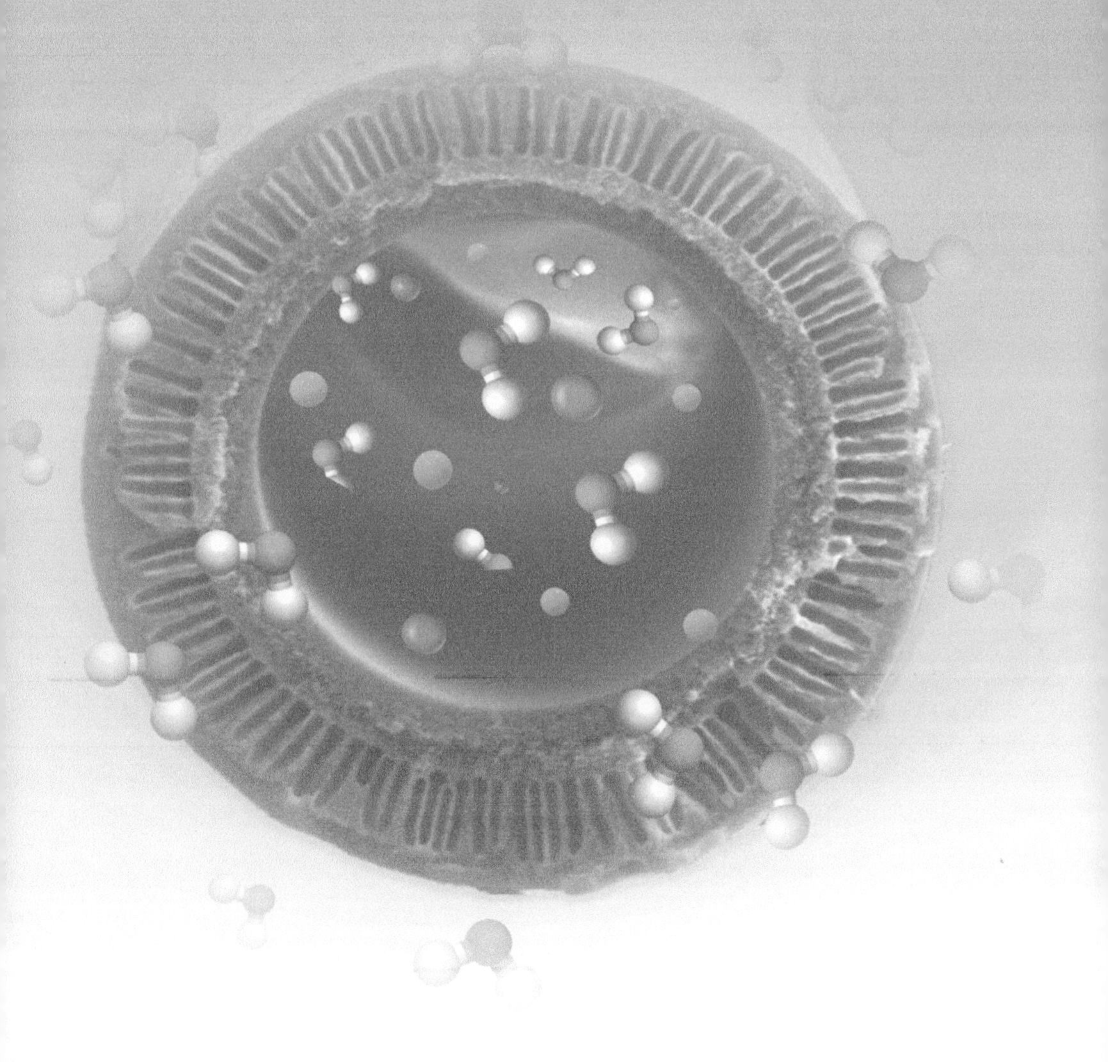

Hollow Fibers and Nanofibers in Membrane Science

Preparation, Characterization, and Applications

edited by

Alberto Figoli

Mir Saeed Seyed Dorraji

Francesco Galiano

JENNY STANFORD
PUBLISHING

Published by

Jenny Stanford Publishing Pte. Ltd.
101 Thomson Road
#06-01, United Square
Singapore 307591

Email: editorial@jennystanford.com
Web: www.jennystanford.com

British Library Cataloguing-in-Publication Data
A catalogue record for this book is available from the British Library.

Hollow Fibers and Nanofibers in Membrane Science: Preparation, Characterization, and Applications

ISBN 978-981-4968-03-4 (Hardcover)
ISBN 978-1-003-25663-2 (eBook)

Contents

Preface

Membrane processes are highly efficient and reliable separation methods. They represent, more than ever, a key technology in several industrial applications including water treatment, gas separation, desalination, food processing, and medical applications.

The wide range of membrane configurations and modules, which are available nowadays, makes membranes a highly versatile and resilient technology able to respond to the different needs of the industry. In this framework, fiber-shaped membranes play a crucial role in separation-based operations, thanks to their specific properties, compared to their flat-sheet counterparts.

The principal objective of this book is to provide a comprehensive review on the state of the art of fiber technologies including the preparation, characterization, and fields of applications. Among the fiber-configuration membranes, hollow fibers are the most diffused geometry and, for this reason, they are discussed to a greater extent in this book.

Following a general introduction to hollow fiber membranes (Chapter 1), the first part of the book discusses the fundamental aspects of fiber preparation and characterization (Chapters 2–5). In particular, Chapter 3 deals with the preparation and characterization of polymeric nanofibers produced by the electrospinning processes, while Chapters 2, 4, and 5 deal with the preparation and characterization of polymeric, inorganic, and mixed matrix hollow fibers, respectively.

The second part of the book focuses on the various fields of application of fibers and their practical uses covering micro- and ultrafiltration processes in the agrofood sector (Chapter 6), reverse osmosis and nanofiltration (Chapter 7), membrane distillation and membrane contactors (Chapter 8), gas separation (Chapter 9), pervaporation (Chapter 10), and finally, biomedical applications (Chapter 11). Concrete examples are discussed throughout the various chapters highlighting the advantages and potential hurdles to the use of fibers in the different applications.

We hope that the present text will provide useful information to the readers interested in membrane science and technology and, in particular, new insights on the state of the art and future trends of fibers.

This book has been written with the precious support and participation of the authors, to whom goes our sincere gratitude and appreciation.

Alberto Figoli
Mir Saeed Seyed Dorraji
Francesco Galiano
March 2022

Chapter 1

An Introduction to Hollow Fiber Membranes

Francesco Galiano and Alberto Figoli
Institute on Membrane Technology,
National Research Council of Italy (ITM-CNR),
via P. Bucci, Cubo 17/C Rende (CS), Italy
f.galiano@itm.cnr.it; a.figoli@itm.cnr.it

Hollow fibers display unique properties, which make them highly attractive in many industrial fields. This chapter introduces the reader to the world of hollow fibers with a brief and general introduction to membranes. Hollow fibers have played a significant role in the field of membranes from the moment of the development of the first module to the present day, in various fields of application. They have totally revolutionized the world of membranes, thanks to the new technological breakthroughs and innovative materials discovered.

Hollow Fibers and Nanofibers in Membrane Science: Preparation, Characterization, and Applications
Edited by Alberto Figoli, Mir Saeed Seyed Dorraji, and Francesco Galiano

ISBN 978-981-4968-03-4 (Hardcover), 978-1-003-25663-2 (eBook)
www.jennystanford.com

1.1 General Consideration on Membranes

Nowadays membranes represent a consolidated technology in different industrial sectors, including desalination, agrofood, gas separation, wastewater treatment, and pharmaceutical and biomedical applications. Membranes are used to separate, fractionate, concentrate, and purify liquid or gaseous mixtures presenting some important advantages over conventional separation technologies such as distillation, adsorption, extraction, and crystallization. Membranes are particularly appreciated, in fact, thanks to their ability to operate the separation under mild conditions of temperature and pressure, having a low environmental footprint, a low energy input requirement, and easiness in scale-up. Depending on the application they are intended for, membranes differ from each other in their morphology, structure, material, and geometry (Fig. 1.1).

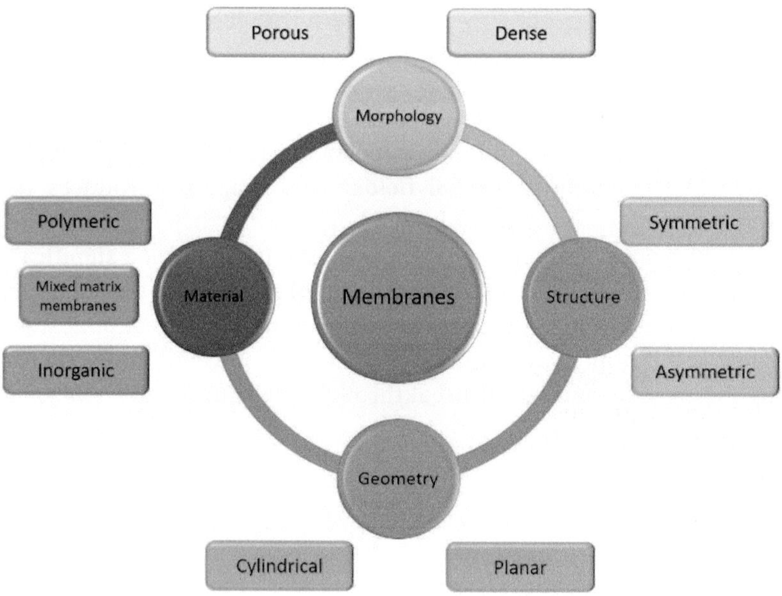

Figure 1.1 Classification of membranes on the basis of their morphology, structure, geometry, and material.

Porous membranes act as a selective barrier toward the permeation of target compounds, which are discriminated on the basis of their molecular size and membrane pore size. In classical membrane filtration operations, such as microfiltration (MF), ultrafiltration (UF), and nanofiltration (NF), the driving force is represented by a pressure difference across the membrane.

In the case of membrane contactors, on the contrary, the membrane does not act as a selective barrier but it has the only function of separating and putting in contact two adjacent phases, avoiding their mixing. The components are transported through the membrane under a concentration or temperature difference or under a chemical potential gradient. The permeation of the components through the membrane occurs by diffusion without the direct control of the membrane.

In non-porous or dense membranes, used in pervaporation (PV) and gas separation, the transport of target compounds is governed by the solution-diffusion model. The diffusion occurs because of a concentration or a chemical potential gradient across the membrane and the separation depends on the different solubility and diffusivity of permeating species within the membrane material.

Membranes can be classified, on the basis of their structure, in two main general categories: symmetric and asymmetric. To the first group belong the membranes with an isotropic morphology, which can be either dense or porous. In this case, the membrane exhibits the same characteristics (e.g., pore size, porosity) along its entire section. The mass transport resistance is mainly determined by the overall membrane thickness. In the second group are the membranes with an anisotropic morphology where the pore size, porosity, and composition differ from the top to the bottom. A typical asymmetric membrane is made of a thin skin layer with a thickness generally less than 1 μm and a very porous support layer (with a thickness of 50–100 μm). In these membranes, the separation is governed by the skin layer, while the underneath support has the only function of mechanical support.

Based on the material the membranes are made of, they can be classified as organic (including polymeric and liquid membranes) and inorganic (including ceramic, metallic, and glass membranes).

Mixed matrix membranes (MMMs) represent an intermediate condition where a solid inorganic dispersed phase is incorporated into a polymer matrix combining the benefits of both polymeric and inorganic materials. Polymeric membranes currently dominate the market, thanks to their huge number of advantages, including the vast availability of different polymers, the possibility to produce membranes in different shapes (tubular, flat sheet) with a wide membrane area, and the possibility of obtaining different membrane structures (porous, dense, charged) using versatile manufacturing processes such as phase-inversion techniques.

Membranes can be finally categorized based on their geometry: planar (such as flat-sheet membranes) or cylindrical (such as hollow fibers, capillaries, or tubes). Depending on their shape, in industrial installations, membranes are assembled in plate-and-frame and spiral-wound modules (in the case of flat-sheet membranes) or in tubular and hollow fiber modules (in the case of membranes with a cylindrical geometry).

This book will be almost fully dedicated to hollow fibers and to all the aspects related to this membrane geometry. The aim of this chapter is to provide a general and brief introduction to hollow fiber membranes, retracing their history and providing an outlook on their future direction.

1.2 Brief History of Hollow Fibers

Since Loeb and Sourirajan [1] conceived in the late 1950s the preparation of asymmetric cellulose acetate membranes for reverse osmosis (RO) using the phase-inversion technique, membranes with different materials, morphologies, and intended for several applications have been developed. The first hollow fiber membrane modules were developed during the 1960s by Mahon at Dow Chemical where a multitude of parallel, thin, and continuous fibers were bundled together within a specific cell [2, 3]. Pioneering were also the works carried out during the same period by DuPont, which introduced the first hollow fiber RO system for water desalination in 1969 with the trade name of Permasep.

In the same decade, at the Western Division of Dow, in San Francisco, the production of hollow fibers for several commercial and industrial applications was launched. Cellulose acetate fibers were manufactured for the brackish water treatment in RO application [4]. In the late 1960s the potentiality of hollow fibers started to be commercially explored for their use in hemodialysis [5].

The company Dupont made the seawater desalination commercially feasible in 1974 with the installation of the Permasep B-9 permeator installing in the following years, a series of modules in the Middle East, including Bahrain, which had the world's largest reverse osmosis plant [6].

From that time, their excellent mass transport properties led hollow fibers to be applied in a variety of applications, including microfiltration (MF), ultrafiltration (UF), nanofiltration (NF), reverse osmosis (RO), pervaporation (PV), gas separation, and dialysis (as reported in Table 1.1).

Table 1.1 Hollow fibers in membrane applications

Application	Reference
Microfiltration	[7–10]
Ultrafiltration	[11, 12]
Nanofiltration	[13, 14]
Reverse osmosis	[15, 16]
Pervaporation	[17, 18]
Gas separation	[19, 20]
Dialysis	[21, 22]

1.3 Hollow Fibers in Membrane Science

The use of membranes in "cylindrical" configuration is rapidly increasing worldwide in many sectors spanning a wide range of applications in water treatment, food, pharmaceutical, chemicals, and dairy industries. Depending on their dimensions, membranes with cylindrical geometry can be distinguished in three main types: tubular membranes (with a diameter higher than 5 mm), capillary membranes (with a diameter between 5 and 0.5 mm), and hollow fiber membranes (with a diameter lower than 0.5 mm) [23].

Tubular membranes are not self-supporting; therefore, they are allocated in porous stainless steel or plastic tubes having the same diameter of the tubular membrane. From 4 to 18 tubes are mounted together within the module, and the stream flows inside the lumen of each tubular membrane, while the permeate is collected into the module housing [23]. Their use is mainly limited to UF processes.

Capillary membranes are used in both "inside-out" and "outside-in" configurations. The smaller diameter of the capillary membranes makes them more prone to be affected by fouling in comparison to tubular ones but, at the same time, allows a higher packing density with a self-supporting structure.

Hollow fibers are the most common type of membranes with a cylindrical geometry. They are self-supporting membranes that can display different cross-sectional shapes such as round, pentagonal, trilobal, etc.

Hollow fibers can be made of polymers or inorganic materials. Organic materials represent, nowadays, the most diffused and preferred choice for the preparation of hollow fibers, thanks to their lower cost, high surface area per unit volume, easy scale-up, and more simple membrane fabrication. In this regard, a plethora of polymers are available and have been used for the fabrication of hollow fiber membranes, including polysulfone (PSf), polyethersulfone (PES), polyimide (PI), polyetherimide (PEI), polyvinylidene fluoride (PVDF), and cellulose acetate (CA).

Several techniques, which will be discussed in detail in Chapter 2, are currently employed for the fabrication of polymeric hollow fibers, such as melt spinning, dry spinning, wet spinning, or dry/wet spinning. The preparation of hollow fibers is more complicated with respect to flat-sheet membranes since it requires a specific spinning setup, and a series of variables, during the spinning process, are involved simultaneously. Different parameters, in fact, can influence the structure and the properties of the hollow fibers, such as the dimension of the spinneret, the viscosity of the solution, the temperature of dope solution and coagulation bath, the extrusion rate, the air gap, the bore fluid flow and composition.

Inorganic materials are also employed for the preparation of hollow fibers, thanks to their high chemical and thermal resistance, high mechanical strength, biocompatibility, and long lifetime. Inorganic hollow fibers are prepared from microporous glasses, silicon carbide, aluminum oxide (Al_2O_3), silicon nitride (Si_3N_4), and metal oxides and alloys [24].

Inorganic hollow fibers are generally prepared following a series of different steps comprising the powder preparation from a precursor material, the shaping of the powder followed by one or more thermal treatments, the layer deposition, and a possible final functionalization [25]. Their preparation will be extensively treated in Chapter 4.

Mixed matrix membranes (MMMs) combine the benefits of both polymeric and inorganic materials and are attracting more and more attention at industrial and academic levels being often considered the new-generation membranes. They basically consist of polymers and inorganic particles such as metal organic frameworks (MOFs), zeolites, nanoparticles, and nanocatalysts. The addition of inorganic domains into polymeric hollow fibers, whose preparation techniques and materials employed will be thoroughly discussed in Chapter 5, aims to improve membrane performance in terms of selectivity [26], antifouling properties [10], antimicrobial activity [27], and permeate flux [28].

The main advantage of hollow fibers is represented by their capacity to combine a large membrane area in a restricted volume, thanks to their high packing density when compared to their flat counterparts. Table 1.2 shows the most diffused membrane modules based on their area/volume ratio.

Table 1.2 Membrane modules on the basis of their area/volume ratio [6, 29]

Membrane Module	Area/Volume Ratio
Plate and frame	100–400 m^2/m^3
Tubular	300 m^2/m^3
Spiral wound	300–1000 m^2/m^3
Hollow fiber	up to 9000 m^2/m^3

Another important characteristic is that hollow fibers are generally mechanically self-supported resulting in a larger space in membrane modules where more fibers can be bundled into.

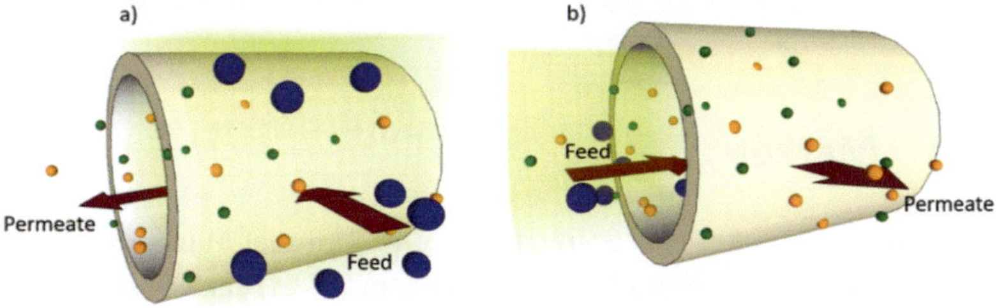

Figure 1.2 Filtration process configuration in hollow fibers: (a) outside-in and (b) inside-out.

Moreover, hollow fibers are flexible membranes since the filtration process can be carried out in two different ways: "outside-in" or "inside-out" configuration (Fig. 1.2). The feed can, in fact, be supplied from the inside or outside wall of the fiber and the permeate can be collected on the opposite side.

The hollow fibers can have a symmetric (Fig. 1.3a) or an asymmetric structure (Fig. 1.3b), and in the latter case, the active skin layer is put in contact with the feed side.

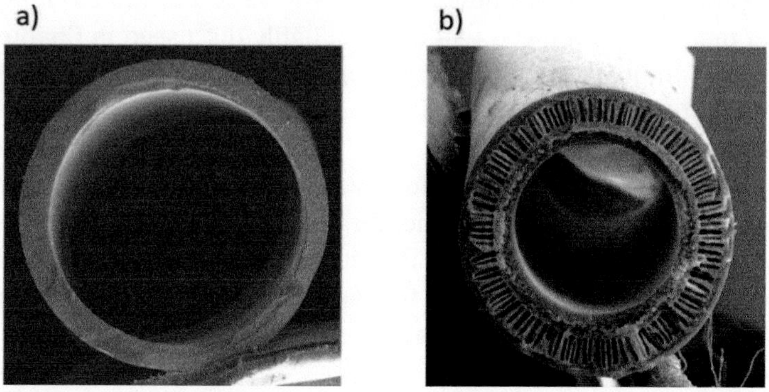

Figure 1.3 SEM morphology of a symmetric (a) and an asymmetric (b) hollow fiber.

Fouling still remains one the main bottlenecks in membrane filtration process in general, and hollow fibers are not relieved by this phenomenon. In some cases, membrane fouling can be more frequent in hollow fibers than other membrane configurations. Another drawback is related to their geometry, which can make them more susceptible to breakage under high strains. Hollow fiber membranes are generally competitive, from the cost point of view, with respect to other membrane systems present on the market, but they may require higher operating costs compared to other configurations.

1.4 Market

According to recent research conducted by Global Market Insights, the hollow fiber membrane market is expected to reach \$6.14 billion in 2025, including different sectors [30]. The most significant growth will be represented by water and wastewater treatment applications driven by the increase in desalination plants, improved demand for food and beverages along with the rising regulations on water preservation, water recovery and reuse in accordance to the process intensification strategy. In these sectors, the hollow fiber membrane market constituted about 46% in 2017 with an annual growth rate of 9%.

The hollow fiber membrane market is composed of large- and small-scale producers. The leading players are Polymem (prominent global manufacturer in hollow fiber membrane and modules), Asahi Kasei, Mitsubishi Chemical, Koch Membranes, Evonik Industries, Toray, Dow-Dupont, and Atec Innovations GmbH.

North and Latin America, the Middle East, Europe, and Asia Pacific are the key regions for the global hollow fiber membrane market. Asia Pacific, in particular, will likely hold the most significant share of the hollow fiber membrane market due to the more and more stringent environmental policies for wastewater treatment adopted by several Asian countries such as China, Thailand, and India. Europe and North America have an important share of the fiber membrane market due to the growing demand for biotech and pharmaceutical products [31].

On the basis of the material, polymers represent the most diffused choice with polyethersulfone (PES) and polysulfone (PS) accounting for the largest share of the polymeric hollow fiber market, thanks to their excellent oxidative, hydrolytic, and thermal stability as well as good mechanical properties.

1.5 Future Outlook

Membrane technology demonstrated during the last decades its huge potential in providing sustainable and concrete solutions in the field of clean water production, energy demand, and health care. In this context, hollow fibers will play a more and more important role. The need for membranes with high performance in terms of flux and selectivity reaffirmed the focus on new membrane materials research. In this regard, the development of mixed matrix hollow fibers is considered a promising approach for producing membranes with superior performance. With the advent of the dual-layer technique [32], it is possible to preferentially locate the inorganic particles on the outer active layer of the fibers considerably decreasing the amount of material required, which results in a lower production cost [33]. Several nanomaterials have been explored so far for their incorporation into hollow fiber membrane matrixes, including different types of zeolites for improving membrane selectivity toward gases [34, 35], carbon nanotubes (CNTs) for enhancing water permeability [36] and antifouling properties [37], graphene oxide for CO_2 capture [38], inorganic catalysts, such as TiO_2, for conferring peculiar catalytic properties [9, 10].

The production of hollow fibers with an ultra-thin selective layer able to reduce mass transfer, improving permeation flux, is still a challenge. This is mainly due to the complexity of the phase-inversion process in producing defects-free ultra-thin layers, which is particularly crucial in membrane processes such as gas separation and pervaporation [39–41]. However, many breakthroughs have been accomplished in this direction with the development of new manufacturing strategies [42], such as co-extrusion technology, and the use of the silicon rubber coating technique to seal the defects on membrane selective layer.

The future of hollow fibers also passes through new manufacturing processes that are sustainable and eco-friendly. Currently, most of the hollow fibers are prepared by phase-inversion techniques where the use of toxic and hazardous solvents, for the environment and human health, such as N-methyl pyrrolidone (NMP), dimethylformamide (DMF), and dimethylacetamide (DMA), threatens the concept of sustainable development [43–45]. Recently, pioneering articles dealing with the replacement of traditional toxic solvents with greener alternatives came into light. New solvents such as Rhodiasolv® PolarClean [46], triethyl phosphate [47], and ionic liquids [48] were investigated and proposed for making hollow fiber membranes following a more sustainable route.

Nowadays, many hollow fiber products succeeded in establishing themselves in different market sectors. However, there are still some open technical and scientific challenges such as the deep understanding of hollow fiber morphology during membrane formation, the effects of material chemistry and of all spinning process parameters, which are waiting for resolution and continue to stimulate the research and progress in this field.

References

1. Loeb, S. and Sourirajan, S. (1963) Sea water demineralization by means of an osmotic membrane. *Adv. Chem.*, **38**, 117–132.
2. Mahon, H. I. (1966) Permeability separatory apparatus and process utilizing hollow fibers. US Patent, 3,228,877, issued 1966.
3. Mahon, H. I. (1966) Permeability separatory apparatus, permeability separatory membrane element, method of making the same and process utilizing the same. US Patent, 3,228,876, issued 1966.
4. Sargent, J. A. (2007) Dialysis in the 1960s and the first hollow fiber dialyzer. *Int. J. Artif. Organs*, **30**(11), 953–963.
5. Stanton, G. W. (1961) An artificial kidney utilizing hollow fibers. *Dow Chemical Company Report* 974, Pittsburg, CA.
6. Khulbe, K. C., Feng, C., Matsuura, T., and Khayet, M. (2006) Polymeric hollow fibers: State of the art review of their preparation, characterization and applications in different research areas. *J. Appl. Membr. Sci. Technol.*, **4**, 53–88.
7. Shimizu, Y., Okuno, Y.-I., Uryu, K., Ohtsubo, S., and Watanabe, A. (1996) Filtration characteristics of hollow fiber microfiltration membranes used in membrane bioreactor for domestic wastewater treatment. *Water Res.*, **30**(10), 2385–2392.
8. Cassano, A., Bentivenga, A., Conidi, C., Galiano, F., Saoncella, O., and Figoli, A. (2019) Membrane-based clarification and fractionation of red wine lees aqueous extracts. *Polymers (Basel).*, **11**(7), 1089.
9. Benhabiles, O., Galiano, F., Marino, T., Mahmoudi, H., Lounici, H., and Figoli, A. (2019) Preparation and characterization of TiO_2-PVDF/PMMA blend membranes using an alternative non-toxic solvent for UF/MF and photocatalytic application. *Molecules*, **24**(4), 724.
10. Galiano, F., Song, X., Marino, T., Boerrigter, M., Saoncella, O., Simone, S., Faccini, M., Chaumette, C., Drioli, E., and Figoli, A. (2018) Novel photocatalytic PVDF/Nano-TiO_2 hollow fibers for environmental remediation. *Polymers (Basel).*, **10**(10), 1134.
11. Kurada, K. V., Tanmay, and De, S. (2018) Modeling of cross flow hollow fiber ultrafiltration for treatment of effluent from Railway Workshop. *J. Memb. Sci.*, **551**, 223–233.
12. Dutta, M., Bhattacharjee, S., and De, S. (2020) Separation of reactive dyes from textile effluent by hydrolyzed polyacrylonitrile hollow fiber ultrafiltration quantifying the transport of multicomponent species

through charged membrane pores. *Sep. Purif. Technol.*, **234**, 116063.

13. Asadi Tashvigh, A. and Chung, T.-S. (2019) Robust polybenzimidazole (PBI) hollow fiber membranes for organic solvent nanofiltration. *J. Memb. Sci.*, **572**, 580–587.
14. Xia, L., Ren, J., and McCutcheon, J. R. (2019) Braid-reinforced thin film composite hollow fiber nanofiltration membranes. *J. Memb. Sci.*, **585**, 109–114.
15. Chen, K., Xiao, C., Liu, H., Ling, H., Chu, Z., and Hu, Z. (2019) Design of robust twisted fiber bundle-reinforced cellulose triacetate hollow fiber reverse osmosis membrane with thin separation layer for seawater desalination. *J. Memb. Sci.*, **578**, 1–9.
16. Usta, M., Morabito, M., Anqi, A., Alrehili, M., Hakim, A., and Oztekin, A. (2018) Twisted hollow fiber membrane modules for reverse osmosis-driven desalination. *Desalination*, **441**, 21–34.
17. Tsai, H. A., Chung, L. H., Lee, K. R., and Lai, J. Y. (2013) The preparation of polyamide/polyacrylonitrile composite hollow fiber membranes for pervaporation. *Appl. Mech. Mater.*, **377**, 246–249.
18. Zhang, Y., Chen, S., Shi, R., Du, P., Qiu, X., and Gu, X. (2018) Pervaporation dehydration of acetic acid through hollow fiber supported DD3R zeolite membrane. *Sep. Purif. Technol.*, **204**, 234–242.
19. Wijiyanti, R., Ubaidillah, A. N., Gunawan, T., Karim, Z. A., Ismail, A. F., Smart, S., Lin, R., and Widiastuti, N. (2019) Polysulfone mixed matrix hollow fiber membranes using zeolite templated carbon as a performance enhancement filler for gas separation. *Chem. Eng. Res. Des.*, **150**, 274–288.
20. Liang, C. Z., Liu, J. T., Lai, J.-Y., and Chung, T.-S. (2018) High-performance multiple-layer PIM composite hollow fiber membranes for gas separation. *J. Memb. Sci.*, **563**, 93–106.
21. Zhang, Q., Lu, X., and Zhao, L. (2014) Preparation of polyvinylidene fluoride (PVDF) hollow fiber hemodialysis membranes. *Membranes (Basel).*, **4**(1), 81–95.
22. Barzin, J. (2004) Characterization of polyethersulfone hemodialysis membrane by ultrafiltration and atomic force microscopy. *J. Memb. Sci.*, **237**(1–2), 77–85.
23. Mulder, M. (1996) *Basic Principles of Membrane Technology*, Second edition, Kluwer Academic Publishers, Dordrecht.
24. Tan, X., Liu, S., and Li, K. (2001) Preparation and characterization of inorganic hollow fiber membranes. *J. Memb. Sci.*, **188**(1), 87–95.
25. Buekenhoudt, A., Kovalevsky, A., Luyten, J., and Snijkers, F. (2010) Basic aspects in inorganic membrane preparation, in *Comprehensive*

Membrane Science and Engineering, Elsevier, pp. 217–252.

26. Jiang, L. Y., Chung, T. S., and Kulprathipanja, S. (2006) Fabrication of mixed matrix hollow fibers with intimate polymer–zeolite interface for gas separation. *AIChE J.*, **52**(8), 2898–2908.

27. Gunawan, P., Guan, C., Song, X., Zhang, Q., Leong, S. S. J., Tang, C., Chen, Y., Chan-Park, M. B., Chang, M. W., Wang, K., and Xu, R. (2011) Hollow fiber membrane decorated with Ag/MWNTs: Toward effective water disinfection and biofouling control. *ACS Nano*, **5**(12), 10033–10040.

28. Wang, K. Y., Foo, S. W., and Chung, T.-S. (2009) Mixed matrix PVDF hollow fiber membranes with nanoscale pores for desalination through direct contact membrane distillation. *Ind. Eng. Chem. Res.*, **48**(9), 4474–4483.

29. Khayet, M. and Matsuura, T. (2011) Formation of hollow fibre MD membranes, in *Membrane Distillation*, Elsevier, Amsterdam, pp. 59–87.

30. Global Market Insights (2017) Hollow Fiber Membranes Market Size By Filtration Type (Microfiltration, Ultrafiltration, Reverse Osmosis), By Membrane Material (Polymer, Ceramic), By End-User (Water & Waste Water Treatment, Food & Beverages, Biotechnology, Pharmaceuticals & Chemicals) R.

31. TMR Research (219AD) Hollow Fiber Membranes Market: Global Industry Analysis, Size, Share, Growth, Trends and Forecast 2018–2028.

32. Jiang, L. Y., Chung, T. S., Cao, C., Huang, Z., and Kulprathipanja, S. (2005) Fundamental understanding of nano-sized zeolite distribution in the formation of the mixed matrix single- and dual-layer asymmetric hollow fiber membranes. *J. Memb. Sci.*, **252**(1–2), 89–100.

33. Peng, N., Widjojo, N., Sukitpaneenit, P., Teoh, M. M., Lipscomb, G. G., Chung, T.-S., and Lai, J.-Y. (2012) Evolution of polymeric hollow fibers as sustainable technologies: Past, present, and future. *Prog. Polym. Sci.*, **37**(10), 1401–1424.

34. Jiang, L., Chung, T., and Kulprathipanja, S. (2006) An investigation to revitalize the separation performance of hollow fibers with a thin mixed matrix composite skin for gas separation. *J. Memb. Sci.*, **276**(1–2), 113–125.

35. Ismail, A., Kusworo, T., and Mustafa, A. (2008) Enhanced gas permeation performance of polyethersulfone mixed matrix hollow fiber membranes using novel Dynasylan Ameo silane agent. *J. Memb. Sci.*, **319**(1–2), 306–312.

36. Wei, G., Chen, S., Fan, X., Quan, X., and Yu, H. (2015) Carbon nanotube

hollow fiber membranes: High-throughput fabrication, structural control and electrochemically improved selectivity. *J. Memb. Sci.*, **493**, 97–105.

37. Yang, Y., Qiao, S., Jin, R., Zhou, J., and Quan, X. (2018) Anti-fouling characteristic of carbon nanotubes hollow fiber membranes by filtering natural organic pollutants. *Korean J. Chem. Eng.*, **35**(4), 964–973.
38. Zhou, F., Tien, H. N., Xu, W. L., Chen, J.-T., Liu, Q., Hicks, E., Fathizadeh, M., Li, S., and Yu, M. (2017) Ultrathin graphene oxide-based hollow fiber membranes with brush-like CO_2-philic agent for highly efficient CO_2 capture. *Nat. Commun.*, **8**(1), 2107.
39. Koros, W. J. and Fleming, G. K. (1993) Membrane-based gas separation. *J. Memb. Sci.*, **83**(1), 1–80.
40. Peng, N., Chung, T.-S., and Lai, J.-Y. (2009) The rheology of Torlon® solutions and its role in the formation of ultra-thin defect-free Torlon® hollow fiber membranes for gas separation. *J. Memb. Sci.*, **326**(2), 608–617.
41. Clausi, D. T. and Koros, W. J. (2000) Formation of defect-free polyimide hollow fiber membranes for gas separations. *J. Memb. Sci.*, **167**(1), 79–89.
42. Peng, N. and Chung, T. S. (2008) The effects of spinneret dimension and hollow fiber dimension on gas separation performance of ultra-thin defect-free Torlon® hollow fiber membranes. *J. Memb. Sci.*, **310**(1–2), 455–465.
43. Marino, T., Galiano, F., Molino, A., and Figoli, A. (2019) New frontiers in sustainable membrane preparation: Cyrene™ as green bioderived solvent. *J. Memb. Sci.*, **580**, 224–234.
44. Russo, F., Galiano, F., Pedace, F., Aricò, F., and Figoli, A. (2020) Dimethyl isosorbide as a green solvent for sustainable ultrafiltration and microfiltration membrane preparation. *ACS Sustain. Chem. Eng.*, **8**(1), 659–668.
45. Marino, T., Galiano, F., Simone, S., and Figoli, A. (2019) DMSO EVOL™ as novel non-toxic solvent for polyethersulfone membrane preparation. *Environ. Sci. Pollut. Res.*, **26**(15), 14774–14785.
46. Hassankiadeh, N. T., Cui, Z., Kim, J. H., Shin, D. W., Lee, S. Y., Sanguineti, A., Arcella, V., Lee, Y. M., and Drioli, E. (2015) Microporous poly(vinylidene fluoride) hollow fiber membranes fabricated with PolarClean as water-soluble green diluent and additives. *J. Memb. Sci.*, **479**, 204–212.
47. Chang, J., Zuo, J., Zhang, L., O'Brien, G. S., and Chung, T.-S. (2017) Using

green solvent, triethyl phosphate (TEP), to fabricate highly porous PVDF hollow fiber membranes for membrane distillation. *J. Memb. Sci.*, **539**, 295–304.

48. Xing, D. Y., Peng, N., and Chung, T.-S. (2010) Formation of cellulose acetate membranes via phase inversion using ionic liquid, [BMIM]SCN, as the solvent. *Ind. Eng. Chem. Res.*, **49**(18), 8761–8769.

Chapter 2

Preparation and Characterization of Polymeric Hollow Fibers by Phase-Inversion Techniques

Francesco Galiano, Francesca Russo, and Alberto Figoli
Institute on Membrane Technology,
National Research Council of Italy (ITM-CNR),
via P. Bucci, Cubo 17/C Rende (CS), Italy
f.galiano@itm.cnr.it; a.figoli@itm.cnr.it

Polymeric hollow fibers are one of the preferred choices of membranes at the industrial level. This chapter will illustrate the different techniques employed, both at small and large scale, for their preparation with particular emphasis on the phase-inversion spinning techniques. The variables entering into play and involved in determining the properties of hollow fibers are multiple, and their role will be thoroughly discussed. A separate section will be dedicated to the polymers employed and the conventional techniques used for characterizing the membranes.

Hollow Fibers and Nanofibers in Membrane Science: Preparation, Characterization, and Applications
Edited by Alberto Figoli, Mir Saeed Seyed Dorraji, and Francesco Galiano

ISBN 978-981-4968-03-4 (Hardcover), 978-1-003-25663-2 (eBook)
www.jennystanford.com

2.1 Introduction

Hollow fiber membranes are of great industrial interest in several membrane-based operations spanning from the treatment of wastewaters, the separation of gases, bio-separations to seawater desalination processes. Hollow fibers are often the preferred geometry on large-scale applications since they can offer a large effective area per unit volume maximizing the overall process, are mechanically self-supporting, and display good flexibility. Hollow fibers, as their flat sheet counterpart, can be prepared by employing traditional phase-inversion techniques even if their preparation can result more challenging. Numerous factors are, in fact, involved during their preparation process, which affect their final morphology and properties in different ways. Hollow fibers are prepared by a spinning process, which consists of the extrusion of a dope polymer solution through a spinneret generally containing two orifices of different dimensions. The following paragraphs will focus on the various factors involved in the preparation of hollow fibers by spinning techniques. The polymeric materials used for their fabrication and the preparation conditions that influence their properties will be thoroughly analyzed and discussed. Moreover, the characterization techniques employed for studying hollow fibers' physical, chemical, and structural characteristics will also be examined.

2.2 Preparation of Hollow Fibers by Spinning Process

The spinning process, employed for the preparation of hollow fibers, consists in five steps [1]:

1. Dope solution preparation;
2. Dope solution extrusion through the spinneret;
3. Possible passage of the nascent fiber into an air-gap region;
4. Coagulation;
5. Post-treatment processes.

In the spinning process of hollow fibers (schematically illustrated in Fig. 2.1a), a dope polymeric solution, prepared at a suitable viscosity, is forced and extruded, by means of a pump or by the pressure exerted by an inert gas, through the annular space of the spinneret while an internal coagulant (called bore fluid) is pumped, at the same time, through the central orifice of the spinneret (Fig. 2.1b). The spinneret dimensions are 0.9–2 mm for the outer diameter and 0.5–1 mm for the inner diameter [2]. The spun polymer solution, after the extrusion, generally enters the air-gap region. This region, representing the distance between the spinneret and the coagulation bath and that can vary from 0 to 1 m, is characterized by a very well-controlled environment in terms of temperature and humidity. Both parameters, in fact, can play a crucial role in determining hollow fiber properties such as pore size, porosity, and morphology. The nascent hollow fibers enter, at this point, a coagulation bath, represented by a non-solvent, responsible for the precipitation and solidification of the fibers. The formed hollow fibers can now be collected, usually by means of guiding wheels, into a reservoir. Post-treatment procedures are generally aimed at removing residual solvent traces, washing away soluble additives and preserving hollow fiber structure by reducing their possible shrinkage and pores collapse.

Adequate comprehension of the spinning process, which considers the effect of several variables occurring during the preparation of hollow fibers, is fundamental to understand the resulting membrane properties fully. There are three principal and different forces acting on the nascent hollow fibers: (1) shear and elongation stresses of the dope solution within the spinneret; (2) gravity forces burdening on the just spun fibers; and (3) stress forces exerted by the collecting system [2]. The shear stress generally occurring at the spinneret walls, since the dope solution is most of the times represented by a non-Newtonian fluid, may greatly influence the molecular orientation of polymer chains located at the outer surface of the nascent membrane. Besides, the polymer macromolecules leaving the spinneret can be subjected to the die swell and relaxation phenomena, which change their orientation. This is particularly significant if there is an air gap between the spinneret and the coagulation bath. For these reasons, shear stress can affect both the outer skin formation of the fibers and their thermomechanical properties [3].

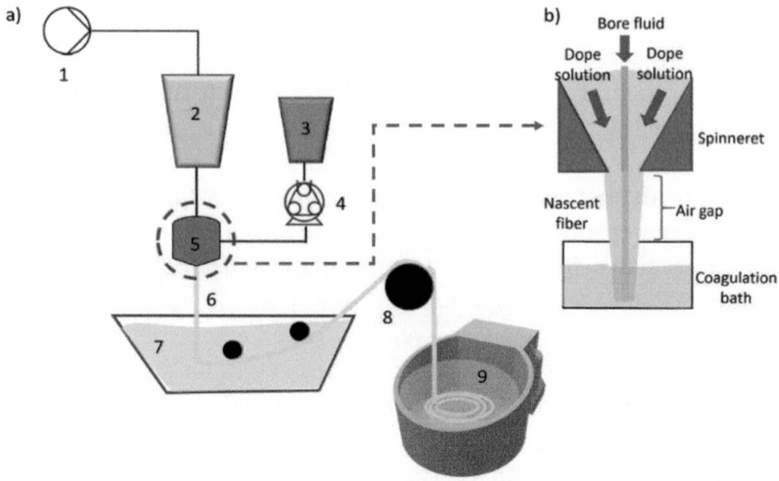

Figure 2.1 (a) Schematic diagram of a hollow fiber spinning setup: (1) pump; (2) dope solution tank; (3) bore fluid tank; (4) pump; (5) spinneret; (6) extruded hollow fiber; (7) coagulation bath; (8) guiding wheels; (9) reservoir. (b) Schematic illustration of hollow fiber formation at spinneret.

Gravity force and stress forces exerted by the collecting unit can also influence the surface structure and cross-sectional morphology of the forming hollow fibers. This is particularly true in the case of high air gaps and fast take-up speed, which can be responsible for the formation of defects in the hollow fibers [4].

In general, the spinning process is influenced by a series of parameters (e.g., temperature, bore fluid and dope solution flow rate, air gap, dimension of the spinneret, fiber take-up speed), which together with dope solution characteristics (e.g., viscosity, composition, concentration of the chemicals, presence of additives) are responsible for the final membrane properties. The most crucial variables will be addressed and discussed in more detail in Section 2.4 of this chapter.

2.3 Techniques Employed for the Preparation of Polymeric Hollow Fibers

Hollow fibers can be prepared following different methods of spinning, which include melt spinning, dry spinning, wet spinning, and dry/wet spinning.

Melt spinning is considered the fastest and most economical way for the production of hollow fibers [5]. The polymer is first heated above its melting point and then extruded from the annular hole of a spinneret, while a bore fluid (which can be represented by nitrogen gas) is injected, at the same time, through the spinneret lumen. This technique is characterized by the fact that the polymer is in the melt form and the formation of the hollow fibers occurs by cooling after their extrusion (thermal precipitation). The micropores are formed on the membrane surface by the mechanical forces acting on the fibers and by the following stretching step.

The best-suited materials for melt spinning are represented by crystalline thermoplastic polymers such as nylon, polyethylene, polyester, polypropylene (PP), and polyethylene terephthalate (PET).

One of the advantages of this approach is that no solvent is needed, which is particularly beneficial in terms of environmental sustainability and safety. Moreover, no additives and solvent diluents are requested. Another advantage is represented by the fact that the hollow fibers prepared with this technique exhibit outstanding mechanical properties [6]. However, the resulting fibers often suffer from low porosity and poor permeability [7].

Huang et al. [6], for instance, fabricated poly(tetrafluoroethylene-co-hexafluoropropylene) (FEP) hollow fiber membranes via the melt spinning technique without the use of any solvent and coagulation bath. $CaCO_3$ and dioctyl phthalate (DOP) were used as a pore-forming agent and plasticizer, respectively. Nitrogen gas was used as a bore fluid. The formation mechanism of membrane pores was studied based on the different draw ratio stretching (0%, 20%, 40%, and 60% at 140°C) to which hollow fibers were subjected and based on the post-treatment protocol carried out using an HCl aqueous solution (5 wt%). The morphology of the hollow fibers revealed the presence of micropores on their surface attributed to $CaCO_3$ particles that occupied the space that led to the formation of the micropores. The subsequent post-treatment with HCl solution promoted the formation of a sponge-like structure. When the hollow fibers were hot stretched, the shape of pores changed from a circular to an elongated one. The effect of the hot stretching was also evident on the mean pore size that increased from 76 nm (0% draw ratio stretching) to 277 nm (60% draw ratio stretching), on the porosity (which increased from 10 to 50%) and on the surface roughness.

Liu et al. [7] studied the preparation of polyurethane (PU) hollow fibers via the melt spinning technique and evaluated the effect of different post-treatments (stretching and hot-air treatment) on their performance and morphology. By increasing the theoretical draw ratio (TDR), the inner pore size of the hollow fibers increased (from 0.11 to 0.29 μm). The increase in hot-air treatment temperature (from 50 to 150°C) was responsible for the increase in pore size on the inner surface of the fibers (Fig. 2.2). The same hot-air treatment, however, caused a shrinkage on the outer surface of the membranes, resulting in a decrease in membrane water flux (from about 200 to about 50 L/m^2 h). Once tested in microfiltration (MF) application for the removal of Direct Sky Blu (DSB) 5B molecule, the hollow fibers showed a rejection toward the dye higher than 95.5%.

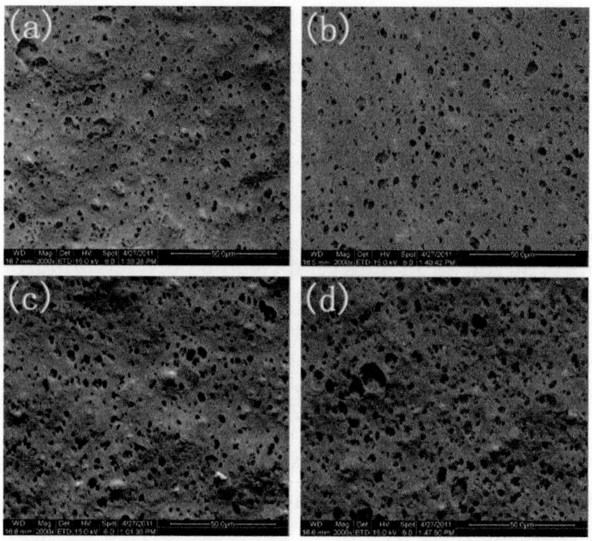

Figure 2.2 Effect of hot-air treatment on the inner surface morphology of PU-based hollow fiber membrane (heat treatment time: 30 min; heat treatment temperature: (a) 50°C, (b) 90°C, (c) 120°C, (d) 150°C). Reprinted from Ref. [7] with permission of Elsevier.

In the dry spinning method, the polymer is dissolved in an appropriate volatile solvent and then extruded through a spinneret into the air. The formation of the hollow fibers occurs by solvent evaporation followed by their stretching to orient the polymer chains in the gas gap direction [2]. Air or inert gas is used instead of a liquid bore fluid.

Berghmans et al. [8] produced poly(2,6-dimethyl-l,4-phenylene ether) (PPE) hollow fibers using cyclohexanol as a solvent by the dry spinning technique with a set of spinneret temperatures ranging from 80 to 150°C. Scanning electron microscopy (SEM) images of fibers prepared with 20 wt% of polymer revealed that the outer layer of the spun hollow fibers was quite dense and compact when high spinneret temperatures were used (150, 120, and 100°C). However, when the temperature was further decreased (80°C), the outer layer became less regular, and some pores started to be visible. The properties of the prepared hollow fibers were then evaluated in gas separation tests. It was found that the permeability for oxygen and nitrogen increased as the spinning temperature decreased; in contrast, the selectivity followed the opposite trend even if in all cases it was characterized by low values (O_2/N_2 selectivity < 3) due to the presence of holes.

A large majority of hollow fibers are currently prepared by wet or dry/wet-spinning techniques. These techniques allow the formation of different structures, due to the many parameters involved, widening the production of hollow fibers for several applications. Herein, the polymer solution, after the extrusion from the spinneret, is coagulated in a coagulation medium represented by a non-solvent. In the coagulation bath, the exchange between solvent and non-solvent occurs (demixing), leading to the precipitation of the polymer and the formation of the final membrane. The region between the spinneret and the coagulation bath represents the air gap, which may vary from 0 (wet spinning) to 1 m (dry/wet spinning).

Qin et al. [9] studied the effect of wet and dry/wet-spinning techniques on the properties of hollow fibers, employing polyethersulfone (PES) and polysulfone (PS) as polymers, various solvents, and additives. From the SEM images, a difference in morphology was revealed: the hollow fibers prepared by wet spinning were characterized by a finger-like morphology; in contrast, the ones prepared by dry/wet spinning were characterized by a sponge-like structure. In all cases, however, the outer surface of the prepared hollow fibers was characterized by a dense skin layer caused by the instantaneous liquid–liquid demixing. Wet spun hollow fibers showed a reduced pore size and lower water permeability compared to the dry/wet spun hollow fibers. These results confirmed the fact that the air gap (even if only of 1 cm) played a crucial role in the formation of the membranes. The shear and the gravity-induced orientation in the air gap, in fact, were key in determining the skin structure of the final membrane. Wet spun hollow fibers, moreover, showed better mechanical properties, in terms of tensile strength, with respect to the dry/wet spun fibers. These results were ascribed to the greater orientation and higher packing of the molecular chains at the outer surface of wet spun hollow fibers.

Alsalhy et al. [10] studied the fabrication of PES hollow fibers via dry/wet-spinning technique but using a steam gap, located immediately after the spinneret, in order to promote the formation of a sponge-like structure. Different variables such as the type of bore fluid, the concentration of polyethylene glycol (PEG) used as a pore former, and bore fluid flow rate were evaluated on membrane structure, permeability, and performance.

The combination of PEG with steam effectively resulted in the formation of a sponge-like architecture visible along the cross section of the fibers. Moreover, the increase in PEG concentration resulted, as expected, in an increase in membrane porosity (from 66 to 76%). However, the increase in PEG concentration was also responsible for a decrease in the mechanical properties of fibers as a consequence of the higher porosity and pore size, which weakened membrane structure. All the membranes presented a pore size in the microfiltration (MF) range (0.32–0.43 μm) with a water permeability up to 236 L/m^2 h bar.

The effect of humidity (RH) in the air gap was also studied by Tasselli et al. [11] for the preparation of hollow fibers made from poly(ether ether ketone) (PEEK-WC) and poly(vinyl pyrrolidone) (PVP) blends using dimethyl formamide (DMF) and dimethyl acetamide (DMA) as solvents. The morphology of the resulting hollow fibers was found to be related to two main aspects: the concentration of PVP and the amount of humidity. Regarding the PVP, the increase in its concentration caused a reduction in the overall porosity and the suppression of macrovoids. The humidity was crucial when it was around 100% (supersaturated conditions). In this case, the contact between the water vapor droplets and the outer surface of the nascent hollow fiber was responsible for the formation of a microporous structure (pores up to 1 μm) (Fig. 2.3) characterized by high water flux (up to 570 L/m^2 h bar).

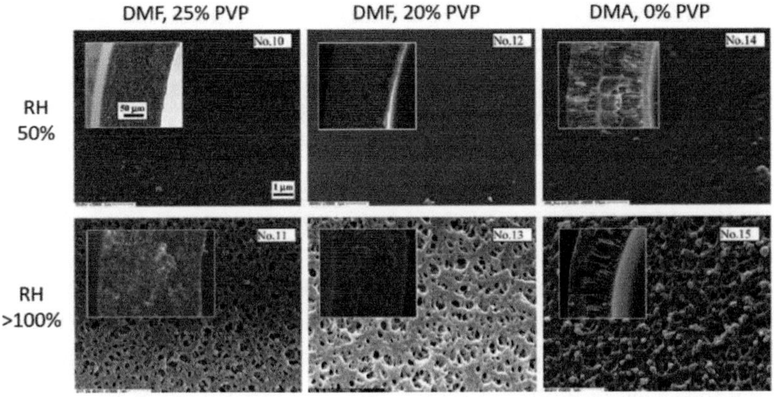

Figure 2.3 SEM image of the outer surface of hollow fibers, prepared at RH 50 ± 10% and in water vapor supersaturated air (>100%). Spinning solutions in DMF with 25% PVP and 20% PVP, and in DMA without PVP. Inserts: detail of the cross section. Same magnification in all membranes. Reprinted from Ref. [11] with permission of Elsevier.

2.4 Dope Solution and Spinning Parameters

As mentioned before, in hollow fibers production, a multitude of variables are involved. This paragraph focuses on the main elements that enter into play during hollow fiber production and formation in wet spinning and dry/wet-spinning processes.

2.4.1 Effect of Dope Solution Composition

The dope solution is represented by a thermodynamically stable solution made of a polymer (or copolymer), a solvent (or a solvent mixture), and possibly a series of additives (e.g., nanoparticles, pore-forming agents, non-solvent). Besides the chemical characteristics of the dope solution, also its physical properties (such as viscosity) affect the morphology of the final hollow fibers.

2.4.1.1 Polymer concentration

Polymer concentration is undoubtedly one of the most critical parameters that have to be carefully taken into consideration before preparing hollow fibers. The polymer concentration requires, in fact, to be within a certain range for the preparation of self-supporting and mechanically stable hollow fibers. Depending on the polymer employed, the minimum concentration required can vary significantly. Usually, high polymer concentrations are requested for the preparation of hollow fibers intended for pervaporation and gas separation applications, where a dense skin layer is necessary. Hollow fibers for water treatment applications, on the contrary, are prepared with lower polymer concentrations since a porous architecture is, in this case, necessary [12]. As a rule of thumb, the increase in polymer concentration is related to a change in hollow fibers morphology (e.g., from finger-like to sponge-like structure), to a decrease in membrane pore size and porosity, and an increase in the mechanical properties and thickness of the walls of hollow fibers.

It was observed that, for instance, in the case of polyvinylidene fluoride (PVDF) hollow fibers, the increase in polymer concentration (from 15 to 20 wt%) resulted in a disappearance of macrovoids in favor of a more dense and compact structure [13]. This was probably the consequence of a faster solidification, occurring at higher polymer concentrations, which hindered the formation of the open macrovoid structure. Pore size and porosity, in this case, were also affected by the increase in polymer concentration in the dope solution. The pore size decreased from 0.013 to 0.008 μm, while the porosity decreased from 22 to 8%, with the increase in PVDF concentration from 15 to 20 wt%.

Liu and Bai [14] prepared hollow fiber membranes through the blending of chitosan (CS) and cellulose acetate (CA), evaluating the effect of both polymers concentrations on fibers properties. The increase in CA concentration (from 12 to 18 wt%) caused a decrease in membrane pore size (from 0.54 to 0.09 μm on the outer surface) and porosity (from 80 to 70%). This was related to the fact that a higher content of CA in the dope solution resulted in the formation of a denser membrane matrix and consequently in lower pore size and porosity. An increase in CA also led to an improvement in the mechanical properties of hollow fibers in terms of tensile stress (from 7.8 to 18.2 MPa). The suppression of pores and the reduction in porosity, which represent weak points in hollow fibers, generally cause an improvement in the mechanical resistance of hollow fibers. The concentration of CS could not exceed 6 wt% within the dope solution due to the high increase in viscosity, making the fibers not spinnable. Its concentration was, therefore, varied from 4 to 6 wt%. As expected, the increase in CS led to a decrease in membrane pore size (from 0.54 to 0.063 μm on the outer surface) as a consequence of the higher viscosity of the dope solution; at the same time, the porosity remained almost unchanged (around 80%) due to the small amount of CS added in the blend solution.

2.4.1.2 Solvent

The selection of the proper solvent to prepare a polymeric dope solution is a key factor in membrane fabrication. The solvent, in fact, is not merely involved in the solubilization of the polymer, but the polymer/solvent and the solvent/non-solvent interactions are crucial in the determination of hollow fiber properties. Hansen solubility parameters are commonly employed to predict the solubility of a polymer in a given solvent understanding the polymer–solvent affinity through the evaluation of their dispersive, hydrogen, and polar bonding contributions [15]. The effect of different solvents such as DMA and a mixture of DMA/trimethyl phosphate (TMP) and DMF/TMP was studied by García-Fernández et al. [16] for the preparation of PVDF hollow fibers. Hollow fibers prepared with the DMF/TMP mixture presented larger macrovoids located in the inner layer compared to the fibers prepared with DMA/TMP. This phenomenon was explained considering that DMF has a lower affinity for PVDF in comparison to DMA but higher affinity for the non-solvent water used as a bore fluid. This favored a higher interdiffusion of water during the membrane formation, leading to the formation of bigger fingers in the inner layer. The hollow fibers prepared with the solvent mixtures (DMF/TMP and DMA/TMP) revealed a narrower pore size (between 65 and 69 nm) in comparison to the ones prepared with the pure solvent DMA (pore size of about 71 nm). When the membranes were tested in direct contact membrane distillation (DCMD), the hollow fibers prepared with DMA showed higher flux (with a salt rejection of 99.8%) than the ones prepared with DMF, thanks to their thicker internal sponge-like structure characterized by a few macrovoids.

The effect of three different solvents (DMF, N-methyl pyrrolidone (NMP), and dimethylsulfoxide (DMSO)) was evaluated by Roy et al. [17] for the preparation of blend cellulose acetate phthalate (CAP) and polyacrylonitrile (PAN) hollow fiber membranes. In all the cases, the membranes prepared with DMSO as a solvent revealed a more porous structure characterized by bigger cavities and macrovoids in comparison to the hollow fibers prepared with NMP and DMF. This was attributed to the fact that DMSO, as revealed by the thermodynamic studies carried out, is, among the three, the worst solvent able to induce a faster liquid/liquid demixing during the membrane formation. Conversely, the hollow fibers prepared with NMP, which was revealed to be the most effective solvent for the CAP/PAN solution, presented a denser structure. For the same reason, the hollow fibers prepared with DMSO showed higher permeability (up to 365 L/m^2 h bar) and greater porosity (up to 75%) in comparison to the membranes prepared with DMF and NMP as solvents.

2.4.1.3 Additives

The addition of inorganic or organic components in a dope solution is often carried out to improve the hollow fibers' properties in terms of permeation rate, selectivity, fouling, and mechanical resistance. The use of many additives such as PEG, PVP, LiCl, $ZnCl_2$, and water has been widely explored by many authors. PVP, for instance, is one of the most common additives used for the preparation of hollow fibers in PSU, PES, and PVDF intended for MF and UF application. PVP, in fact, is able to tune membrane pore size and pore-size distribution, to adjust membrane porosity, to increase the permeability, and to improve membrane resistance to fouling. Simone et al. [18], for instance, studied the effect of PVP (0–15 wt%) for the preparation of PVDF microporous hollow fibers for membrane distillation (MD) applications. It was found that small amounts of PVP (up to 9 wt%) promoted the formation of bigger cavities and voids on the outer surface of the produced hollow fibers. However, further increases in PVP concentration (up to 15 wt%) suppressed the formation of macrovoids in favor of a sponge-like structure, which improved the mechanical properties of membranes and slightly decreased the porosity. PVP is recognized to foster the formation of finger-like and macrovoid structures [19] but up to a certain concentration. Big cavities are the result of a fast solvent/non-solvent demixing boosted by the presence of PVP that causes an increase in thermodynamic instability of the solution during the phase-inversion process. However, at high concentrations, PVP increases solution viscosity, thus hindering the liquid demixing, leading to the suppression of macrovoids.

The effect of PEG on CA membranes was studied by Chou et al. [20]. The increase in PEG concentration (from 7 to 21 wt%) led to an increase in water permeability (from 5.6 to 18 L/m^2 h bar) due to the formation of more pores and a decrease in membrane retention toward dextrans. The increase in PEG molecular weight (MW), from 1,000 to 40,000, on the contrary, was responsible for a decrease in water permeability due to the lower solubility of PEG and to the decrease in the number of micropores.

In the work of Peng et al. [21], four different additives (water, methanol, ethanol, and tetrahydrofuran (THF)), acting as non-solvents, were investigated for the preparation of Torlon® poly(amide imide) hollow fibers for gas separation. THF and methanol, due to their similar low viscosity, enhanced the mutual solvent/non-solvent exchange favoring the formation of macrovoids visible in fibers cross section. Water and ethanol, on the contrary, as a consequence of their higher viscosities and of their higher solubility toward the polymer Torlon®, retarded the non-solvent intrusion into the forming membrane, hindering the formation of macrovoids. Through the adjustment of dope solution composition and spinning conditions, water was revealed to be the most effective additive for the preparation of defect-free Torlon® hollow fibers with an extra-thin selective layer of 488 Å and showing an O_2/N_2 selectivity of 8.55.

Pluronic is the trade name for tri-block copolymers of poly(ethylene oxide) (PEO) and poly(propylene oxide) (PPO). It has been considered an additive in membrane preparation for its ability to improve surface hydrophilicity and pores formation. Loh et al. [22] worked on the preparation of PES hollow fibers using different types of Pluronic differing for their molecular architecture. All the hollow fibers prepared with the different Pluronic presented a morphology characterized by finger-like macrovoids much larger than the hollow fibers prepared without additives. The addition of Pluronic in the dope solution, in fact, reduced its thermodynamic stability that, together with the hydrophilic nature of PEO chains in Pluronic molecules, accelerated the intrusion of non-solvent during the membrane formation. In general, in comparison to the pristine PES membranes, the addition of Pluronic led to (1) a reduction in the mechanical properties of hollow fibers in terms of Young's modulus, tensile stress at break, and strain at break as a consequence of the more porous membrane structure; (2) an increase in water permeability (up to 130 L/m^2 h bar); (3) an increase in the molecular weight cut-off (MWCO) (up to 218 KDa). However, important differences were found on the basis of the different types of Pluronic investigated. Pluronic 127 and Pluoronic 108, for instance, displayed high water permeability values despite their low MWCO. It was supposed that a layer of PEO chains was formed on the surface of membrane internal pores reducing the apparent pore size of the membrane leading to a decrease in MWCO (Fig. 2.4).

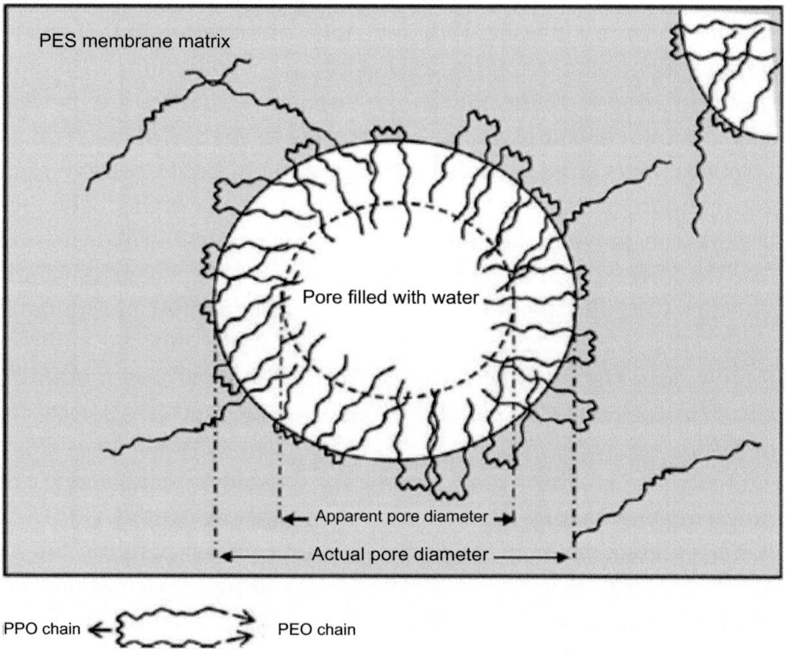

Figure 2.4 PEO layer formation on the internal pores located on the membrane surface. Reprinted from Ref. [22] with permission of Elsevier.

2.4.2 Effect of Dope Solution Viscosity

The dope solution viscosity is a key parameter in hollow fiber preparation since it can regulate, from a kinetic point of view, the exchange between solvent and non-solvent, allowing the production of membranes with different properties. Dope solution viscosity can be basically modified by acting on two variables: (1) the polymer concentration; (2) the presence and concentration of additives. The effect of polymer concentration has already been discussed in Section 2.4.1, so this section will focus on the effect of additives on solution viscosity. Generally, the increase in dope solution viscosity is responsible for decreasing the mass exchange process during membrane formation. This leads to a delay in the kinetics of polymer precipitation, having an effect on precipitation rate and the stretch ratio of hollow fibers spinning [23]. As suggested by Chou and Young [24], the solution viscosity has a significant influence also on the shape and the number of pores. The delay in polymer precipitation, in fact, causes an increase in pore size as the dope viscosity increases.

Liu et al. [25] studied the effect of viscosity on PES hollow fiber membranes by varying the concentration of PEG400. By increasing the concentration of PEG in the dope solution (from 37 to 64 wt%), a consequent increase in the solution viscosity was obtained (from 28,000 to 72,000 cP). At high PEG concentrations and in the presence of water, used as a non-solvent, the suppression of macrovoids was observed and a sponge-like structure was obtained.

Ohya et al. [26] studied the effect of dope solution viscosity in the preparation of polysulfone (PSF) hollow fibers. The viscosity was tuned by adding PEG of different molecular weights (from 600 to 150,000 Da). Different dope solutions were, thus, prepared with viscosity values ranging from 5.7 to 53.8 Pa*s. From the SEM cross-sectional analyses, only the fibers prepared with PEG 600 Da showed a finger-like structure, while the other membranes presented a sponge-like morphology. By looking at the SEM images of the hollow fibers' outer surface (Fig. 2.5), it can be clearly seen that, as the viscosity of the solution increased, the pore size and the porosity of the membranes also increased. In particular, the pore size increased from 0.19 μm (when PEG 600 was used) to 0.98 μm (when PEG 150,000 was used), while the porosity increased from 18.2 to 43.9%.

The reason why the increase in PEG molecular weight caused an increase in membrane pore size and porosity was explained based on the different dope solution viscosities. As the viscosity increased, in fact, the diffusion rate between the solvent (NMP) and the non-solvent (water), during membrane formation, slowed down. This phenomenon led to a delayed time of coagulation favoring the development of a polymer-lean phase responsible for pores formation.

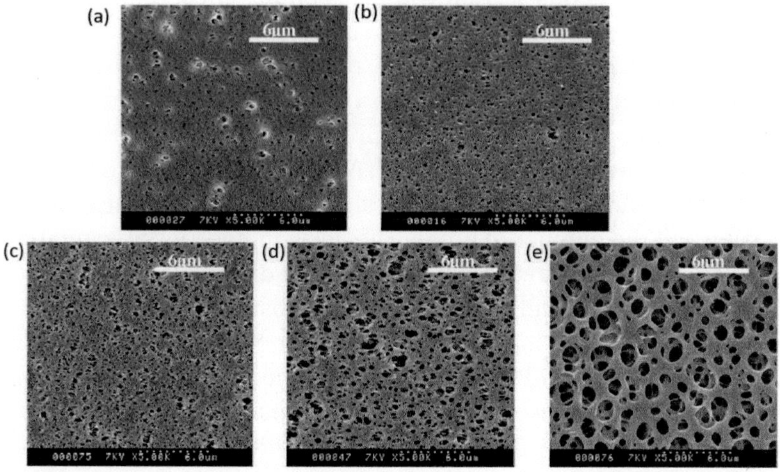

Figure 2.5 SEM of the outer surface of hollow fibers prepared with PEG at different molecular weights: (a) 600 Da; (b) 6000 Da; (c) 20,000 Da, (d) 35,000 Da; (e) 150,000 Da. Reprinted from Ref. [26] with permission of Elsevier.

The effect of dope solution viscosity was also evaluated by Agra Dias et al. [23] for the preparation of PES hollow fiber membranes using different concentrations of PVP and clay as additives. The use of clay as a unique additive (at concentrations above 5 wt%) drastically increased the viscosity of the spinning solution preventing membrane formation. The concomitant addition of PVP with clay decreased the viscosity of the solution, when the polymer concentration was kept constant (20 wt%), making the dope solutions spinnable. The decrease in dope viscosity (range 1245–3274 MPa·s) allowed the preparation of membranes with a finger-like structure, thanks to the high precipitation rate of the polymer. High viscosity values (about 16,000–30,000 MPa·s) on the contrary, delayed the precipitation rate of the polymer, leading to the suppression of the fingers and the formation of more spherical pores.

2.4.3 Effect of Spinning Parameters

In dry/wet spinning, the polymer solution, under gas pressure, is extruded from a spinneret and, at the same time, a bore fluid, acting as an internal coagulant, is driven to the central orifice of the spinneret. The dope solution generally passes a certain distance (air gap) before entering the coagulation bath, where the hollow fibers form and are finally collected. During this process, several variables such as the polymer solution rate, the bore fluid rate and composition, the air gap, the coagulation bath, and the collection speed enter into play and can have a big impact on hollow fiber formation and final properties.

2.4.3.1 Polymer solution flow rate

The extrusion rate of the dope solution from the spinneret has been found, by many authors, to have an important impact on hollow fiber features. On an industrial scale, it is important to keep a high extrusion rate of hollow fibers to improve productivity. It is always a challenge to scale up the hollow fiber production keeping the same membrane properties as before. The pressurized dope solution, in fact, can undergo several stresses, which can influence the molecular orientation and relaxation affecting fibers formation and characteristics. Aptel et al. [27], for instance, realized that the increase in the dope solution flow rate causes an increase in the shear stress, resulting in hollow fibers with a lower water permeability. Shilton et al. [28] studied the effect of dope solution extrusion rate in the preparation of PSF gas selective hollow fibers. The extrusion rate did not affect the morphology and the mechanical properties of the hollow fibers prepared. When tested for gas separation (with a CO_2/CH_4 mixture), however, the hollow fibers were very sensitive to dope solution extrusion rate. Both permeability and selectivity, in fact, increased as the extrusion rate increased from 0.7 to 2.0 cm^3/cm. The increase in permeability was attributed to an increase in surface pore area as the extrusion rate was increased. The increase in selectivity was related to the fact that as the extrusion rate was increased, the polymer molecules became more aligned, enhancing the orientation in the skin of the spun hollow fibers.

Qin et al. [29] studied the effect of dope solution flow rate (from 0.71 to 6.02 cm^3/cm) on the morphology, mechanical properties, and separation properties of PES hollow fiber membranes. The increase in the extrusion rate, as evidenced by SEM images, resulted in thicker and denser morphologies. This was explained by considering the rheological behavior of the dope solution. At higher dope rates, in fact, the polymer molecular chains aligned much better and packed closer to each other leading to a denser structure. This resulted in a decrease in hollow fiber pore size, as evidenced by the decrease in water permeability and an increase in the separation performance (higher rejection). Mechanical properties of the hollow fibers were enhanced at higher dope flow rates, as a consequence of the greater molecular orientation, with an improvement in both Young's modulus and tensile strength.

2.4.3.2 Bore fluid type and flow rate

The bore fluid is generally represented by a liquid that is injected through the inner orifice of the spinneret at the same time as the polymer solution extrusion. The bore fluid is one of the main parameters that has to be controlled during the preparation of the hollow fibers. It is generally represented by a non-solvent, but often a mixture of solvent/non-solvent is used in order to retard the demixing process. The composition of the bore fluid, in fact, influences the rate of polymer precipitation affecting the final structure of the hollow fibers. Tasselli and Drioli [30], for instance, studied the effect of different R-OH bore fluids (with R = H; CH_3; C_2H_5; n-C_3H_7, and n-C_4H_9) on the morphology and transport properties of PEEK-WC hollow fibers prepared using DMF as a solvent. From SEM images, the hollow fibers, prepared with the different bore fluids, did not display important differences in terms of the morphology of the outer layer, which was characterized by large pores. On the contrary, the lumen of the fibers was affected by the bore fluid type. A denser structure was observed for the fibers prepared using water as a bore fluid. Among the various bore fluids, in fact, water is the strongest non-solvent responsible for a fast phase separation, which does not allow significant growth in the pores. The fibers prepared with water as a bore fluid also showed the lowest values of water permeability (about 158 L/m^{2}hbar) with the highest rejection for dextrans (92% for dextran of 480 KDa). By increasing the apolarity of the bore fluid (methanol, ethanol, isopropanol, butanol), the water permeability increased (from 644 to 1209 L/m^{2}hbar), while the dextran rejection decreased (from 13 to 6%).

The effect of the two bore fluids on the formation of polyetherimide (PEI) hollow fiber membranes was studied by Bakeri et al. [31]. NMP and water were evaluated as bore fluids representing a solvent and a non-solvent for the polymer. The formation of a skin layer was observed in the inner and outer surfaces for the membrane prepared with water as bore fluid; while for the membrane prepared with NMP as a bore fluid, the skin layer was visible just on the outer surface of the fiber (Fig. 2.6). The solvent NMP, in fact, penetrated through the nascent fibers from the lumen side, leading to the formation of a finger-like structure. The use of NMP, moreover, improved both the pore size of the fibers, the effective surface porosity, and gas permeance when the membranes were tested in gas–liquid contacting process for CO_2 adsorption. This was the result of the skin layer suppression when NMP was employed.

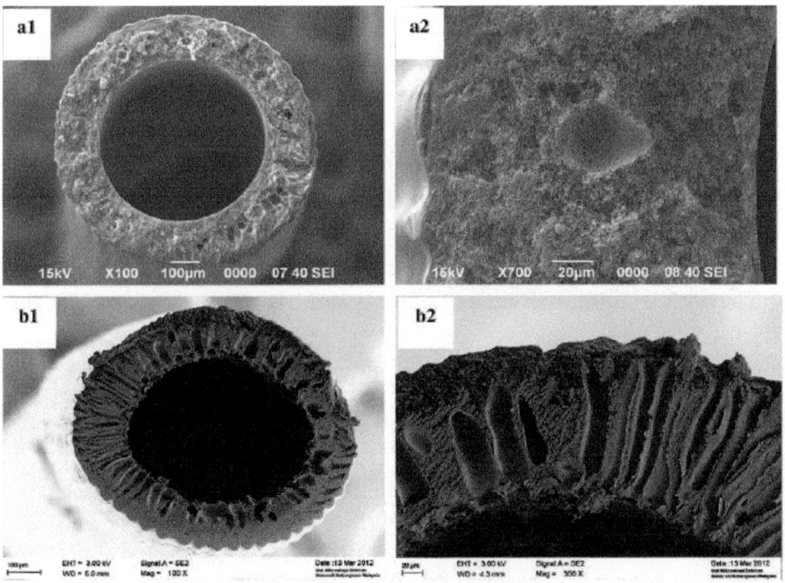

Figure 2.6 SEM cross section and magnification of the hollow fibers prepared with water (a1, a2) and NMP (b1, b2) as bore fluids. Reprinted from Ref. [31] with permission of Elsevier.

For PVDF hollow fibers, Khayet et al. [32] observed that when a strong internal coagulant, such as water, is used, longer finger macrovoids can be observed in the proximity of the inner surface with a sponge-like structure between the inner and outer surfaces. When the composition of the bore fluid was changed to water/ethanol (50:50), however, the inner layer was reduced. Water, for PVDF, is a strong non-solvent and, therefore, responsible for the fast coagulation of the polymer. Generally, finger-like macrovoids are, in fact, characteristics of a fast coagulation process. The addition of ethanol in the bore fluid, on the contrary, delays the coagulation process favoring the formation of a sponge structure.

In addition to bore fluid composition, also the flow rate can play a significant role in the preparation of the hollow fibers. Simone et al. [33], for instance, evaluated the effect of four different bore fluids (DMF/water 45/55) flow rates (9, 12, 15, and 18 ml/min) for the preparation of PVDF hollow fibers. It was found that the inner diameter of the fibers was strongly influenced by the bore fluid flow rates. In particular, the dimension of the hollow fibers lumen increased when the bore fluid flow rate was increased (from 1.4 to 1.74 mm). The same phenomenon was also observed by Alobaidy et al. [34] in the preparation of polyvinyl chloride (PVC) hollow fibers, when the bore fluid flow rate was increased from 2.6 to 13 ml/min. In this case, a reduction in more than 300% of membrane thickness was observed. As a consequence, the hollow fibers spun at the highest bore fluid flow rate exhibited an increase in water permeability 17 times higher than the ones prepared at lower flow rates with a slight decrease in bovine serum albumin (BSA) rejection (from 93.4 to 90.4).

2.4.3.3 Air gap

The air gap represents the distance between the spinneret and the coagulation bath, which can be varied from 0 to 1 m. In the dry/wet-spinning technique, the effect of the air-gap length on hollow fibers has been investigated by several authors. The influence of air gap on the internal and external morphology of PVDF hollow fibers was, for instance, studied by Khayet [35]. The extruded polymer solution entered the coagulation bath after passing a distance varied from 1 to 80 cm. It was observed that the inner and outer diameters of hollow fibers increased when the air gap was increased. This effect was probably related to the fact that higher air gaps introduced elongational stress on spun hollow fibers due to the gravity force. The fibers spun with an air gap higher than 45 cm showed a closer structure, characterized by a lower free volume, in comparison to the fibers spun at lower air gaps. When the air gap was lower, in fact, the fibers entered more rapidly in the coagulation bath where a higher amount of solvent and non-solvent was trapped in the polymer chains resulting in a more porous structure. The water permeability of the prepared hollow fibers decreased when the air-gap distance was increased due to the decrease in their MWCO. This was explained considering that the increase in the air-gap distance may induce a bigger orientation and tighter molecular packing of the polymer chains leading to the formation of membranes with reduced pore size.

Chung and Hu [36] studied the effect of the air gap (from 0 to 14.4 cm) on the preparation of PES hollow fiber membranes. The different air gaps were found to play a significant role when the membranes were tested for the permeance of different gases. In particular, as the air gap was increased, the permeance toward the three different gases evaluated (O_2, N_2, and He) decreased. The hollow fibers spun with 0 air-gap distance showed a morphology characterized by two arrays of finger-like voids closed at the inner and outer surfaces; while the hollow fibers spun with 14.4 cm of air-gap distance displayed an array of finger-like voids closed at the inner surface. The double finger-like layer created for the hollow fibers spun at 0 air-gap distance was the result of a convective-type coagulation occurring at the two sides of the nascent fiber. The hollow fibers prepared with the highest air gap, on the contrary, went first in contact with the air where the moisture contained in it diffused through the external surface by a non-convective process. The relation of the gas permeance and the air-gap distance was explained considering that, likely, the hollow fibers spun with 14.4 cm of air gap presented a tighter molecular packing that decreased the gas permeation; while the wet spun hollow fibers (0 air gap) showed a random and unoriented polymer chain entanglement characterized by a looser external structure responsible for the higher permeance to gas species.

2.4.4 Effect of the Coagulation Bath

In the phase separation process induced by non-solvent, the effect of the coagulation bath on membrane formation is a critical point. The coagulation medium is, most of the time, represented by water, which acts as a non-solvent for the polymer, allowing its precipitation. The miscibility of the solvent, employed for the solubilization of the polymer, with the non-solvent is an important aspect to be considered since their exchange rate, during membrane formation, influences the structure of the membrane. Different amounts of solvent can be, sometimes, added to the coagulation bath to slow down the phase separation process for producing membranes with a closed structure. Even the temperature of the coagulation bath can impact the structure of the final membrane. Generally, higher temperatures favor the demixing process leading to the formation of membranes with a more open structure.

The effect of the coagulation bath composition (0–50% of ethanol concentration in an aqueous coagulation bath) was investigated by Deshmukh and Li [37] for the preparation of PVDF hollow fibers using DMA as a solvent. From the SEM pictures, it was observed that as the ethanol concentration in the coagulation bath was increased from 18 to 50%, the morphology of the prepared hollow fibers in the outer wall gradually changed from a finger-like to a sponge-like structure (Fig. 2.7).

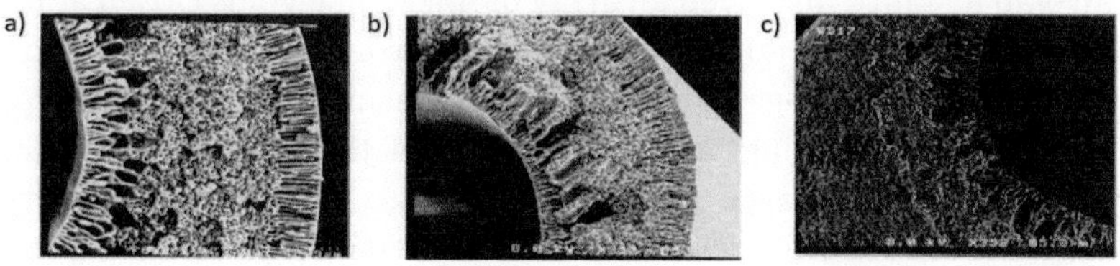

Figure 2.7 SEM cross section of hollow fibers coagulated in (a) 18:85 of ethanol: water bath; (b) 30:70 of ethanol: water bath; (c) 50:50 of ethanol: water bath. Adapted and reprinted from Ref. [37] with permission of Elsevier.

Based on the diffusivity values, the presence of ethanol in the coagulation bath slowed down the precipitation rate of the polymer leading to the formation of a closer membrane structure. In agreement with SEM pictures, the porosity of the prepared hollow fibers decreased when the concentration of ethanol in the coagulation bath increased as a consequence of the formation of a more closed structure.

Yu et al. [38] studied the effect on the morphology, mechanical and permeation properties of PAN hollow fibers of the addition of

DMF solvent (from 0 to 60%) in the coagulation bath. By increasing the concentration of DMF in the coagulation bath, the outer and inner diameters of the hollow fibers were reduced as a consequence of a denser morphology. This was related to a retarded solidification of the polymer due to the presence of DMF solvent in the coagulation bath. The increase in the DMF content improved the tensile strength of the prepared hollow fibers as a result of the decrease in membrane porosity. The closed hollow fiber structure, at higher DMF contents, was also related to higher solute retention (dextran of 100.2 KDa) of the hollow fibers.

The effect of coagulation bath temperature was investigated by Choi et al. [39] for the preparation of PVDF membranes by a wet-spinning technique using DMA as a solvent. Two different water coagulation bath temperatures (14 and 50°C) were evaluated on dope solutions prepared with 27 and 30 wt% of PVDF. The morphology of the hollow fibers prepared did not show significant changes. However, by measuring the gas properties of the produced hollow fibers, a decrease in permeance and an increase in the perm-selectivity were observed when the temperature of the coagulation bath was higher.

At 14°C, the slower solid–liquid demixing was responsible for the formation of more porous and, therefore, less selective membranes. At 50°C, the liquid–liquid demixing mechanism was dominant in the phase-inversion process, favoring the formation of a selective dense skin layer.

The difference in permeance at 50°C between the membranes prepared at 27 or 30 wt% of polymer concentration was not significant and within the experimental error.

2.4.5 Effect of the Post-Treatments

Once prepared, the hollow fibers are generally subjected to different post-treatment processes aiming at removing solvent traces, washing away additives, preserving or modifying the morphology of the prepared membranes [40–42].

Tsai et al. [43], for instance, applied two different post-treatment processes, including a water bath at different temperatures for 4 h and a heat treatment in the oven at 120°C for 12 h, during the preparation of PAN hollow fibers via wet-spinning technique. PEGs of different molecular weights were employed during the fabrication of hollow fibers, leading to the formation of membranes with a porous structure. In order to achieve the goal of producing hollow fibers with a dense skin layer for possible applications in pervaporation, the PAN hollow fibers were then subjected to a heat treatment at 120°C for 12 h. The treatment resulted in the formation of hollow fibers with a denser skin layer as a consequence of macrovoids suppression typical of the not heat-treated membranes. Hollow fibers prepared with PEG 200 were also subjected to a second alternative post-treatment consisting of a water bath at different temperatures. The aim of this post-treatment was to provide an efficient protocol for the complete removal of PEG 200 from PAN hollow fibers. Increasing the water bath temperature from 25 to 40°C, the outer skin dense layer of the fibers became more porous. This effect was related to a better removal, at higher temperatures, of the pore-forming additive PEG 200. However, when the temperature was further increased to 60 and 80°C, the skin layer of the PAN hollow fibers was denser than the one obtained at 40°C. This phenomenon was explained by considering the annealing effect on the morphology of PAN hollow fibers, which was greater when the water bath temperatures were higher.

The post-treatment of PVDF hollow fibers with ethanol was investigated by Tan et al. [44] for the production of membranes applied in the removal of ammonia from water. Immersion of the produced fibers in an ethanol bath for 5 min, before drying, was found to improve the hydrophobicity of the hollow fibers. The ethanol post-treatment, moreover, led to an increase in the effective surface porosity favoring the removal of ammonia.

The process of annealing as post-treatment was evaluated by Kim et al. [45] for the preparation of PAN hollow fibers with the aim of reducing their pore size. After the formation of the hollow fibers, the annealing was carried out in hot water (70–96°C) and then the annealed membranes were hydrolyzed with NaOH and CH_3ONa. PAN has a glass transition temperature of 120°C, which means that a rod-like and regular crystalline structure of the polymer can be created when the annealing temperature is increased. In other words, the annealing temperature improves the packing of the PAN molecules reducing membrane pore size. By evaluating the properties of annealed membranes in terms of flux and rejection with a 1,000 ppm PEG 600 solution (Fig. 2.8), a decrease in flux and an increase in rejection were observed as the annealing temperature was increased. This effect was explained considering that the annealing affected the polymer chains, causing a constriction of membrane pores.

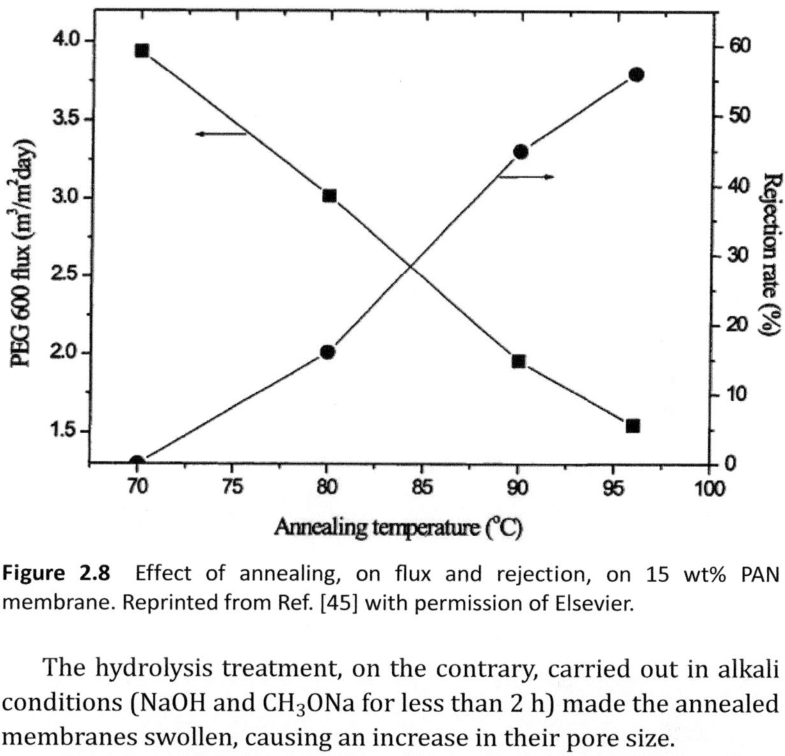

Figure 2.8 Effect of annealing, on flux and rejection, on 15 wt% PAN membrane. Reprinted from Ref. [45] with permission of Elsevier.

The hydrolysis treatment, on the contrary, carried out in alkali conditions (NaOH and CH_3ONa for less than 2 h) made the annealed membranes swollen, causing an increase in their pore size.

To preserve the porous architecture of hollow fiber membranes, an aqueous solution of glycerol is often employed. Glycerol, in fact, penetrates the porous structure of the membrane and, thanks to its very low vapor pressure, does not evaporate, allowing the preservation of the hollow fiber structure for a longer time [46]. Before using the membranes, glycerol can be easily flushed away with water.

2.4.6 Effect of Hollow Fiber Shape

The cross-sectional geometry of hollow fibers is a key factor in numerous membrane processes spanning from desalination to biomedical applications [47]. The most common geometry in hollow fibers is the circular one; however, different other cross-sectional structures (X-shaped, trilobal, polygonal, sinusoidal) can be designed and produced. Irregular structures present, in fact, numerous advantages such as the induction, into the fibers lumen or outer shell, of turbulence phenomena and secondary flows, which can limit the formation of concentration polarization and fouling phenomena, the enhancement of the mass and heat transport rates, the reduction in pressure drop, and the improvement in the surface area. The formation of these irregular structures can be achieved by modulating the spinning condition parameters (mainly the bore fluid composition and flow rate) or by using spinnerets of different shapes.

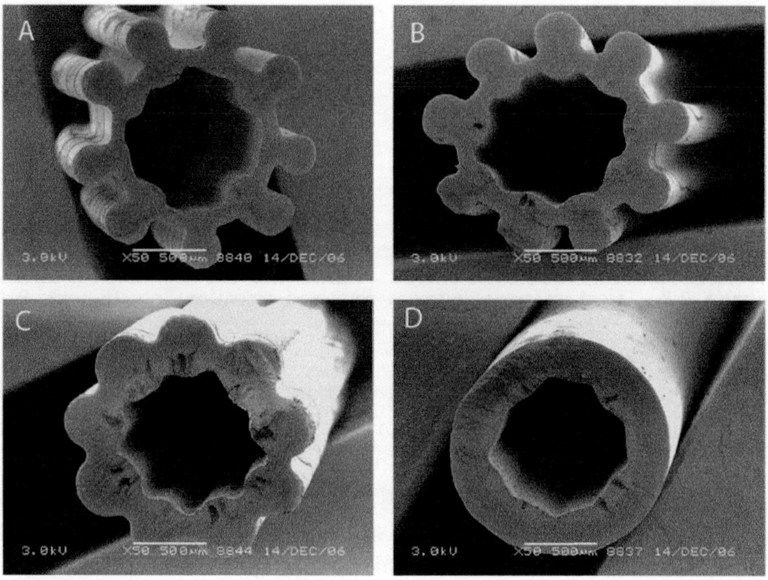

Figure 2.9 SEM cross section of hollow fibers spun at different air gaps of (a) 5 mm; (b) 12 mm; (c) 32 mm, and (d) 58 mm. Reprinted from Ref. [48] with permission of Elsevier.

A microstructured outer hollow fiber geometry was investigated, for instance, by Çulfaz et al. [48], from a PES/PVP blend. An insert with a microstructured opening, located immediately after the spinneret, was responsible for the structure modification of the outer surface of the hollow fibers. By increasing the air gap (from 5 to 58 mm), the residence time of the hollow fibers in the air increased and the microstructure gradually faded in favor of a round morphology as a consequence of a surface energy reduction (Fig. 2.9).

By comparing the performance of microstructured hollow fibers with the circular ones, it was demonstrated that the first exhibited similar values of water permeability, MWCO, and pore-size distribution with respect to the traditional circular hollow fibers. However, due to the higher surface area, their water permeability was enhanced while keeping the same separation properties of circular hollow fibers.

Le et al. [47] studied the shape evolution of the lumen of hollow fibers prepared with PEI and different diethylene glycol (DEG)/NMP ratios. Basically, by changing the bore fluid flow rate (from 2 to 8 ml/min) and composition (water/ethylene glycol ratio), it was possible to obtain several lumen geometries from circular to polygonal with a different number of sides (N = 2 to 8), as shown in Fig. 2.10.

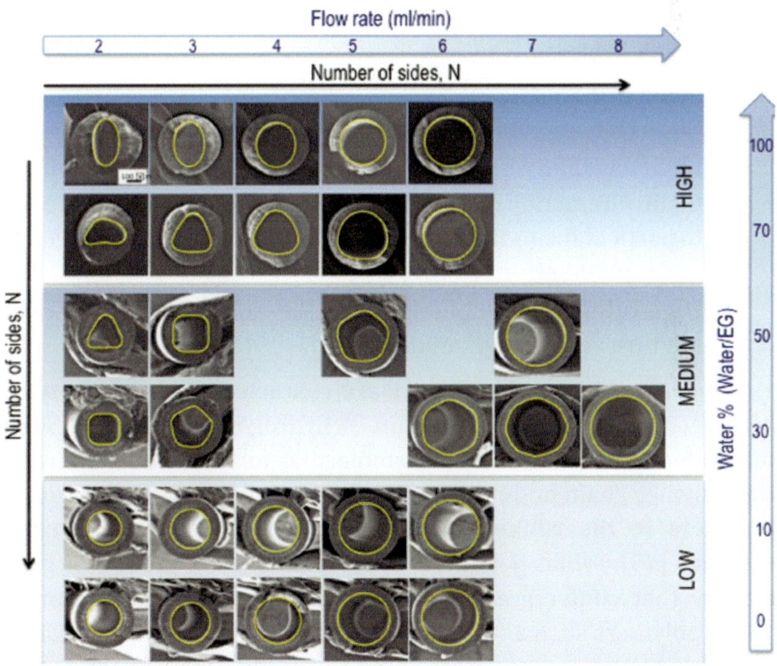

Figure 2.10 SEM cross section of PEI hollow fibers prepared by varying bore fluids composition and flow rate. The lumen shape is circled in yellow [47] (Open Access).

As shown in Fig. 2.10, the lumen shape varies according to the bore fluid composition and flow rate. By increasing the concentration of ethylene glycol in the bore fluid (at a flow rate of 3 ml/min), the number of sides increased from 2 to 5, becoming finally circular at low water content. In Fig. 2.10, the effect of the bore fluid flow rate on the lumen's fibers shape can also be clearly seen. At high water content, the increase in flow rate did not lead to an increase in the number of sides. At medium water content, on the contrary, the increase in flow rate led to an increase in the number of sides ($N = 8$ at a flow rate of 7 ml/min). At low water content, the lumen was circular independently from the flow rate considered.

2.5 Polymers Used

The choice of the suitable polymer for the preparation of hollow fibers is based on polymer specific requirements that influence the structure (symmetric, asymmetric, etc.) and performance of the membranes, such as:

1. Molecular weight;
2. Glass transition temperature;
3. Chain flexibility in the amorphous or semicrystalline state;
4. Chain interaction for thermal, chemical stability and resistance to particular conditions (pH, temperature, or organic solvents) of the polymer;
5. High affinity and solubility for the solvent or the solvent mixture at a certain temperature [1, 49].

Linear or branched polymers are characterized by three types of molecular bonds: dispersion forces, hydrogen bonding forces and dipole forces, or secondary intermolecular forces. The flexibility of the polymer chain and its mobility depend on the type of polymer: polymers in the rubbery state present a high degree of chain mobility [49], while polymers in the glassy state present a great rigidity that confer great thermal and chemical stability. For 30 years, polymers such as PVDF, PS, PES, and polyimide (PI), obtained from petroleum sources, dominated the membrane market, thanks to their high chemical and thermal stability with good processability for spinning [50, 51]. In the way of the circular economy, which is based on the paradigm of recycling, reducing CO_2 emissions and re-using of natural resources, there is an increasing necessity and attention toward the production and utilization of bio-based polymers as alternatives to the traditional fossil-based ones, now considered to be hazardous for the environment as a consequence of their synthesis and disposal [51, 52]. Some bio-based polymers have already started to be explored for hollow fibers preparation, such as CA and CS. They present, in fact, a series of different advantages:

- Low carbon impact;
- Bio-compostability, biodegradability, biocompatibility;
- Support of the sustainable industrial production and competitiveness.

Table 2.1 summarizes the most common polymers and biopolymers used for the preparation of hollow fibers.

Fluoropolymers (PVDF, PTFE, etc.) represent a major class of polymers for a wide range of applications in membranes. Their choice is related to their interesting properties such as thermal stability, chemical resistance, and low surface tension due to the high electronegativity of the fluorinate atom and the strong C–F bond (485 kJ/mol) [53, 54]. PVDF polymer, in a semicrystalline state, is widely used for hollow fiber preparation. Its degree of crystallinity is in the range of 35–70% with a melting point of 155–192°C. It presents five crystal polymorphs: α, β, γ, δ, and ε. Generally, the α and β phases are dominant in PVDF membranes. They have a different polarity and stability. The α phase is non-polar and kinetically favorable, while the β phase is more polar and thermodynamically more stable [55]. The TIPS and NIPS are techniques applied for producing PVDF hollow fibers [56]. The mechanism of phase separation is influenced by the interaction and solubility of PVDF with the selected solvent. PVDF has good miscibility with common aprotic solvents such as NMP, DMA, DMSO, triethyl phosphate (TEP), and Polarclean® [57–59] that influence the structure and the performance of the membranes produced by phase inversion. The properties of PVDF membranes can be modulated acting on dope solution composition, temperature of casting, phase-inversion conditions, and rheology of the material. One of the drawbacks of PVDF membranes is the low wetting resistance in the mass-transfer-based processes; in fact, the contact angle (CA) of PVDF polymer is less than 90° [60]. The strategy to improve the hydrophobicity of PVDF membranes is generally carried out through the addition of salts such as $CaCO_3$ or LiCl [55]. Fosi-Kofal et al. [61] and Wang D. et al. [62], for instance, produced asymmetric PVDF hollow fibers with good hydrophobicity induced through the incorporation of these additives. The same results were confirmed by Huang et al. [63], who studied the effects of $CaCO_3$ nanoparticles using a bore fluid containing LiCl. The prepared hollow fibers showed high porosity with a water permeation of 580 L/m^2h. Another important strategy is blending the PVDF membranes with other perfluorinated polymers such as PTFE or PVDF copolymers. The PTFE and PVDF copolymers (hexafluoropropylene (HFP) and tetrafluoroethylene (TFE)) display high hydrophobicity with respect to the PVDF due to their higher fluorinate content.

Table 2.1 Most common polymers and biopolymers used to prepare membranes in hollow fiber configuration

Polymers	Acronym
Polyvinilidene fluoride	PVDF
Poly(tetrafluoroethylene) PTFE	PTFE
Polypropylene	PP
Polyether ketone	PEK
Polyether ether ketone	PEEK-WC
Polysulfone	PSf
Poly(ethersulfone)	PES
Polyacrylonitrile	PAN
Polyaniline	PANI
Polydimethylsiloxane	PDMS
Poly(4-methyl-1-pentene)	PMP
Poly(4-vinylpyridine)	P4VP
Polybenzimidiazole	PBI
Amides/Imides	
Matrimid®5218	
3.3′.4.4′-benzophenone tetracarboxylicdianhydride and 80% methylphenylene-diamine + 20% methylene diamine co-polyimide precursor	BTDA–TDI/MDI, P84

Polymers	Acronym
Poly(1,5-naphthalene-2.2′-bis(3.4-phthalic) hexafluoropropanediimide/polyethersulfone	6FDANDA/PES
Polyetherimide	PEI
Ultem®	
Biopolymers	**Code**
Cellulose acetate	CA
Chitosan	CS
Poly(vinyl alcohol)	PVA
Ethylene vinyl alcohol copolymer	EVOH
Poly(lactic acid)	PLA
Polyhydroxyalkanoates	PHA
Polyurethane	PU
Poly(ε-caprolactone)	PCL

Sulfone polymers are a class of thermoplastic polymers (PS, PES, etc.) exhibiting high processability, thanks to the presence of a unit of diphenyl sulfone in their chain. Being characterized by good mechanical and thermal stability, they are employed in the preparation of hollow fibers for water treatment, gas separation, and biomedical applications [64]. PES has the highest hydrophilicity [40] with good compatibility with different additives such as PVP, PEG [65], propionic acid, glycerol, and TiO_2 [66]. For improving the pore size of PES hollow fibers, Ochoa et al. [67] studied the effect of different amounts of PVP in a system composed of PES/NMP/PVP. The resulting membrane pore size was in the range of ultrafiltration application. Interesting results have been reported by Simone et al. [40] for the preparation of nanocomposite PES/TiO_2 hollow fiber membranes in combination with Pluronic® and water as additives. They found an optimal TiO_2 concentration around 0.3%, with membranes characterized by a symmetric finger-like structure. In the optic of improving the sustainability in membrane preparation, Ursino et al. [58], for the first time, prepared PES hollow fibers using Polarclean® as a green solvent. Two variables were investigated: bore fluid composition and polymer concentration. The PES membranes showed a sponge-like structure with narrow pore-size distribution in the range of MF.

Noteworthy are also other polymers used for the preparation of hollow fibers such as PEI [68], PEEK-WC [69], PAN [70], and PI [71]. PEI, for instance, has high thermal resistance, high dielectric strength, and stiffness and good chemical resistance. Thanks to its properties, it is well used in CO_2 absorption processes [72]. PAN presents a high resistance to degradation (above 300°C); PI has a low free volume with high mechanical resistance, and it finds applications for membranes applied in gas separation [73]. PEEK-WC is characterized by a lactone group. It is a semicrystalline and biocompatible material. It was employed by Tasselli et al. [69] for the preparation of hollow fibers by the dry/wet-spinning technique for ultrafiltration applications.

2.6 Characterization of Hollow Fiber Membranes

Membrane characterization is crucial for three important aspects, related to each other, such as (1) the thermodynamics and kinetics of membrane preparation; (2) the structure of the membranes; and (3) the performance in application processes. Characterization tests can be divided into dynamic and static techniques with the possibility to destroy or preserving the samples. Static techniques include characterization tests oriented in acquiring information about the morphology and the physical and chemical properties; while dynamic techniques give information about the performance of the membranes [74]. Table 2.2 summarizes the main typologies and methods of characterizations.

Table 2.2 Main typologies and methods of characterizations

Characterizations	Typology	Method
Pore size	Dynamic	Bubble point method
Pore-size distribution	Dynamic	Permporometry
Pore-size distribution	Dynamic	Gas and liquid displacement methods (GLDP-LLDP)
Pore-size distribution	Static	Mercury porosimetry (MP)
Pore-size distribution	Static	Gas adsorption/ desorption (GAD)

Characterizations	Typology	Method
Morphology, thickness, pore-size distribution, qualitative structure analysis	Static	Scanning electron microscopy (SEM), Transmission electron microscopy (TEM)
Surface porosity, roughness	Static	Atomic force microscopy (AFM)
Surface studies, functional group analysis	Static	Infrared spectroscopy (FT-IR, ATR, photoacoustic)
Surface studies, chemical analysis	Static	Energy dispersive X-rays spectroscopy (EDX or EDS), wavelength dispersive X-ray spectroscopy (WDS)
Thermal analysis	Dynamic	Differential scanning calorimetry (DSC)
Wettability	Static and dynamic	Contact angle measurement
Mechanical properties	Static and dynamic	Mechanical resistance, stress–strain measurements
Chemical stability	Static and dynamic	Swelling and contact angle

Note: Adapted from Ref. [74].

2.6.1 Microscopy Techniques

The first approach to be considered for studying membrane properties is generally represented by the investigation of membrane structure by means of microscopy techniques such as transmission (TEM), scanning (SEM), and field emission systems, based on the interaction of an electron beam with the membrane sample. Atomic force microscopy (AFM) is also used to study the topography and roughness of the hollow fibers. The difference among the techniques is represented by their resolution scale, as follows [49]:

- SEM: up 0.5 nm
- Field emission electron microscopy: 0.6–0.7 nm
- TEM: 0.4–0.5 nm
- AFM: ca. 1 nm.

The aforementioned microscopic techniques are powerful tools for studying the morphology and topography of the hollow fibers. SEM, for instance, can be used for determining the thickness of the hollow fiber walls, the dimension of the lumen, and the presence of possible defects. The images acquired give direct evidence of the structure of the membrane and its symmetry. By SEM analyses, it is possible to detect the presence and the thickness of a dense selective layer and to get an indication of membrane pore size and porosity degree. TEM can be used for studying the distribution and possible agglomeration of nanoparticles in the case of MMM hollow fibers.

AFM allows having a 3D image of membrane surface, giving information on its topography. AFM measures the roughness of the membrane, and it is also able to determine the pore-size distribution.

2.6.2 Pore Size and Pore-Size Distribution

The determination of pore size and pore-size distribution can be studied using different methodologies such as bubble point method, liquid/gas displacement, and liquid–liquid porometry. The membrane sample is generally soaked in a wetting liquid (phase 1) with a low surface tension. An increasing pressure, represented by air or another immiscible liquid (phase 2), is then applied to expel the wetting liquid from the membrane pores. The pressure exerted is in relation to the pore dimensions of the membrane. The measured pressure can be used for determining the maximum pore size, the mean pore size, and pore-size distribution.

2.6.2.1 Bubble point method

In the bubble point method, the theoretical base is the Young–Laplace equation [75] where ΔP is the transmembrane pressure (Pa) required to extrude the liquid from the pores and that can be calculated with the following equation (2.1):

$$\Delta P = \frac{4\gamma L \cos\theta}{2r(p, \max)} \tag{2.1}$$

where $r(p, \max)$ is the maximim pore radius (m), γL is the surface tension of the wetting liquid (N/m), and θ is the contact angle between the liquid and the membrane surface (°).

2.6.2.2 Mercury intrusion porosimetry method

Mercury intrusion porosimetry requires air as phase 1 and mercury as phase 2. This method is based on the same equation as the bubble point method. However, this technique requires high pressure to force mercury to intrude into the membrane's pores. The capability of pore size measurement (dp) is about 3 nm with respect to the bubble point method, where the dp is about 13 nm [76].

2.6.2.3 Liquid–liquid porometry method

In this method, both phases (1 and 2) are liquids, immiscible together. Generally, it is possible to use water–alcohol at low pressure (around 10 bar) with a dp of 1 nm. The pressure depends on the interfacial tension between the two liquids. The following information can be acquired: bubble point, pore size, pore-size distribution, and pore density.

Bubble point is one of the preliminary measures detected. It is defined as the pressure at which the first gas bubbles are revealed and correspond to the maximum pore size.

Pore size (D) is determined by the Washburn equation (Eq. 2.2):

$$D = 4\,\gamma\cos\theta/p \tag{2.2}$$

where γ is the surface tension of the wetting liquid, θ is the contact angle of the liquid and the membrane, and p is the differential gas pressure. Generally, the pore diameter is assumed to be that of a cylindrical pore [77].

The maximum pore size diameter (μm) is expressed by using Eq. (2.3):

$$d = C\,\gamma/p \tag{2.3}$$

where C is a constant (0.0286 bar), p is differential gas pressure across the wetting liquid in the pore, and γ is the surface tension of the wetting liquid.

Pore density can be identified in terms of fractional pore number distribution (f_N) and represents the percentage of the pore in the indicated diameter range.

2.6.3 Differential Scanning Calorimetry

Differential scanning calorimetry (DSC) is the most common thermal analysis allowing the detection of thermal transitions such as melts, glass transitions, phase changes, or curing. The technique allows determining the physical transition of the material under thermal treatment. This information is pivotal for understanding the capacity, enthalpy, and efficiency of the material for different application processes [78]. During a temperature change, DSC measures the amount of heat radiated or absorbed by the sample based on a temperature difference between the sample and a reference material.

2.6.4 Fourier Transform Infrared

Fourier transform infrared (FT-IR) is a static surface study spectroscopy, which is a qualitative analysis for the identification of the chemical nature of an organic substance providing information on chemical bonds, molecular orientations, molecular energy levels, and molecular interactions by using specific absorption peaks for particular groups [79]. FT-IR, in membranes, can be used for the identification of macromolecules adsorbed at the membrane surface (fouling), the detection of functional groups of specific polymers or additives included in the membrane matrix, and the detection of the presence of residual solvent traces.

2.6.5 Contact Angle

The hydrophobicity/hydrophilicity, or wettability, of hollow fiber membranes, can be determined by contact angle measurements. The contact angle of water is the angle formed between a water droplet and membrane surface, and it is mainly influenced by the chemistry of the membrane surface and by its roughness or smoothness.

The contact angle behavior (advancing, receding, and hysteresis) can be determined by Young's, Wenzel's, and Cassie's equations [80, 81]. Young's formula or tangent line method (Fig. 2.11a) is the simplest method used for measuring the static contact angle ($\cos\theta$):

$$\cos\theta = \frac{\gamma_{\mathrm{sv}} - \gamma_{\mathrm{sl}}}{\gamma_{\mathrm{lv}}} \tag{2.4}$$

where γ_{sv}, γ_{sl}, and γ_{lv} are the surface tensions (or surface free energy) of the solid–vapor, solid–liquid, and liquid–vapor pairs. Water contact angle $< 90°$ is typical of a surface that is considered hydrophilic, while contact angle > 90° is characteristic of a surface considered hydrophobic [82, 83]. When a surface is roughened, the solid–liquid interactions are amplified, and the contact angle can be determined using Wenzel's equation (Fig. 2.11b) as follows:

$$\cos \theta^* = r\, x \cos \theta \tag{2.5}$$

where θ^* is the apparent contact angle on the roughened surface, θ is the contact angle on a smooth surface, and r is the roughness value.

The Cassie–Baxter equation (Eq. 2.6) considers the air pockets trapped between the solid surface of a rough material and the liquid (Fig. 2.11c), resulting in a more hydrophobic surface:

$$\cos \theta^* = f_1 \cos \theta_1 + f_2 \cos \theta_2 \tag{2.6}$$

where θ_1 is the contact angle of the smooth solid surface and θ_2 represents the contact angle for air.

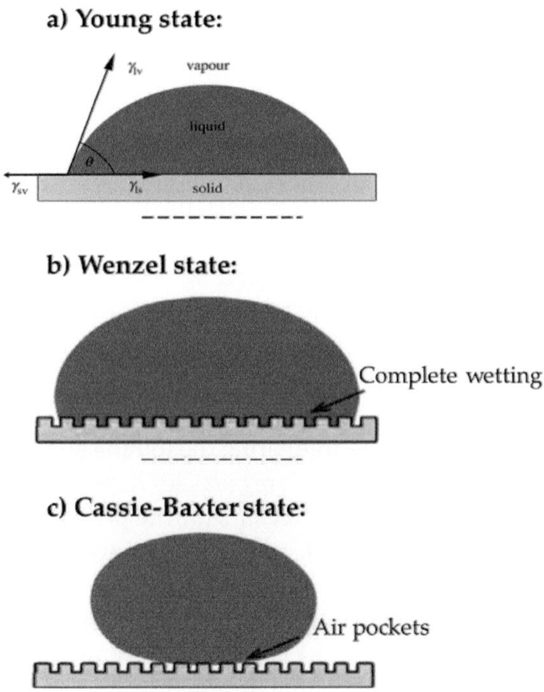

Figure 2.11 The contact angle method in (a) Young, (b) Wenzel, and (c) Cassie–Baxter state. Adapted from Ref. [84].

2.6.6 Mechanical Properties

Good mechanical properties, in particular for membranes in hollow fiber configuration, are crucial for having membranes able to withstand the applied pressure and to exhibit a certain degree of durability in time. The hollow fiber is generally subjected to a strain at a controllable strain rate, measuring the corresponding required load. Bending test, uniaxial test, dynamic mechanical analysis, nanoindentation, and bursting tests are the most employed characterization techniques employed for investigating the mechanical behavior of polymeric membranes [85]. Tensile strength (stress at break), elongation at break (strain at break), and Young's modulus are the parameters generally measured.

2.7 Conclusion and Future Perspectives

This chapter summarizes the principles of hollow fiber preparation by spinning techniques elucidating the variables that dictate the morphology and the properties of the final membranes. Hollow fibers are already the preferred choice in large-scale membrane applications, and many breakthroughs, during the last years, have been accomplished in terms of understanding their formation mechanism. Dry/wet and wet-spinning techniques are still the major procedures employed in hollow fiber fabrication.

Novel hollow fiber designs, as in the case of microstructured and microshaped structures, demonstrated to have some important benefits in enhancing hollow fiber performance in terms of improving surface area and mitigating the fouling and concentration polarization phenomena. Pioneering articles have recently proposed the use of green and nontoxic solvents for the preparation of hollow fibers to make the entire fabrication process more sustainable and safer.

To improve the productivity and the performance of membranes in several applications such as gas separation, water treatment as well as pervaporation and biofuel production, hollow fibers with an ultrathin selective layer are highly desired. Even if the realization of very thin selective layers without defects is quite challenging, many advances in this direction have been made with the development, for instance, of dual-layer hollow fibers [86].

A series of challenges and opportunities in hollow fiber production and applications are still open. The future of hollow fibers will be strictly related to developing new performing materials, new formulations, and innovative and more efficient preparation conditions. The basic research will be fundamental in assisting this process by elucidating even more the mechanisms underlaying the hollow fiber formation process. Moreover, modeling and simulation [87, 88], which can correlate and predict the relationship between preparation conditions and membrane structure, will be powerful tools to design the next generation of membranes.

References

1. Feng, C. Y., Khulbe, K. C., Matsuura, T., and Ismail, A. F. (2013) Recent progresses in polymeric hollow fiber membrane preparation, characterization and applications. *Sep. Purif. Technol.*, **111**, 43–71.
2. Khayet, M. and Matsuura, T. (2011) Formation of hollow fibre MD membranes, in *Membrane Distillation*, Elsevier, Amsterdam, pp. 59–87.
3. Qin, J.-J., Wang, R., and Chung, T.-S. (2000) Investigation of shear stress effect within a spinneret on flux, separation and thermomechanical properties of hollow fiber ultrafiltration membranes. *J. Memb. Sci.*, **175**(2), 197–213.
4. Peng, N., Chung, T.-S., and Wang, K. Y. (2008) Macrovoid evolution and critical factors to form macrovoid-free hollow fiber membranes. *J. Memb. Sci.*, **318**(1–2), 363–372.
5. De Rovère, A. and Shambaugh, R. L. (2001) Melt-spun hollow fibers: Modeling and experiment. *Polym. Eng. Sci.*, **41**(7), 1206–1219.
6. Huang, Y., Liu, H., Xiao, C., Wang, C., Chen, D., and Sun, K. (2019) Robust preparation and multiple pore structure design of poly (tetrafluoroethylene-co-hexafluoropropylene) hollow fiber membrane by melt spinning and post-treatment. *J. Taiwan Inst. Chem. Eng.*, **97**, 441–449.
7. Liu, H., Xiao, C., Hu, X., and Liu, M. (2013) Post-treatment effect on morphology and performance of polyurethane-based hollow fiber membranes through melt-spinning method. *J. Memb. Sci.*, **427**, 326–335.
8. Berghmans, S. (1996) Spinning of hollow porous fibres via the TIPS mechanism. *J. Memb. Sci.*, **116**(2), 171–189.

9. Qin, J. (2001) Effect of wet and dry-jet wet spinning on the shear-induced orientation during the formation of ultrafiltration hollow fiber membranes. *J. Memb. Sci.*, **182**(1–2), 57–75.

10. Alsalhy, Q. F., Salih, H. A., Simone, S., Zablouk, M., Drioli, E., and Figoli, A. (2014) Poly(ether sulfone) (PES) hollow-fiber membranes prepared from various spinning parameters. *Desalination*, **345**, 21–35.

11. Tasselli, F., Jansen, J., Sidari, F., and Drioli, E. (2005) Morphology and transport property control of modified poly(ether ketone) (PEEKWC) hollow fiber membranes prepared from PEEKWC/PVP blends: Influence of the relative humidity in the air gap. *J. Memb. Sci.*, **255**(1–2), 13–22.

12. Peng, N., Widjojo, N., Sukitpaneenit, P., Teoh, M. M., Lipscomb, G. G., Chung, T.-S., and Lai, J.-Y. (2012) Evolution of polymeric hollow fibers as sustainable technologies: Past, present, and future. *Prog. Polym. Sci.*, **37**(10), 1401–1424.

13. Wu, B., Li, K., and Teo, W. K. (2007) Preparation and characterization of poly(vinylidene fluoride) hollow fiber membranes for vacuum membrane distillation. *J. Appl. Polym. Sci.*, **106**(3), 1482–1495.

14. Liu, C. and Bai, R. (2006) Preparing highly porous chitosan/cellulose acetate blend hollow fibers as adsorptive membranes: Effect of polymer concentrations and coagulant compositions. *J. Memb. Sci.*, **279**(1–2), 336–346.

15. Russo, F., Castro-Muñoz, R., Galiano, F., and Figoli, A. (2019) Unprecedented preparation of porous Matrimid® 5218 membranes. *J. Memb. Sci.*, **585**, 166–174.

16. García-Fernández, L., García-Payo, M. C., and Khayet, M. (2014) Effects of mixed solvents on the structural morphology and membrane distillation performance of PVDF-HFP hollow fiber membranes. *J. Memb. Sci.*, **468**, 324–338.

17. Roy, A., Bhunia, P., and De, S. (2017) Solvent effect and macrovoid formation in cellulose acetate phthalate (CAP)-polyacrylonitrile (PAN) blend hollow fiber membranes. *J. Appl. Polym. Sci.*, **134**(1).

18. Simone, S., Figoli, A., Criscuoli, A., Carnevale, M. C., Rosselli, A., and Drioli, E. (2010) Preparation of hollow fibre membranes from PVDF/PVP blends and their application in VMD. *J. Memb. Sci.*, **364**(1–2), 219–232.

19. Smolders, C. A., Reuvers, A. J., Boom, R. M., and Wienk, I. M. (1992) Microstructures in phase-inversion membranes. Part 1. Formation of macrovoids. *J. Memb. Sci.*, **73**(2–3), 259–275.

20. Chou, W.-L., Yu, D.-G., Yang, M.-C., and Jou, C.-H. (2007) Effect of molecular weight and concentration of PEG additives on morphology and permeation performance of cellulose acetate hollow fibers. *Sep. Purif. Technol.*, **57**(2), 209–219.

21. Peng, N., Chung, T.-S., and Li, K. Y. (2009) The role of additives on dope rheology and membrane formation of defect-free Torlon® hollow fibers for gas separation. *J. Memb. Sci.*, **343**(1–2), 62–72.

22. Loh, C. H., Wang, R., Shi, L., and Fane, A. G. (2011) Fabrication of high performance polyethersulfone UF hollow fiber membranes using amphiphilic Pluronic block copolymers as pore-forming additives. *J. Memb. Sci.*, **380**(1–2), 114–123.

23. Dias, R. A., Medeiros, V. da N., Silva, B. I. A., Araújo, E. M., and Lira, H. de L. (2019) Study of the influence of viscosity on the morphology of polyethersulfone hollow fiber membranes/additives. *Mater. Res.*, **22**.

24. Chou, W. and Yang, M. (2005) Effect of take-up speed on physical properties and permeation performance of cellulose acetate hollow fibers. *J. Memb. Sci.*, **250**(1–2), 259–267.

25. Liu, Y., Koops, G., and Strathmann, H. (2003) Characterization of morphology controlled polyethersulfone hollow fiber membranes by the addition of polyethylene glycol to the dope and bore liquid solution. *J. Memb. Sci.*, **223**(1–2), 187–199.

26. Ohya, H., Shiki, S., and Kawakami, H. (2009) Fabrication study of polysulfone hollow-fiber microfiltration membranes: Optimal dope viscosity for nucleation and growth. *J. Memb. Sci.*, **326**(2), 293–302.

27. Aptel, P., Abidine, N., Ivaldi, F., and Lafaille, J. P. (1985) Polysulfone hollow fibers: Effect of spinning conditions on ultrafiltration properties. *J. Memb. Sci.*, **22**(2–3), 199–215.

28. Shilton, S. J., Bell, G., and Ferguson, J. (1994) The rheology of fibre spinning and the properties of hollow-fibre membranes for gas separation. *Polymer (Guildf).*, **35**(24), 5327–5335.

29. Qin, J. (1999) Effect of dope flow rate on the morphology, separation performance, thermal and mechanical properties of ultrafiltration hollow fibre membranes. *J. Memb. Sci.*, **157**(1), 35–51.

30. Tasselli, F. and Drioli, E. (2007) Tuning of hollow fiber membrane properties using different bore fluids. *J. Memb. Sci.*, **301**(1–2), 11–18.

31. Bakeri, G., Ismail, A. F., Rahimnejad, M., Matsuura, T., and Rana, D. (2012) The effect of bore fluid type on the structure and performance of polyetherimide hollow fiber membrane in gas–liquid contacting processes. *Sep. Purif. Technol.*, **98**, 262–269.

32. Khayet, M., Feng, C. Y., Khulbe, K. C., and Matsuura, T. (2002) Preparation and characterization of polyvinylidene fluoride hollow fiber membranes for ultrafiltration. *Polymer (Guildf).*, **43**(14), 3879–3890.

33. Simone, S., Conidi, C., Ursino, C., Cassano, A., and Figoli, A. (2016) Clarification of orange press liquors by PVDF hollow fiber membranes. *Membranes (Basel).*, **6**(1), 9.

34. Alobaidy, A. A., Sherhan, B. Y., Barood, A. D., and Alsalhy, Q. F. (2017) Effect of bore fluid flow rate on formation and properties of hollow fibers. *Appl. Water Sci.*, **7**(8), 4387–4398.

35. Khayet, M. (2003) The effects of air gap length on the internal and external morphology of hollow fiber membranes. *Chem. Eng. Sci.*, **58**(14), 3091–3104.

36. Chung, T.-S. and Hu, X. (1997) Effect of air-gap distance on the morphology and thermal properties of polyethersulfone hollow fibers. *J. Appl. Polym. Sci.*, **66**(6), 1067–1077.

37. Deshmukh, S. P. and Li, K. (1998) Effect of ethanol composition in water coagulation bath on morphology of PVDF hollow fibre membranes. *J. Memb. Sci.*, **150**(1), 75–85.

38. Yu, D.-G., Chou, W.-L., and Yang, M.-C. (2006) Effect of draw ratio and coagulant composition on polyacrylonitrile hollow fiber membranes. *Sep. Purif. Technol.*, **52**(2), 380–387.

39. Choi, S.-H., Tasselli, F., Jansen, J. C., Barbieri, G., and Drioli, E. (2010) Effect of the preparation conditions on the formation of asymmetric poly(vinylidene fluoride) hollow fibre membranes with a dense skin. *Eur. Polym. J.*, **46**(8), 1713–1725.

40. Simone, S., Galiano, F., Faccini, M., Boerrigter, M., Chaumette, C., Drioli, E., and Figoli, A. (2017) Preparation and characterization of polymeric-hybrid PES/TiO_2 hollow fiber membranes for potential applications in water treatment. *Fibers*, **5**(2), 14.

41. Galiano, F., Song, X., Marino, T., Boerrigter, M., Saoncella, O., Simone, S., Faccini, M., Chaumette, C., Drioli, E., and Figoli, A. (2018) Novel photocatalytic PVDF/Nano-TiO_2 hollow fibers for environmental remediation. *Polymers (Basel).*, **10**(10), 1134.

42. Galiano, F., Figoli, A., Conidi, C., Menichini, F., Bonesi, M., Loizzo, M., Cassano, A., and Tundis, R. (2016) Functional properties of *Punica granatum L.* juice clarified by hollow fiber membranes. *Processes*, **4**(3), 21.

43. Tsai, H.-A., Ma, L.-C., Yuan, F., Lee, K.-R., and Lai, J.-Y. (2008) Investigation of post-treatment effect on morphology and pervaporation performance of PEG added PAN hollow fiber membranes. *Desalination*, **234**(1–3), 232–243.

44. Tan, X., Tan, S. P., Teo, W. K., and Li, K. (2006) Polyvinylidene fluoride (PVDF) hollow fibre membranes for ammonia removal from water. *J. Memb. Sci.*, **271**(1–2), 59–68.

45. Kim, I.-C., Yun, H.-G., and Lee, K.-H. (2002) Preparation of asymmetric polyacrylonitrile membrane with small pore size by phase inversion and post-treatment process. *J. Memb. Sci.*, **199**(1–2), 75–84.

46. Bildyukevich, A. V. and Usosky, V. V. (2014) Prevention of the capillary contraction of polysulfone based hollow fiber membranes. *Pet. Chem.*, **54**(8), 652–658.

47. Le, N. L., Kim, D., and Nunes, S. P. (2017) Evolution of regular geometrical shapes in fiber lumens. *Sci. Rep.*, **7**(1), 9171.

48. Çulfaz, P. Z., Rolevink, E., van Rijn, C., Lammertink, R. G. H., and Wessling, M. (2010) Microstructured hollow fibers for ultrafiltration. *J. Memb. Sci.*, **347**(1–2), 32–41.

49. Strathmann, H., Giorno, L., and Drioli, E. (2010) Basic aspects in polymeric membrane preparation. *Compr. Membr. Sci. Eng.*, **1**, 1570.

50. Abdul Khalil, H. P. S., Saurabh, C. K., Adnan, A. S., Nurul Fazita, M. R., Syakir, M. I., Davoudpour, Y., Rafatullah, M., Abdullah, C. K., Haafiz, M. K. M., and Dungani, R. (2016) A review on chitosan-cellulose blends and nanocellulose reinforced chitosan biocomposites: Properties and their applications. *Carbohydr. Polym.*, **150**, 216–226.

51. Russo, F., Galiano, F., Iulianelli, A., Basile, A., and Figoli, A. (2020) Biopolymers for sustainable membranes in CO_2 separation: A review. *Fuel Process. Technol.*, 106643.

52. Galiano, F., Briceño, K., Marino, T., Molino, A., Christensen, K. V., and Figoli, A. (2018) Advances in biopolymer-based membrane preparation and applications. *J. Memb. Sci.*, **564** (July), 562–586.

53. Drobny, J. G. (2008) *Technology of Fluoropolymers*, Second edition, CRC Press.

54. Holding, S. (2010) Technology of fluoropolymers. *Chromatographia*, **71**, 173.

55. Matsuyama, H., Rajabzadeh, S., Karkhanechi, H., and Jeon, S. (2017) 1.7 PVDF hollow fibers membranes, in *Comprehensive Membrane Science and Engineering*, vol. 1, Elsevier, pp. 137–189.

56. Figoli, A., Simone, S., and Drioli, E. (2015) Polymeric membranes, in *Membrane Fabrication* (eds. Hilal, N., Ismail, A. F., and Wright, C.), Taylor & Francis Group, Boca Raton, FL, USA, pp. 1–42.

57. Russo, F., Ursino, C., Avruscio, E., Desiderio, G., Perrone, A., Santoro, S., Galiano, F., and Figoli, A. (2020) Innovative poly (vinylidene fluoride) (PVDF) electrospun nanofiber membrane preparation using DMSO as a low toxicity solvent. *Membranes (Basel).*, **10**(3), 36.

58. Ursino, C., Russo, F., Ferrari, R. M., De Santo, M. P., Di Nicolò, E., He, T., Galiano, F., and Figoli, A. (2020) Polyethersulfone hollow fiber membranes prepared with Polarclean® as a more sustainable solvent. *J. Memb. Sci.*, **608**, 118216.

59. Hassankiadeh, N. T., Cui, Z., Kim, J. H., Shin, D. W., Lee, S. Y., Sanguineti, A., Arcella, V., Lee, Y. M., and Drioli, E. (2015) Microporous poly(vinylidene fluoride) hollow fiber membranes fabricated with PolarClean as water-soluble green diluent and additives. *J. Memb. Sci.*, **479**, 204–212.

60. Feng, X. and Jiang, L. (2006) Design and creation of superwetting/antiwetting surfaces. *Adv. Mater.*, **18**(23), 3063–3078.

61. Fosi-Kofal, M., Mustafa, A., Ismail, A. F., Rezaei-DashtArzhandi, M., and Matsuura, T. (2016) PVDF/$CaCO_3$ composite hollow fiber membrane for CO_2 absorption in gas–liquid membrane contactor. *J. Nat. Gas Sci. Eng.*, **31**, 428–436.

62. Wang, D. (2000) Porous PVDF asymmetric hollow fiber membranes prepared with the use of small molecular additives. *J. Memb. Sci.*, **178**(1–2), 13–23.

63. Huang, D.-xi, Wang, L., Meng, X. rong, Wang, X. dong, and Zhao, L. (2014) Preparation of PVDF hollow fiber ultrafiltration membrane via phase inversion/chemical treatment method. *Desalin. Water Treat.*, **52**(25–27), 5076–5083.

64. Khulbe, K. C., Feng, C., Matsuura, T., and Khayet, M. (2006) Polymeric hollow fibers: State of the art review of their preparation, characterization and applications in different research areas. *J. Appl. Membr. Sci. Technol.*, **4**, 53–88.

65. Xu, Z.-L. and Alsalhy Qusay, F. (2004) Polyethersulfone (PES) hollow fiber ultrafiltration membranes prepared by PES/non-solvent/NMP solution. *J. Memb. Sci.*, **233**(1–2), 101–111.

66. Rahbari-Sisakht, M., Ismail, A. F., and Matsuura, T. (2012) Development of asymmetric polysulfone hollow fiber membrane contactor for CO_2 absorption. *Sep. Purif. Technol.*, **86**, 215–220.

67. Ameri, A., Gholami, M., Nasseri, N., and Matsuura, T. (2004) Modification of polyether sulfone (PES) hollow fiber membranes characteristics for more efficient water treatment process. *Iran. J. Public Health.*, **33**(2), 49–55.

68. Gao, J., Wang, K. Y., and Chung, T. S. (2020) Design of nanofiltration (NF) hollow fiber membranes made from functionalized bore fluids containing polyethyleneimine (PEI) for heavy metal removal. *J. Memb. Sci.*, **603**, 118022.

69. Tasselli, F., Jansen, J. C., and Drioli, E. (2004) PEEKWC ultrafiltration hollow-fiber membranes: Preparation, morphology, and transport properties. *J. Appl. Polym. Sci.*, **91**(2), 841–853.

70. Morelli, S., Piscioneri, A., Salerno, S., Tasselli, F., Di Vito, A., Giusi, G., Canonaco, M., Drioli, E., and De Bartolo, L. (2012) PAN hollow fiber membranes elicit functional hippocampal neuronal network. *J. Mater. Sci. Mater. Med.*, **23**(1), 149–156.

71. Xu, S., Wang, Z. Y., Li, S., Tian, L., and Su, B. (2020) Fabrication of polyimide-based hollow fiber membrane by synergetic covalent-crosslinking strategy for organic solvent nanofiltration (OSN) application. *Sep. Purif. Technol.*, **241**, 116751.

72. Naim, R. and Ismail, A. F. (2013) Effect of polymer concentration on the structure and performance of PEI hollow fiber membrane contactor for CO_2 stripping. *J. Hazard. Mater.*, **250–251**, 354–361.

73. Russo, F., Castro-Muñoz, R., Galiano, F., and Figoli, A. (2019) Unprecedented preparation of porous Matrimid® 5218 membranes. *J. Memb. Sci.*, **585**, 10.1016/j.memsci.2019.05.036.

74. Tylkowski, B. and Tsibranska, I. (2015) Overview of main techniques used for membrane characterization. *J. Chem. Technol. Metall.*, **50**.

75. Drioli, E. and Giorno, L. (2010) *Comprehensive Membrane Science and Engineering*, Elsevier.

76. Tung, K. L., Chang, K. S., Wu, T. T., Lin, N. J., Lee, K. R., and Lai, J. Y. (2014) Recent advances in the characterization of membrane morphology. *Curr. Opin. Chem. Eng.*, **4**, 121–127.

77. Jena, A. and Gupta, K. (2010) Advances in pore structure evaluation by porometry. *Chem. Eng. Technol.*, **33**(8), 1241–1250.

78. Marino, T., Galiano, F., Molino, A., and Figoli, A. (2019) New frontiers in sustainable membrane preparation: Cyrene™ as green bioderived solvent. *J. Memb. Sci.*, **580**, 224–234.

79. Vinodhini, P. A., Kirubanandam, S., Gomathi, T., P.N., S., Jayachandran, V., and Sukumaran, A. (2017) FTIR, XRD and DSC studies of nanochitosan,

cellulose acetate and polyethylene glycol blend ultrafiltration membranes. *Int. J. Biol. Macromol.*, **104**, 1721–1729.

80. Gao, L. and McCarthy, T. J. (2007) How Wenzel and Cassie were wrong. *Langmuir*, **23**(7), 3762–3765.

81. Edalatpour, M., Liu, L., Jacobi, A. M., Eid, K. F., and Sommers, A. D. (2018) Managing water on heat transfer surfaces: A critical review of techniques to modify surface wettability for applications with condensation or evaporation. *Appl. Energy.*, **222**, 10.1016/j.apenergy.2018.03.178.

82. Srinivasarao, M. (2017) Biomimetics: Bioinspired Hierarchical-Structured Surfaces for Green Science and Technology, Bharat Bhushan, Springer, 2016 (2nd ed.), 591 p, $279.00, ISBN 978-3-31. *Phys. Today*, **70**(9), 60–61.

83. He, Y., Xiao, S., Wu, J., and Fang, H. (2019) Influence of multiple factors on the wettability and surface free energy of leaf surface. *Appl. Sci.*, **9**, 593, 10.3390/app9030593.

84. Brown, P. S. and Bhushan, B. (2016) Bioinspired materials for water supply and management: Water collection, water purification and separation of water from oil. *Philos. Trans. R. Soc. A Math. Phys. Eng. Sci.*, **374**(2073), 20160135.

85. Wang, K., Abdala, A. A., Hilal, N., and Khraisheh, M. K. (2017) Mechanical characterization of membranes, in *Membrane Characterization* (eds. Nidal, H., Ismail, A. F., Matsuura, T., and Oatley-Radcliffe, D.), Elsevier, pp. 259–306.

86. Bonyadi, S. and Chung, T. S. (2007) Flux enhancement in membrane distillation by fabrication of dual layer hydrophilic–hydrophobic hollow fiber membranes. *J. Memb. Sci.*, **306**(1–2), 134–146.

87. Aroon, M. A., Ismail, A. F., Montazer-Rahmati, M. M., and Matsuura, T. (2010) A mathematical analysis of hollow fiber spinning: Bore and dope velocity profiles in the air gap. *J. Memb. Sci.*, **348**(1–2), 13–20.

88. Balasubramanian-Rauckhorst, H. A., Lloyd, D. R., and Lipscomb, G. G. (2009) Predicting extent of anisotropy in anisotropic hollow fiber membrane formation. *J. Memb. Sci.*, **339**(1–2), 250–260.

Chapter 3

Preparation and Characterization of Polymeric Nanofibers by Electrospinning

Sergio Santoro,[a] Javier Aragón,[b,c] and Efrem Curcio[a]
[a]*Department of Environmental Engineering (DIAm), University of Calabria, via P. Bucci CUBO 45A, Rende (CS), 87036, Italy*
[b]*Department of Chemical Engineering and Environmental Technologies, Aragon Nanoscience Institute, University of Zaragoza, Zaragoza, Spain*
[c]*Aragon Health Research Institute (IIS Aragon), Zaragoza, Spain*
sergio.santoro@unical.it

Nowadays electrospinning plays an influential role in the scenario of membrane technologies where it reached an active and dynamic stage. Electrospinning has emerged as an innovative, straightforward, and cost-effective method for the preparation of nanofibrous polymeric membranes characterized by outstanding porosity and surface area. The recent scientific and technological advancements have enabled the feasibility of electrospinning at industrial scale, preserving a precise control of the membrane morphology at a sub-micrometric scale leading to immense opportunities for a practical application

Hollow Fibers and Nanofibers in Membrane Science: Preparation, Characterization, and Applications
Edited by Alberto Figoli, Mir Saeed Seyed Dorraji, and Francesco Galiano

ISBN 978-981-4968-03-4 (Hardcover), 978-1-003-25663-2 (eBook)
www.jennystanford.com

of electrospun nanofibers. The first part of the chapter focuses on the fundamentals of the electrospinning process and the effect of the operating conditions on the morphology of the nanofibers. The second part gives an overview of the physicochemical properties of electrospun membranes and the methodologies for their characterization with a special focus on strategies for tailoring the key features of the membrane.

3.1 Introduction

The electrospinning technique has gained popularity in the last two decades for the development of fibrous membranes. Electrospun membranes are obtained from different precursor materials offering a wide variety of properties in terms of compositions, shapes, and structures exploited in many applications. The diameter of the fibers usually varies from only a few nanometers to tens of microns [1–3], and the thickness of the membrane typically ranges from hundreds of micrometers (considered as 2D structures) to some centimeters (considered as 3D structures) [4].

Recently the electrospinning technique has injected new vitality in biomedical research [5] and separative processes [6] gaining more ground in emerging areas of nanotechnology [7].

Electrospinning has roots in the 16th century when William Gilbert noticed an interesting phenomenon observing a water droplet displaced near an electrically charged amber: the droplet changed its shape, forming first a cone and spraying small droplets from the cone tip, showing an electrospraying process (Fig. 3.1c) [8, 9]. Only in the 20th century, the theoretical aspects of the phenomena were revealed by the mathematical description of a fluid cone formed (Taylor cone) by an electrical field [10], opening up an interesting perspective for innovative and practical applications [11]. In the early 1990s, the electrospinning process was developed by Reneker et al. based on the electrical spinning process employed to produce nanofibers from polymers using high voltage [12, 13].

To date, electrospinning is considered a cost-effective process for the large-scale production of fibrous membranes characterized by a random distribution (Fig. 3.1f) or a partial alignment of the nanofibers (Fig. 3.1e). Several natural polymers (hyaluronic acid, collagen, alginate, cellulose, chitosan, gelatin, pullulan, zein, silk fibroin) and synthetic polymers (polyamide, polycaprolactone, polylactic acid, poly (lactic-co-glycolic acid), polyvinyl alcohol, polyurethane, poly (ethylene-co-vinyl alcohol), polystyrene, etc.) have been successfully electrospun and employed in membrane processes [2, 14].

The electrospinning setups (horizontal or vertical) are usually formed of a needle (single, coaxial, or triaxial) [15], or a multiple needles system [16], collector (flat or drum) [17], high-voltage power supply connected to the needle [18, 19], and a solution feeding system [20]. Some commercial electrospinning equipment includes a robotic arm to facilitate the movement of the needle, multiples sensors to measure and control the environmental parameters (temperature and humidity) [4].

In this chapter, we discuss the fundamentals of the electrospinning technique, elucidating about the effects of the operative parameters and electrospinning configurations on the specific structures of the membranes with a special focus on the leap from lab to large scale. The key features of electrospun membranes and the methodologies employed for the characterization have been highlighted as well as their exploitation in separative applications.

3.2 Electrospinning Process

Usually, the precursor of the electrospinning process consists of a polymeric solution prepared by the dispersion of a polymer amount (or a polymers mixture) in a solvent (or a solvents blend) under vigorous stirring until the complete dissolution of the polymer. The polymer solution is then transferred to the syringe and fed by a pump to the needle at a certain flow rate. Simultaneously, the needle tip is also connected to a positive high-voltage power supply and fixed at a specific distance over the collector coupled to a negative high-voltage power supply or grounded. As a consequence, a Taylor cone is ejected from the needle tip as soon as the repulsive electrical forces prevail over the surface tension forces on the pending droplet, thanks to the voltage difference applied to the system (Fig. 3.1b) [2, 11, 13, 21, 22]. Subsequently, the solution jet is expelled from the deformed cone and the solvent evaporation leads to the deposition of solid fibers on the collector [2, 3, 13, 23–26].

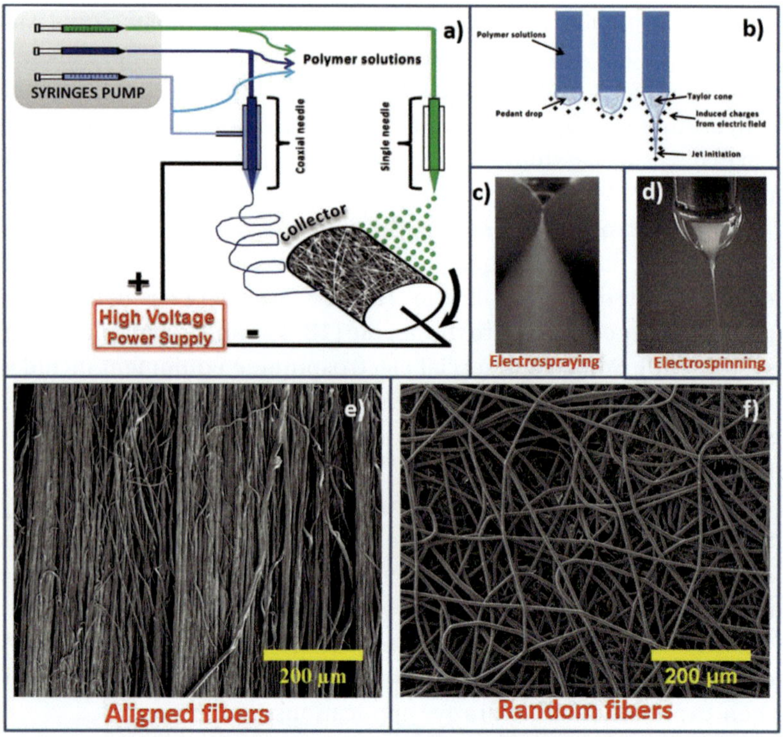

Figure 3.1 (a) Scheme of a vertical electrospinning setup with drum rotating collector, (b) Taylor cone formation, (c) electrospraying, (d) electrospinning with a coaxial needle to obtain a core–shell structure, (e) electrospun membrane with aligned fibers, and (f) electrospun membrane with random fibers distribution. Image courtesy of Dr. Javier Aragón Fernández, Institute for Nanoscience of Aragon, University of Zaragoza, Spain.

Usually, the morphology of electrospun membranes (flat ribbons, hollow, with wrinkled surfaces, porous fibers, variable diameters, presence of beads, etc.) is governed by the operative parameters of the electrospinning process [27–29]. In fact, the electrospinning process requires the fine control of many variables in order to stabilize the Taylor cone over the production time. Once the parameters are optimized by acting on the main components of the electrospinning setup, the preparation of electrospun membranes looks a simple and easy process. The key parameters are the needle-collector distance, the flow rate of polymer solutions, the humidity, and the temperature inside the equipment. Definitely, special attention must also be given to the minimum voltage difference able to guarantee the repulsive electrical forces required to overcome the surface tension forces on the pending droplet solution and to eject a polymer jet from the Taylor cone.

On the other hand, the physicochemical properties of the polymeric solutions play an important role in the electrospinning process. The molecular weight and concentration of the polymers affect the structures and the diameters of the fibers. Polymer properties do not solely define the solvent required for its solubilization, but they also modulate the intrinsic viscosity, the surface tension, and the conductivity of the polymeric solution, causing a significant morphological variation on the electrospun membranes [9]. Table 3.1 highlights the basic effects of the aforementioned parameters on the morphology of electrospun membranes.

Table 3.1 Effects and results of different parameters on electrospun morphology [1, 2, 9, 13, 24, 30–32]

Parameters	Linked with	Effect on Fiber Morphology
	Solution Parameters	
Polymer concentration	Viscosity, surface tension, solvent polarity	An increase in polymer concentration enhances the fiber diameter.
Molecular weight (Mw) of polymer	Conductivity, surface tension	The number of defects (beads and droplets) is minimized by high Mw.
Conductivity/ surface charge density	Solvent conductivity	The decrease in conductivity induces an increase in fiber diameter.
Viscosity	Polymer concentration, molecular weight, solvent	An optimal viscosity range is essential to produce fibers. Viscosity is strongly affected by the polymer concentration.
Surface tension	Polymer concentration, molecular weight	Instability of jets is observed at high surface.

(Contd.)

Table 3.1 (*Contd.*)

Parameters	Linked with	Effect on Fiber Morphology
	Processing Parameters	
Applied voltage	Conductivity	The augment of voltage causes a thinning of the fiber.
Feed rate/Flow rate	Polymer concentration	Thinner fibers obtained by reducing the flow rate. Particles can be obtained at very low flow rate.
Tip to collector distance	Solvent volatility	Optimal distance required to avoid the formation of beads. A minimum distance is necessary to ensure the solvent evaporation obtaining homogeneous fibers.
	Ambient Parameters	
Humidity	Solvent volatility	Maintained at value <40%. High humidity favors the formation of beads and small circular pores on the surface of the fiber.
Temperature	Boiling point of solvent	An increase in temperature causes a decrease in fiber diameter.

It is worth to be mentioned that rigorous protocols guarantee stability and reproducibility to the membrane fabrication process, but every electrospinning setup presents distinctive properties requiring consequently the identification of apposite operative parameters [30].

Finally, the process leads to a partial evaporation of the solvent during the flight from the needle to the collector, making the nascent electrospun membranes incompatible with the final application. Consequently, a drying post-treatment (in a desiccator or a dry oven with or without a vacuum) is necessary by controlling the rate of evaporation, the temperature (below the boiling point of the solvent), and the humidity to prevent the formation of defects or pores in the fibers [33].

3.3 Electrospinning Approaches for Electrospun Membrane Fabrication

Currently, many different electrospinning approaches have been extensively developed and successfully employed to produce electrospun membranes. In the last decade, process intensification and sustainability have driven the electrospinning process from proof of concept to industrial production, opening new horizons in membrane technologies.

3.3.1 Single Spinneret Electrospinning

The single spinneret configuration is the simplest setup, presenting just a spinneret/nozzle (single, coaxial, or triaxial needle) fed by a peristaltic pump. This configuration is very flexible and ensures to fabricate electrospun membranes and scaffolds with good reproducibility at lab scale [34]. For that reason, single spinneret electrospinning is commonly employed by scientific research groups involved in the development of innovative nanofibers and membrane for benchmark experiments [33].

3.3.2 Coaxial and Triaxial Electrospinning

Electrospun membranes formed by fibers with a core–shell structure have unpreceded opportunities, mainly for biomedical and pharmaceutical applications and separation processes. This structure is obtained using two or three concentric needles coaxially assembled in the same spinneret, enabling the encapsulation of active ingredients, the preparation of hollow fibers and polymer blending, as shown in Fig. 3.1d [31, 33, 35, 36].

In fact, coaxial and triaxial electrospinning is an excellent electrospinning approach for the polymers with poor solubility or electrospinnability. However, the identification of the appropriate electrospinning parameters to obtain functional electrospun membranes formed by fibers with a core–shell structure is tremendously problematic because of miscibility issues and the individual response of the polymers to the variables of the process [15, 19, 33, 37].

3.3.3 Nanofiber Alignment

The random orientation of formed fibers is an intrinsic feature of the electrospinning process, but it is possible to align the fibers using a specially designed collector (Fig. 3.2).

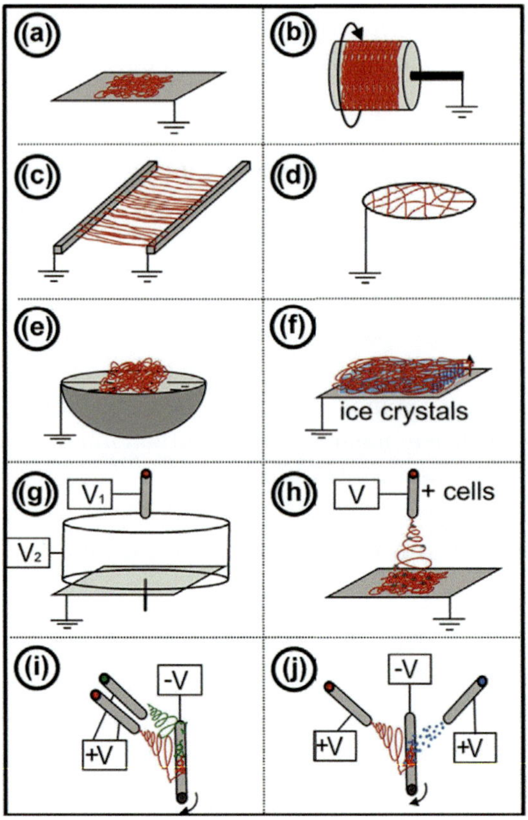

Figure 3.2 Examples of various collector types for electrospinning: (a) single collector; (b) rotating collector; (c) dual collectors; (d) ring collector; (e) cell electrospinning; (f) collecting onto ice crystals; (g) needle and dome collector; (h) simultaneous electrospinning and electrospraying onto a rotating collector; (i) direct writing; and (j) electrospinning onto another scaffold. Reprinted from Ref. [34] with permission of Elsevier.

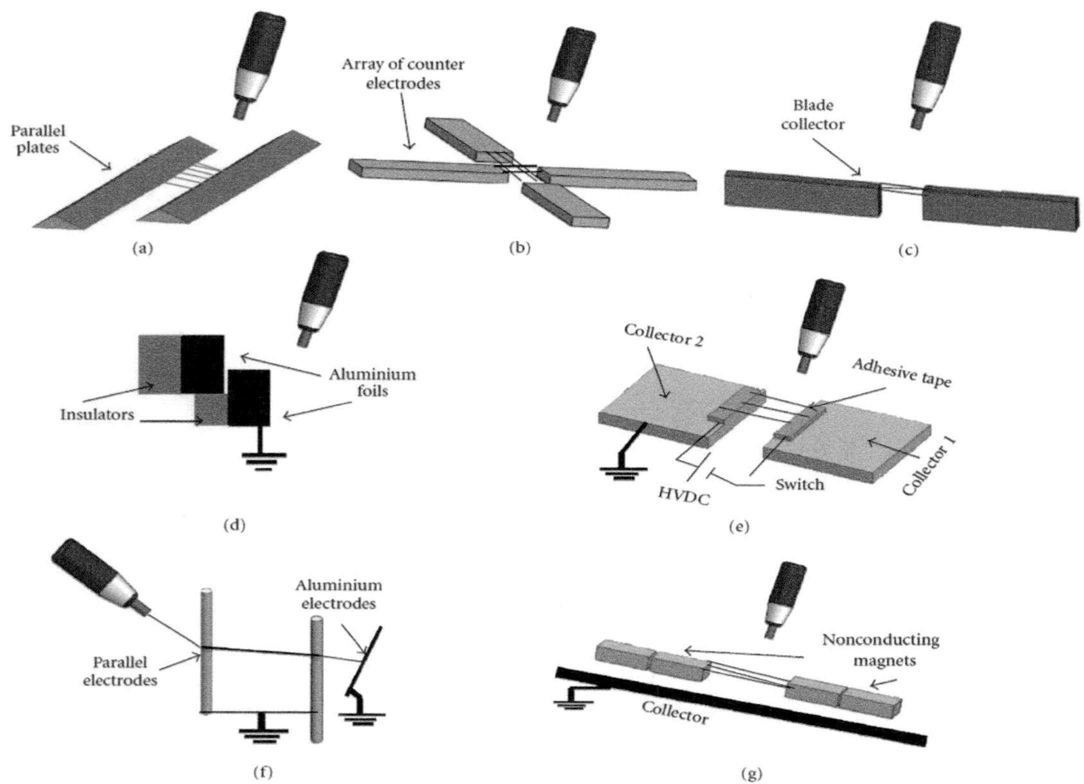

Figure 3.3 Different configurations of the electrode collectors to get aligned nanofibers by electrospinning technique [45]. Available under a Creative Commons Attribution License.

A rotating collector is the most common alternative to the flat configuration to get aligned fibers [38–42]. The rotation speed of the drum effectively controls the fiber diameter and the degree of alignment of the electrospun membrane. Concisely, the centrifugal force induced by high rotational speeds causes the alignment and the elongation of the fiber before the collection on the drum. However, if the rotational speed is too high, the fiber jet may collapse.

An alternative to the rotating collector is the employment of static parallel electrodes to collect the fibers able to ensure a high degree of alignment [43, 44]. This configuration setup is very simple: the employment of two metallic electrodes as collector generates a residual electrostatic repulsion resulting in a facilitated alignment of the fibers, as shown in Fig. 3.3 [45]. Canejo et al., with the help of a high-speed camera, showed that the fibers adhere first to one electrode and then are oriented toward the second electrode, thus achieving a high alignment of the fibers in the electrospun membranes [46, 47].

3.3.4 Multi-Spinneret Electrospinning

The electrospinning setup with a multi-spinneret nozzle system (Fig. 3.4) ensures the process intensification and the scale-up of an optimized electrospinning process [33, 48]. In fact, spinnerets placed side by side along one direction or arranged circumferentially have been used to obtain a large area of electrospun membranes (Fig. 3.4) [49].

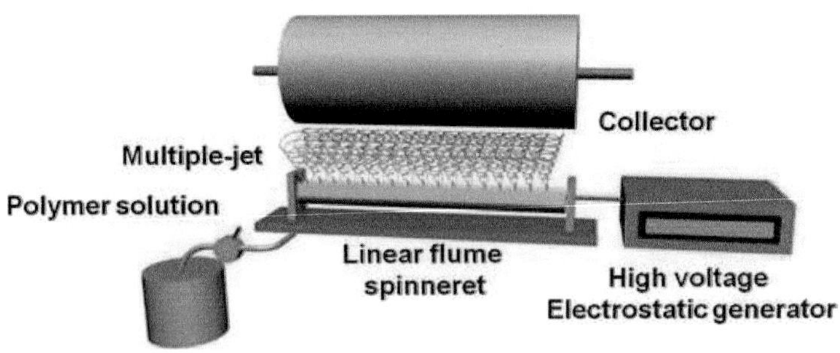

Figure 3.4 Multi-spinneret electrospinning via a linear flume spinneret [49]. Available under a Creative Common CC BY license.

Interestingly, a setup with two spinnerets (single or multi-jet), focused on a single rotating collector, enables combination of polymers [50], inorganic compounds [51], and cells in the same electrospun membrane [32]. Recently, the same electrospinning setup (single rotating collector and two spinnerets) has been used at the University of Zaragoza to develop membranes with different structures for different applications [18, 19]: an asymmetric membrane with different hydrophilicity of both faces loaded with essential oil for wound dressing applications and a scaffold for tissue engineering (i.e., bone tissue) where the polymeric fibers filled with inorganic nanoparticles were mixed with microparticles obtained by electrospraying with the help of the rotation collector, as shown in Fig. 3.5.

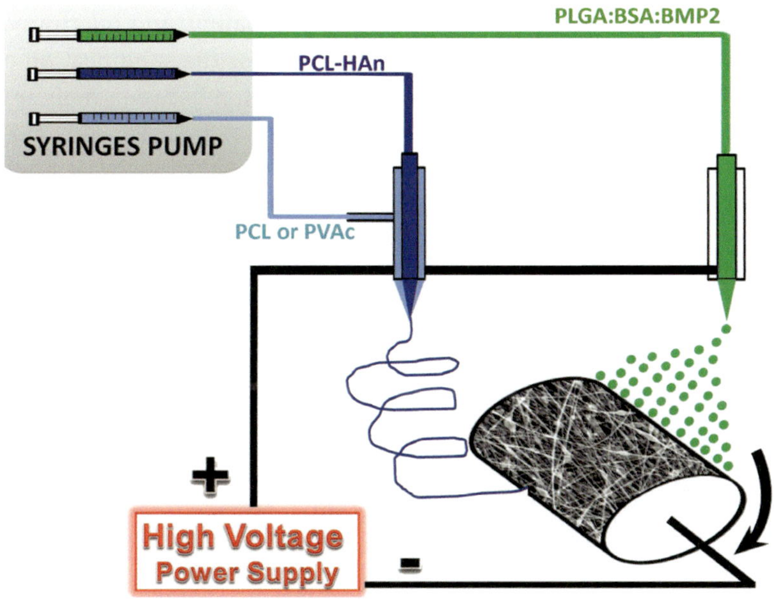

Figure 3.5 Scheme of working of a coaxial and single electrospinning for the fabrication of PCL-HAn or PCL-HAn/PVAc fibers decorated with PLGA-BMP2 particles. Reprinted from Ref. [19] with permission of Elsevier.

3.3.5 Melt Electrospinning

Organic solvents used in the electrospinning process compromise the economic and environmental sustainability of the process [52].

Besides, some common polymers such as polypropylene [33], polyethylene [31], and polytetrafluoroethylene [53] are not easy to be dissolved. Hence, the melt electrospinning process has been developed under the concept of the green production (eco-friendly process) and the process intensification, showing advantageous potentials to fabricate membranes for tissue engineering, wound dressings, textiles, and filtration [33].

Basically, the polymers are melt before the electrospinning process, presenting superior viscosities in comparison with conventional solvent-based solutions. For this reason, melt electrospinning leads to uniform fibers with a larger diameter (around 1 micron). The technical requirements for melt electrospinning are: a heat source needed to melt the polymer (electrical heating, circulating fluids, heated air, heating by a laser source, fire heating, and microwave heating) [54, 55] and a cooling system to favor the solidification of the nascent fibers (Fig. 3.6) [56]. A comparison between melt and solution electrospinning is shown in Table 3.2 [33, 52, 57–60].

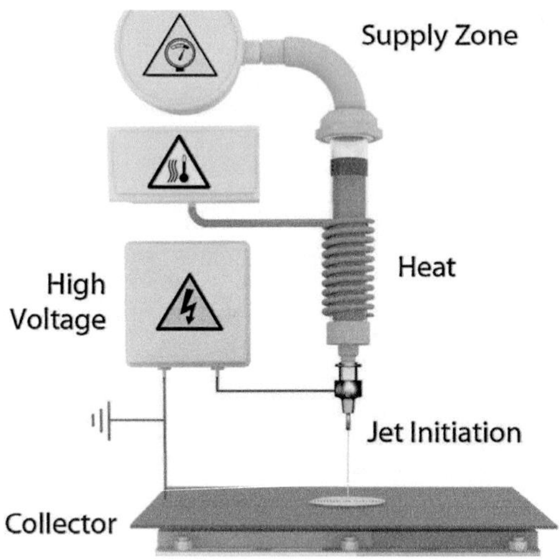

Figure 3.6 Schematic of a typical melt electrospinning system. Reprinted from Ref. [56] with permission of Elsevier.

Table 3.2 Comparison between solution and melt electrospinning

	Method	
	Solution electrospinning	**Melt electrospinning**
Fiber diameter	Dozens to thousands of nanometers.	≥500 nm, about 200 nm partially.
Main factors	Solution solubility, solution system properties, static, voltage, collection distance.	Melt viscosity, static voltage, ambient temperature.
Advantages	Inexpensive, simple device, easy to operate.	Eco-friendly, safety, biomedical applications.
Disadvantages	Environmental pollution, solvent recovery.	Complicated device, expensive, large diameter.

Note: Reprinted (adapted) from Ref. [33] with permission of Elsevier.

3.3.6 Three-Dimensional Electrospun Membrane

Several electrospinning approaches have been developed to produce electrospun 3D membranes for special applications such as scaffolds for tissue engineering, wound dressing, pollutants absorber, or even for solar cell applications as membrane electrodes (Fig. 3.7) [61]. The easiest way to produce 3D electrospun membranes is by piling electrospun membranes on top of each other or via a long-time electrospinning process. Indeed, adequate flow rates and polymer concentrations ensure the preparation of membranes with significant thicknesses after 30 min of spinning.

The layers can be made up of similar or different polymers, with different structures (hollow fibers, core–shell fibers, particles, etc.). Recently, the gas foaming technique has been combined with the electrospinning technique to reorganize the fibers formed through in situ hydrogen gas generation [33, 62].

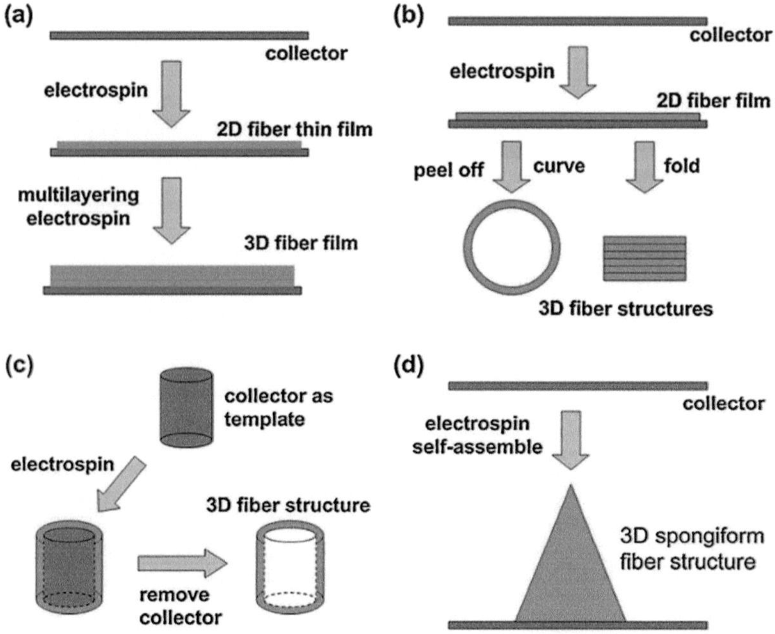

Figure 3.7 Schematic representation of the different approaches to produce 3D nanofiber structure. Reprinted from Ref. [61] with permission of Elsevier.

3.3.7 Inorganic and Mixed Matrix Nanofiber Membranes

Rarely the polymers present functional properties or highly selective interactions. Nevertheless, the electrospinning process guaranties the effortless functionalization of the polymeric fibers with various active molecules or nanoparticles [63]. In fact, nanomaterials are usually directly dispersed into the polymeric solution subsequently submitted to the electrospinning process obtaining nanocomposite nanofibers.

Remarkably, the electrospinning of polymeric solution containing nanomaterials with anisotropic morphology, such as carbon nanotubes, favors the alignment of NPs within the fibers [64]. In fact, the uniaxial stretch of the polymeric solution consists of a straightforward and elegant force for the alignment and immobilization of nanorods [65]. From a practical point of view,

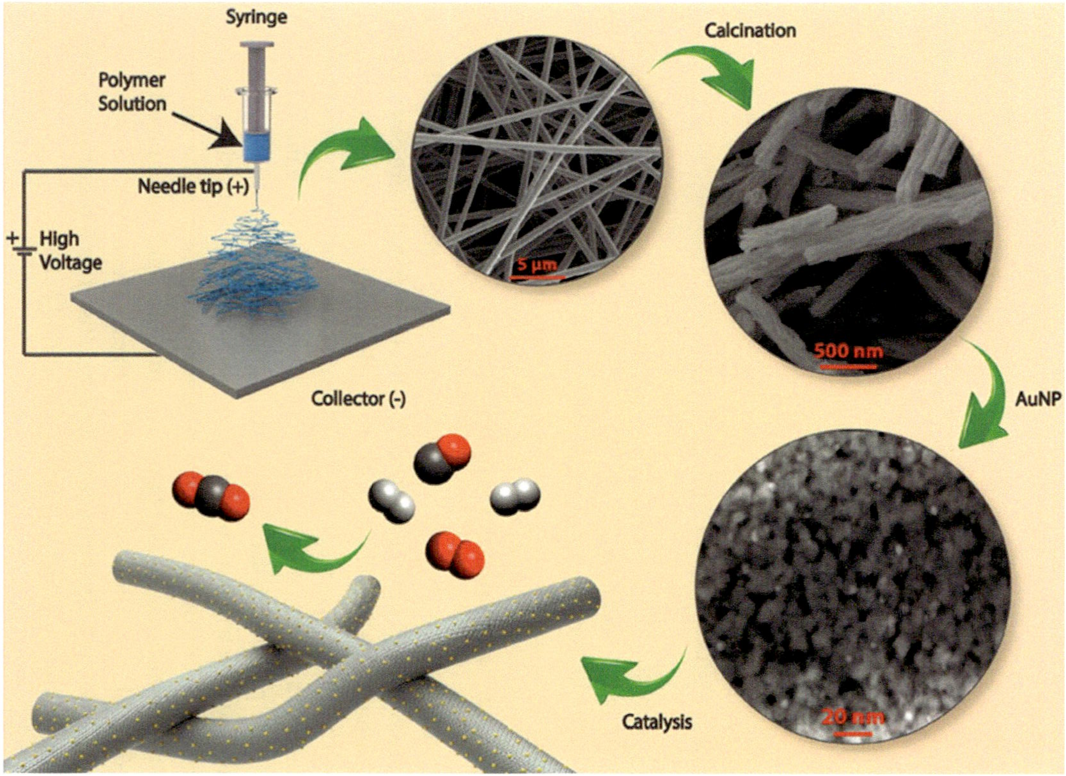

Figure 3.8 Production of ceramic nanofiber for catalysis application via electrospinning process and subsequent calcination. Reprinted from Ref. [63] with permission of Elsevier.

electrospinning is considered a self-assembly method able to stabilize the nanoparticles within the fibers or on their surfaces, overcoming the limitations of the conventional techniques characterized by weak interactions prone to environmental changes [64]. However, more than 100 nanofibers exclusively composed of ceramics in the shape of fibers, porous nanofibers, core–shell structure, nanoparticles, hollow fibers, side by side, etc. have been reported in the literature [33, 66].

In these cases, inorganic solutions (sol-gel) with different precursors are directly electrospun and then submitted to calcination. Unfortunately, the poor electrospinnability of inorganic solutions makes their large-scale application still allusive (Fig. 3.8) [33, 67].

3.3.8 Post-Treatment Techniques

Post-treatment of electrospun membrane is a crucial step to guarantee mechanical stability to the mat and to tailor the physiochemical properties of the fibers. Herein, the most relevant post-treatment methods of electrospun membranes are highlighted.

3.3.8.1 Heat-press treatment

Heat-press treatment is a common post-treatment consisting of exposing the electrospun materials to a certain pressure at high temperature. This method leads to a decrease in the pore size coupled with an improvement in the physical integrity of the electrospun membranes by promoting the junction of adjacent nanofibers [33]. These properties are highly desired for some potential applications of the electrospun membranes, such as gas diffusion, fuel cell, chemical recovery, pollutant adsorption, membrane distillation, forward osmosis, water treatment by micro, ultra, and nanofiltration [68–71]. Additionally, the heat-press treatment improves the sensitivity of electrospun membranes used as biosensing as well as the conductivity for fuel cell application [72].

3.3.8.2 High-energy irradiation (plasma, gamma ray, electron beam, and ultraviolet)

Among the various physical methods, high-energy irradiation has been considered a straightforward strategy able to modulate the behavior of the pores and to introduce functional groups onto the polymer. In general, CF_4 plasma enhances the hydrophobicity of the polymeric network, whereas the introduction of functional groups such as amines, carboxyls, hydroxyls, or anhydrides via atmospheric plasma increases the surface free energy, enhancing the hydrophilic character of the fibers as a consequence [73, 74].

Ultraviolet (UV) light, gamma ray, plasma, and electron beam irradiation were found to be beneficial for the immobilization of biomolecules, sensitivity, cell biocompatibility, hydrophilicity, thermal resistance, the capacity of retention, mechanical property, and antibacterial activity [75–77].

3.3.8.3 Chemical depositions

The hydrophilicity of electrospun membranes can be tailored by using chemical reactions on the surface of the fibers with the help of semicrystalline polymers and inorganic materials through a chemical deposition approach [78]. Usually, polymeric nanofibers rich in hyperbranched structures are chemically functionalized via grafting, exploiting the reaction of their functional groups with monomers [79].

3.3.8.4 Coating

The coating process is an easy modification route provoking dramatic changes on the surface of the electrospun membrane (ionic conductivity, hydrophilicity, bioactivity, etc.) [80, 81]. The coating process consists of immersing the electrospun membranes into a coating solution or using an interface to immobilize the functional materials on the top of electrospun membranes by adhesion [82–84].

3.4 Scale-Up for Industrial Production

The low production rate of electrospinning with a single needle (a few grams of electrospun membrane per hour of electrospinning process) dramatically hampered the large-scale production [42, 85, 86]. Currently, multi-spinneret systems that can eject multiple Taylor cones have been introduced to increase the production rate [87, 88]. MECC Co. Ltd. (Japan), Fluidnatek (Spain), 4SPIN (Czech Republic),

SPUR (Czech Republic), and Inovenso (Turkey) are some of the companies that currently commercialized these types of equipment.

Several traditional polymer-processing techniques have been combined with a multi-spinneret system to tackle the productivity issue [42, 89, 90], such as twin-screw extrusion, including polymer melts [91].

On the other hand, free liquid surface electrospinning has been developed for large-scale production of the electrospun membrane without the employment of nozzles, needles, or spinnerets, thanks to the employment of a rotating electrospinning head [83, 92–95]. This new electrospinning process is based on the fact that the Taylor cone and subsequent fiber jets can be created from a thin film of polymers solution, enabling also the employment of high-viscosity solutions, when a high electrical field is applied. This new configuration has been commercialized by Elmarco's Nanospider technology and Stellenbosch Nanofiber Company's ball electrospinning technology [45].

3.5 Features of Electrospun Membranes and Surface Modifications

Electrospun membranes offer a broad array of attractive tailored features, such as superior porosity (exceeding 90%, ca. 10–20% higher than conventional membranes), narrow pore size, and large surface-area-to-volume ratio (generally ranging from 10 m^2/g to 100 m^2/g, two orders of magnitude larger than flat films) [96–98]. Consequently, the morphology of the electrospun membranes made up of unaccountable nanofibers and interconnected pores maximizes the interactions of the membranes with the feed solutions, whereas their superior porosity minimizes the resistance to mass transport facilitating the diffusion of the permeant molecules [99].

Furthermore, the selectivity depends not solely on the physicochemical properties of the membranes, but also on their morphology. Thus, the separation can take advantage of the steric hindrance of the target molecules and the nanometric size of the nanofibers organized in hierarchical structures with some special size effects for the single-molecule recognition [100].

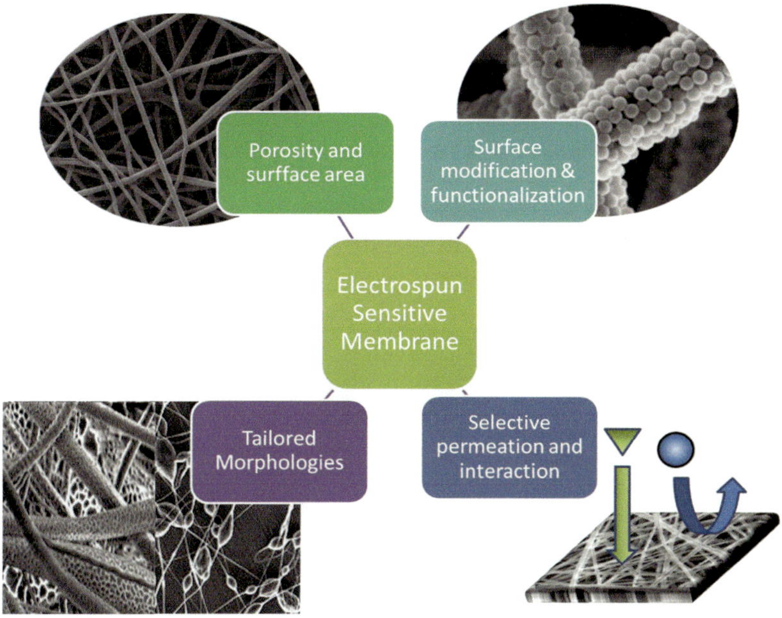

Figure 3.9 Key features of electrospun membranes.

Notably, the tremendous advantage of electrospun fibers concerning other one-dimensional sub-micrometric structures, such as nanorods or nanotubes, is their length and high axial strength, which can guarantee the production of self-standing flexible structures.

These fascinating properties have been exploited in different separative processes, such as air filtration [101], microfiltration [102], ultrafiltration [103], nanofiltration [104], forward osmosis [105], and membrane distillation [106]. Moreover, the hierarchical structures of electrospun membranes have attracted increasing attention for the fabrication of biomimetic materials employed in biomedical applications. Additionally, the synergic exploitation of the changes in the chemical–physical properties of the membranes induced by environmental conditions and of the selective interactions with permeant molecules have led to unprecedented opportunities for practical real-time and in situ sensing of chemical species and physical parameters [107, 108].

Overall, the advancements in the development of electrospun membranes have opened up exciting perspectives for actual surface engineering exploited in molecular recognition and selective permeation, whereas the outstanding specific surface area and porosity of electrospun nanomaterials ensure superior permeabilities and an enormous number of interactions of the target molecules or the environment with the nanofibers.

3.6 Electrospun Membrane Characterization

3.6.1 Morphological Characterization

The appropriate characterization of electrospun membranes at a micrometric level requires advanced techniques. The morphology of the membranes is usually observed with scanning electron microscopes (SEMs) producing images by examining the membrane with a high-energy beam of electrons able to penetrate the sample for a few microns generating different emissions from the surface (e.g., secondary and backscattered electrons, X-rays). The secondary electrons reveal the surface morphology through pictures of capital importance since they provide precious feedback in tailoring the morphology of the membrane, pursuing a proper electrospinning precursor, and optimizing the parameters of the process [109]. Moreover, SEM pictures qualitatively describe the fundamental properties of the membranes, such as surface pore size, pore density, and surface porosity. Additionally, X-rays are employed for the chemical characterization of the membranes useful for the estimation of the surface composition of nanofibers or the dispersion of nanofillers since they are generated by the inelastic collision of the exciting beam with electrons in orbitals of the atoms of the sample.

In the case of the transition electron microscopy (TEM), the highly energetic beam of electrons is transmitted through a sample, producing an image that details the hidden structure inside the nanofiber (e.g., embedded nanomaterials, core–shell structure).

Atomic force microscopy (AFM) is a vital characterization method for defining the surface texture of the membranes by quantifying its roughness, expressed as an arithmetic mean or root mean square roughness [110]. In fact, AFM reproduces the 3D topography of the electrospun mat by scanning a sharp tip with a typical end diameter of 5–10 nm over the membrane surface [110].

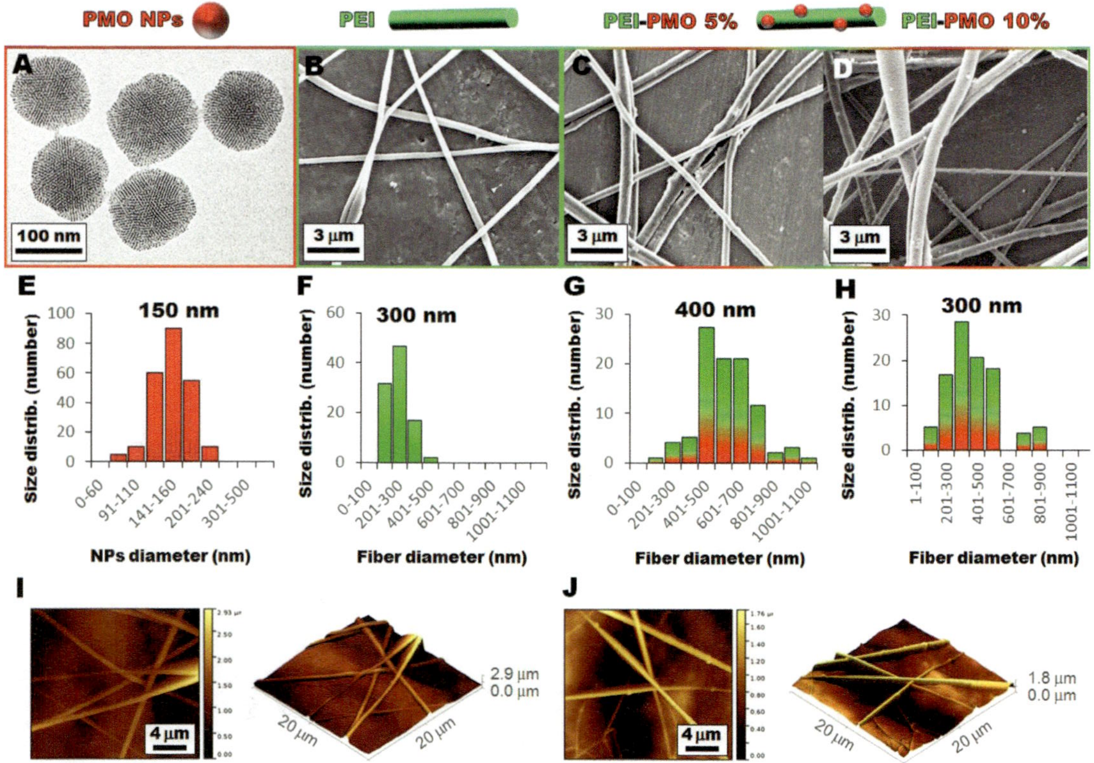

Figure 3.10 TEM pictures of PMO NPs (A). SEM images of PEI (B), and PEI–PMO nanofibers with 5% (C) and 10% (D) of PMO NPs. Statistical analyses of the diameter distributions of PMO NPs (E), PEI (F), PEI–PMO nanofibers with 5% loading (G), and PEI–PMO nanofibers with 10% loading (H) of PMO NPs. AFM pictures of PEI–PMO nanofibers with 5% (I) and 10% (J) of PMO NPs. Reprinted (adapted) from Ref. [111] with permission of American Chemical Society.

The synergic exploitation of these physical methods results in a complete picture of the morphology of electrospun materials. For example, SEM, TEM, and AFM observations were crucial for the development of hydrophobic poly(ether imide) (PEI) nanofibers doped with mesoporous organosilica (PMO) NPs [111], as shown in Fig. 3.10. A systematic study showed that the embodiment of the particles (confirmed by TEM) increased the diameter of the nanofibers (observed by SEM) and enhanced the roughness of the fiber (measured by AFM) [111]. Moreover, the morphological analyses evidenced that a high concentration of NPs (20 wt%) caused the formation of heterogeneous fibers [111]. The dispersion of the NPs into PEI fibers enhanced both the hydrophobic character and the roughness of the electrospun membrane, optimizing its performance in membrane distillation (flux of 31 Lm^2/h) coupled with antifouling properties (reduction of 70% of the fouling caused by the bacteria) [111].

3.6.2 Structural Properties

As already pointed out, high porosity and narrow pore size are the distinctive features of electrospun membranes. The porosity is defined as the ratio between the void space and the total volume of the membrane, and it is usually estimated using a gravimetric method consisting of weighting membrane in dry and wet conditions. Hence, the porosity (ε) is calculated following this formula [112]:

$$\varepsilon(\%) = \left\{ \frac{(W_{\mathrm{w}} - W_{\mathrm{d}}) / \rho_{\mathrm{i}}}{(W_{\mathrm{w}} - W_{\mathrm{d}}) / \rho_{\mathrm{i}} + \dfrac{W_{\mathrm{d}}}{\rho_{\mathrm{P}}}} \right\} \tag{3.1}$$

where W_{W} is the weight of the wet membrane, W_{d} is the weight of the dry membrane, ρ_{i} the density of the liquid wetting the membrane, and ρ_{P} is the density employed to prepare the membrane.

The measurements of pore size usually are carried out according to the wet/dry method [113], based on the measurements of the gas flow across a wet membrane (liquid filled) as a function of the applied gas pressure. The method is grounded on the Laplace equation giving the pressure (p) required to replace one fluid by another through a pore diameter (d_p) as:

$$p = \frac{4\gamma\cos\theta}{d_p} \tag{3.2}$$

where γ is the interfacial tension of the air–liquid interface and θ is the contact angle of the liquid with the solid matrix of the membrane [114]. At the bubble point, the applied pressure is enough to remove the liquid from the largest pore allowing gas permeation. Then the gas pressure is slightly increased, observing an improvement in the gas flow across the sample due to the removal of the residual liquid from smaller pores. Otherwise, a rough estimation of the mean pore size can be carried out knowing the fiber diameter (d) and the porosity (ε) according to the following formula [115]:

$$d_p = \frac{\sqrt{\pi}}{4}\left(\frac{\pi}{2\log\left(\frac{1}{\varepsilon}\right)} - 1\right)d \tag{3.3}$$

The surface area of porous materials, including electrospun membranes, is evaluated using the Brunauer, Emmett, and Teller (BET) technique, which can derive the surface areas from measurements of gas adsorption isotherms. In fact, the isotherm is described as:

$$\frac{p}{p_0} = \frac{1}{a_m C} + \frac{p_0(C-1)}{a_m C} \tag{3.4}$$

where p is the partial pressure of the gas, p_0 is the tension of saturated vapor at a certain temperature, a_m is the absorbed gas per mass of sample (mol/g), and C is a constant, including adsorption and condensation heat. From isotherm measurements, the specific surface (A) area is determined as:

$$A = a_m N_A \sigma \tag{3.5}$$

where N_A is the Avogadro constant and σ is the area of the sample covered by one molecule of the adsorbing gas.

In pursuit of high permeable membranes, high porosity guarantees superior mass transfer but compromises the solidity of the material. Indeed, the mechanical properties are crucial to preserve the shape and size of the pores, guaranteeing long-term stability. Based on the minimal thickness of the electrospun membrane, uniaxial tensile testing is the most simple and common method to evaluate their mechanical response. The mechanical characterization consists of gripping the opposite sides of the membranes within a load frame of an instrument able to constantly stretch the sample until its failure. The test provides a large amount of information describing the mechanical response of the membranes: The combination of the measured load (F), displacement (Δl), the original specimen cross-sectional area (S_0), and initial length (l_0) allow to calculate the engineering/nominal stress ($\sigma_{eng} = F/S_0$) and strain ($\varepsilon_{eng} = \Delta l/l_0$) [116]. The analysis of the stress–strain curve directly determines Young's modulus (E), yield stress (σ_y) and yield strain (ε_y), post-yield strain-softening and strain-hardening characteristics, stress at break (σ_R) and strain at break/elongation at break (ε_R), maximum stress (σ_B), and strain at maximum stress (ε_B), whereas the integral of the stress–strain curve determines the material toughness [116].

The structural properties are essential in membrane processes, especially in pressure-driven liquid filtration since a balance between the permeability of the membrane and mechanical stability is necessary. Unfortunately, the applications of electrospun membranes in pressure-driven processes are limited because of their poor mechanical resistance attributed to the minimal fiber diameters and the highly porous structure combined with weak bonding between the fibers. Besides the employment of fillers to enhance the mechanical properties of the fibers, an interesting approach is the exposure of the membrane to high temperature or vapor solvent promoting the bonding at junction points in the fiber [117]. For instance, a welding method was developed to enhance the mechanical strength of electrospun-nanofibers-covered hollow fiber membrane (N-HFM) by exposing the pristine membrane to vapor of N,N-dimethylacetamide (DMAC) [118]. The N-HFM was produced by electrospinning polyvinylidenefluoride-*co*-hexafluoropropylene (PVDF-HFP) on the surface of a commercial woven tube (average pore diameter of 0.17 mm, external diameter of 1.85 mm, inner diameter of 1.0 mm) [118]. Table 3.3 highlights the benefits of the post-treatment: improvement of 105% in the strain at break (ε_R) and 149% in Young's modulus (E) in the case of N-HFMs post-treated for 60 min (T3) compared with those of untreated ones (M0) [118].

This improvement is essentially due to solvent-induced fusion at the fiber junction.

On the other hand, post-treatment induced a partial collapse of the mat, causing a dramatic decline in the pore size and the porosity of the membranes [118]. The structural properties of the membranes affected their performance in membraned distillation since the decrease in the porosity limited the mass transport leading to a reduction in the flux from 15.1 L/m^{2}h observed for M0 to 13.2 L/m^{2}h in the case of T2 (post-treated for 40 min) [118]. Nevertheless, the poor stability of the M0 membrane limited its employment in membrane distillation processes since the long-term exposure to the feed compromised the integrity of the membrane, causing ultimately a dramatical drop in the salt rejection (*R*) from 99% to 69% (after 250 min of the test) [118]. Conversely, T2 preserved a salt rejection of 99% during experimental studies, thanks to the effective junction of the nanofibers [118].

Table 3.3 Effect of the DMAC post-treatment on the structural properties and the performance of N-HFM [30]

Membrane	Treatment (min)*	ε_R (%)	*E* (MPa)	d_p (μ)	ε (%)	*J* (L/m^{2}h)	*R*** (%)
M0	0	58	8.5	1.91	83	15.1	69
T1	20	90	9.1	1.87	81	14.8	78
T2	40	104	16.2	1.64	74	13.2	99
T3	60	119	21.1	1.01	42	x	x

*DMAC vapor; ** after 250 min of MD

This study proved the capital importance of structural characterizations and post-treatment of electrospun membranes useful to find a balance between the membrane structure requirements able to guarantee high permeation flux and the mechanical resistance ensuring long-term stability.

3.6.3 Hydrophilic/Hydrophobic Character

From a practical point of view, hydrophilicity is a key parameter in membrane processes since it governs the liquid flux rate and the fouling tendency. The sessile drop method is the standard procedure for optical measurements of the angle at the interface where water, air, and membrane surface meet, known as contact angle. Low contact angle implies the tendency of the water to spread on the surface (hydrophilicity), whereas high contact angle values indicate a hydrophobic character [119]. Thus, the contact angle (θ) depends on chemical–physical interactions between the membrane and the liquid that modulate the equilibrium of interfacial tensions as given by Young's equation:

$$\cos(\theta) = \frac{\gamma_{SG} - \gamma_{SL}}{\gamma_{LG}} \tag{3.6}$$

where γ_{LG}, γ_{SG}, and γ_{SL} represent the liquid–vapor, solid–vapor, and solid–liquid interfacial tensions, respectively [120]. Nevertheless, this equation is based on the hypothesis of flat, inert, homogeneous, insoluble, and smooth surface, whereas the measured contact angle (apparent contact angle, θ^*) is dramatically affected by the surface roughness of the membrane. A phenomenological model was proposed long ago by Wenzel [121], for understanding how the roughness (r) affects the apparent contact angle (θ^*) according to the following equation:

$$\cos(\theta^*) = r\cos(\theta) \tag{3.7}$$

where θ is Young's angle.

Hence, the surface topography of the membrane is intimately related to its wettability. In particular, the contact angle dramatically increases with the improvement in the surface roughness, a typical property of the electrospun membranes.

Fluoropolymers are widely employed in the preparation of hydrophobic membranes leading tendentially to the development of smooth surfaces. Typically, PVDF contact angle is ca. 85°C (Virgin membrane), easily improvable to a value of 120°C by controlling the phase inversion process (nanocomposite structure) [122]. Electrospun PVDF membranes guarantee superior surface roughness and contact angle higher than 130° (nanofibrous structure) consequently. As shown in Fig. 3.11, the embodiment of graphene nanosheets into PVDF nanofibers changed their morphology, enhanced the surface roughness, and improved the hydrophobicity, reaching water contact angle of about 160° (hierarchical structure) [122].

Indeed, the design of the surface topography represents an effective strategy to induce superhydrophobicity.

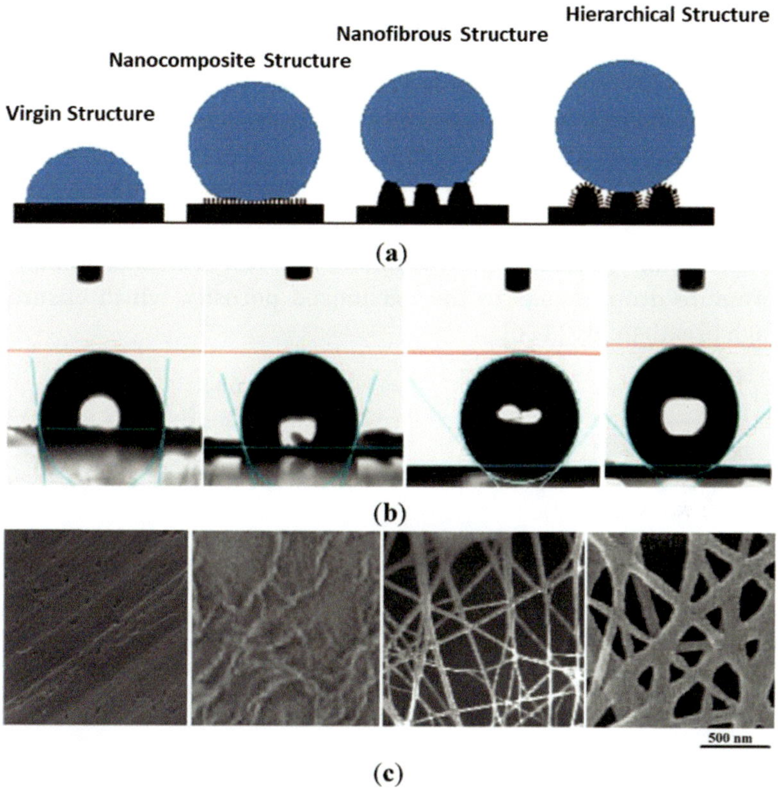

Figure 3.11 (a) Effect of the surface morphology (black colored) on the water contact angle (blue colored). (b) Resulting images (black circles) and angles (green lines) of water contact angle measurements. (c) SEM micrographs of the membranes [122]. Available under a Creative Commons Attribution (CC BY 4.0) license.

3.7 Applications of Electrospun Membrane

3.7.1 Electrospun Membrane in Air and Water Filtration

Nowadays, industries continuously release huge amounts of dust and particles into the atmosphere, extremely dangerous for human health if their size is below 2.5 μm ($PM_{2.5}$) [123]. Moreover, a large number of natural contaminants (e.g., pollen, bacteria, and some microorganisms and virus-carrying aerosol) are responsible for allergies or respiratory infectious diseases [124].

Conventional fiber filters usually suffer from the disadvantages of large pore size and low porosity due to their microsized diameters, whereas electrospun nanofibers integrate small and tortuous pores with highly adjustable porosity. For this reason, nanofiber membranes have gained more attention as high-efficiency particulate air (HEPA) filters characterized by outstanding filtering properties without the sacrifice of air permeability [125]. A tremendous advantage of the employment of electrospun nanofibers as HEPE is their minimal pressure drop, thanks to the maximized porosity, which ensures high breathability [126].

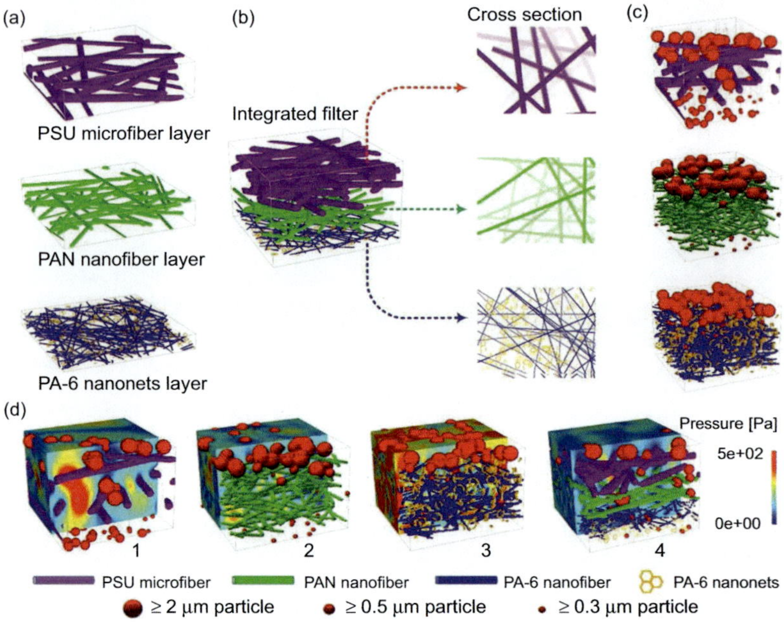

Figure 3.12 (a) Morphologies of PSU, PAN, and PA-6 layers. (b) Morphology of PSU/PAN/PA-6 integrated filter. (c) Mechanism of action of PSU, PAN, and PA-6 layers. (d) Air flow through the different PSU, PAN, and PA-6 layers and PSU/PAN/PA-6 integrated filter. Reprinted (adapted) from Ref. [127] with permission of American Chemical Society.

Moreover, the flexibility of the electrospinning process enables not only fine control of the morphology of the mat, but also production of multilevel composite structures. For instance, PSU/PAN/PA-6 hybrid fibrous membranes (Fig. 3.12) were prepared via sequential electrospinning process, showing surprising performance in capturing airborne particles [127].

Each layer presents a characteristic cut-off due to changes in the diameter of the fibers (1 μm for PSU, 200 nm for PAN, and 20 nm for PA-6) leading to a gradual decrease in the pore size (2.2 μm for PSU, 0.6 nm for PAN, and 0.27 nm for PA-6) [127]. This strategy allowed to efficiently remove (>99.99%) sub-micrometric particles (300 nm) at the low pressure drop (ca. 120 Pa), offering a fascinating strategy to fabricate filters for various applications such as individual and environmental protection, and industrial security to environmental governance [127].

Nevertheless, electrospun membranes have been widely employed in the removal of pollutants (from micron to sub-micron size) from liquid solutions via pressure-driven or sieving-based filtration processes. The attention gained by electrospun membranes is related to the opportunity to develop membranes with high flux and tailored cut-off acting on the diameter of the fibers: A general rule is that the mean pore size is about 3 ± 1 times the mean fiber diameter [128, 129].

In general, water-borne microorganisms are larger than 0.2 μm [130], so efforts have been focused on improving the water permeability of the membrane combined with high rejection fractions of particles of 0.2 μm from water. For instance, in the case of PAN nanofibers, a decrease in the average fiber diameter from 450 nm to 100 nm causes a drop in the mean pore size from 1.8 ± 0.2 μm to 0.22 ± 0.01 μm [131]. The electrospun PAN microfiltration membrane with a pore size of 0.22 μm performed significantly better in terms of flux (2–3 times) over the commercial membrane (Millipore GSWP 0.22 μm) maintaining similar rejection [131].

PAN nanofibers consist of an interesting platform since the active nitrile group in the polymer is available for further chemical modifications widely explored to overcome the biofouling, the bottleneck of hydrophilic membranes employed in MF. For example, nitrile groups on the PAN were functionalized with polyhexamethylene guanidine hydrochloride (PHGH) to provide antibacterial properties [132]. Similar results were observed by immersing electrospun PAN membrane in an aqueous solution of 1-(1-vinylimidazolium)ethyl-3-vinylimidazolium dibromide (VEVIMIBr) or ethoxylated trimethyl propane triacrylate (EOTMTA) and a thermal initiator. The composite VEVIMIBr/PAN with a mean pore size of 0.21 μm was considered an excellent candidate for the removal of bacteria by size exclusion and viruses by adsorption with a retention capacity of 99.99% [133].

A different strategy is the decoration of PAN with Ag nanoparticles reached by treating the electrospun membrane in a hydroxylamine (NH_2OH) aqueous solution to form $-C(NH_2)N-OH$ groups used for the functionalization of the membrane with Ag^+ ions converted to nanoparticles [134]. The electrospun Ag-coated PAN nanofibers showed a 100% reduction in biofouling with an effective filtration rate of 8.0 $mLcm^2/min$ and a rejection of 60% of latex beads of 50 nm [134].

Generally, electrospun membranes emerged as filtration media because they guarantee interconnectivity and uniformity of the pores. Moreover, the superior surface area of nanofibers favors the functionalization of the fibers conferring antifouling properties. Electrospinning has remarkably influenced the investigation of water treatment and air filtering, with a growing potential for industrialization and new avenues open for research [135].

3.7.2 Electrospun Membrane in Desalination

Thin-film composite (TFC) membranes specifically designed for reverse osmosis (RO) consist of three layers: (1) an ultrathin dense polyamide (PA) material as the top hydrophilic selective layer; (2) an interlayer with pores 5–50 nm in size; and (3) a nonwoven substrate ensuring mechanical resistance.

Based on their minimal resistance to mass transport, electrospun membranes have been considered ideal candidates for interlayers aimed to support the active coatings. Rarely, electrospun layers have been explored as a selective coating due to the fact that the surface roughness compromises the hydrophilicity of the surface and enhances the fouling tendency [136]. For example, the contact angle (CA) of polysulfone (PSf) ENMs is ca. 122°, whereas that of PSf flat membrane is ca 82° [137].

On the other hand, hydrophobicity is the fundamental prerequisite in membraned distillation (MD), whereas high porosities reduce the heat conduction through the membrane ensuring high driving force. Unfortunately, electrospun membranes are limited by instability and short shelf-life due to their poor mechanical strength and wetting favored by their large pore size. These issues are generally overcome by heat pressing post-treatment consisting of exposing the membrane under certain pressures and temperatures [127].

PVDF has been widely employed in the fabrication of MD membranes because of its hydrophobic character and solubility in a wide range of solvent, ensuring the preparation of membranes via phase inversion and electrospinning. A post-treatment protocol based on hot pressing at 170°C allowed melting the nanofibers together, enhancing their tensile strength and tensile modulus and ensuring performance stability in the MD process [138, 139]. However, the post-treatment sacrifices the porosity of the membranes (decreasing the thickness from 71 μm to 42 μm) and the super-hydrophobic character by smoothing the membrane surface (reduction in the contact angle from 141° to 120°C) [138, 139]. Despite these side effects, PVDF electrospun membranes guaranteed outstanding water permeation flux (15–21 kg/m^2h) with stable performance in log-term experiments and a 100% rejection of dissolved solutes [138, 139].

Furthermore, the flexibility of the electrospinning process ensured the development of membrane with advanced functionalities, thanks to the employment of the nanotechnologies. In fact, clay nanocomposite and fluorosilane-coated TiO_2 were embedded in PVDF nanofibers to improve the membrane hydrophobicity [79]. In a recent pioneering study, PVDF has been electrospun with (1,10-phenanthroline)-ruthenium(II) ($Ru(phen)_3^{2+}$), a fluorescent molecular probe sensitive toward the temperature [139, 140]. The molecular probe has been employed for the online and in situ monitoring of the MD process by mapping the temperature on the membrane's surface [139, 140]. This strategy has allowed to study the temperature polarization in the membrane module, which is considered a serious limitation of the MD process hindering its industrialization [139, 140]. A possible solution to the polarization phenomenon consists in the employment of photothermal materials, able to convert the absorbed light to heat, offering the opportunity to locally harvest the heat to the surface of the membrane, reversing the temperature polarization and improving the energy efficiency of the MD process ultimately [141, 142]. Recently, photothermal nanoparticles (AgNPs) have been embedded in PVDF nanofibers conferring to the membrane self-heating property upon light exposure [143]. This photothermal behavior was efficiently exploited to promote the evaporation of the water on the membrane surface, thus generating an effective driving force for the MD process [143].

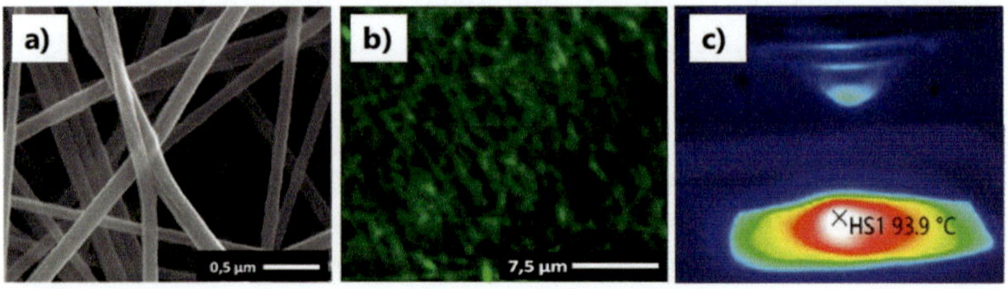

Figure 3.13 (a) Melt PVDF fibers because of hot pressing. Reprinted from Ref. [139] with permission of Elsevier; (b) PVDF electrospun fluorescent membrane doped with Ru(phen)$_3$. Reprinted from Ref. [139] with permission of Elsevier; (c) Photothermal electrospun membrane load with AgNPs exposed to UV light. Reprinted from Ref. [143] with permission of American Chemical Society.

Overall, the bottleneck for scaling the production of electrospun nanofibers in membrane distillation is their inclination to wetting due to their large pore size [135]. Impressive signs of progress have been achieved, thanks to the synergic exploitation of nanotechnology and chemical functionalization of the fibers, paving the way for the exploitation of electrospun membranes in desalination.

3.8 Conclusion

Electrospinning has invigorated the attention of the membrane community and opened an unprecedented horizon in the development of porous membrane with outstanding porosity and surface area. Technological advancements combined with a deep knowledge of the process have guaranteed the fine control of the properties of the fibers.

The synergic exploitation of nanofibers with a large range of nanotechnology dramatically enhanced the basic performance of the membrane and/or conferred to electrospun membranes functional properties such as antifouling, sensitivity, and self-heating. Strategies of post-treatment consist of efficient and effortless alternative routes for the assessment of the nanofibers. Nevertheless, a rational characterization is vital at every stage of the development of electrospun membranes from the design in the laboratory to the quality control in industrial applications.

Nowadays, electrospun membranes have emerged in several applications, with unique features that foresee unprecedented perspectives in membrane technologies.

References

1. Bhattacharjee, P. K. and Rutledge, G. C. (2011) Electrospinning and polymer nanofibers: Process fundamentals, in *Comprehensive Biomaterials* (eds. Ducheyne, P., Healy, K. E., Hutmacher, D. W., Grainger, D. W., and Kirkpatrick, C. J.), Elsevier, pp. 497–512.
2. Bhardwaj, N. and Kundu, S. C. (2010) Electrospinning: A fascinating fiber fabrication technique. *Biotechnol. Adv.*, **28**(3), 325–347.
3. Kijeńska, E. and Swieszkowski, W. (2017) General requirements of electrospun materials for tissue engineering: Setups and strategy for successful electrospinning in laboratory and industry, in *Electrospun Materials for Tissue Engineering and Biomedical Applications: Research, Design and Commercialization* (eds. Uyar, T. and Kny, E.), Woodhead Publishing, pp. 43–56.
4. Liu, Y., Hao, M., Chen, Z., Liu, L., Liu, Y., Yang, W., and Ramakrishna, S. (2020) A review on recent advances in application of electrospun nanofiber materials as biosensors. *Curr. Opin. Biomed. Eng.*, **13**, 174–189.
5. Sill, T. J. and von Recum, H. A. (2008) Electrospinning: Applications in drug delivery and tissue engineering. *Biomaterials*, **29**(13), 1989–2006.
6. Liu, Q., Chen, Z., Pei, X., Guo, C., Teng, K., Hu, Y., Xu, Z., and Qian, X. (2020) Review: applications, effects and the prospects for electrospun nanofibrous mats in membrane separation. *J. Mater. Sci.*, **55**(3), 893–924.
7. Ramakrishna, S., Fujihara, K., Teo, W.-E., Yong, T., Ma, Z., and Ramaseshan, R. (2006) Electrospun nanofibers: Solving global issues. *Mater. Today*, **9**(3), 40–50.

8. Jain, R., Shetty, S., and Yadav, K. S. (2020) Unfolding the electrospinning potential of biopolymers for preparation of nanofibers. *J. Drug Deliv. Sci. Technol.*, **57**, 101604.

9. Peijs, T. (2018) Electrospun polymer nanofibers and their composites, in *Comprehensive Composite Materials II*, vol. 6, Elsevier, pp. 162–200.

10. Taylor, G. I. (1964) Disintegration of water drops in an electric field. *Proc. R. Soc. London. Ser. A. Math. Phys. Sci.*, **280**(1382), 383–397.

11. Taylor, G. I. (1969) Electrically driven jets. *Proc. R. Soc. London. A. Math. Phys. Eng. Sci.*, **313**(1515), 453–475.

12. Reneker, D. H. and Chun, I. (1996) Nanometre diameter fibres of polymer, produced by electrospinning. *Nanotechnology*, **7**(3), 216–223.

13. Doshi, J. and Reneker, D. H. (1995) Electrospinning process and applications of electrospun fibers. *J. Electrostat.*, **35**(2–3), 151–160.

14. Miguel, S. P., Figueira, D. R., Simões, D., Ribeiro, M. P., Coutinho, P., Ferreira, P., and Correia, I. J. (2018) Electrospun polymeric nanofibres as wound dressings: A review. *Colloid. Surfaces B*, **169**, 60–71.

15. Aragon, J., Navascues, N., Mendoza, G., and Irusta, S. (2017) Laser-treated electrospun fibers loaded with nano-hydroxyapatite for bone tissue engineering. *Int. J. Pharm.*, **525**(1), 112–122.

16. Yan, J., Kang, W., and Cheng, B. (2020) Guiding parameters for electrospinning process, in *Advanced Nanofibrous Materials Manufacture Technology Based on Electrospinning*, CRC Press, pp. 82–111.

17. Ge, J., Zong, D., Jin, Q., Yu, J., and Ding, B. (2018) Biomimetic and superwettable nanofibrous skins for highly efficient separation of oil-in-water emulsions. *Adv. Funct. Mater.*, **28**(10), 1705051.

18. Aragón, J., Costa, C., Coelhoso, I., Mendoza, G., Aguiar-Ricardo, A., and Irusta, S. (2019) Electrospun asymmetric membranes for wound dressing applications. *Mater. Sci. Eng. C*, **103**, 109822.

19. Aragón, J., Salerno, S., De Bartolo, L., Irusta, S., and Mendoza, G. (2018) Polymeric electrospun scaffolds for bone morphogenetic protein 2 delivery in bone tissue engineering. *J. Colloid Interface Sci.*, **531**, 126–137.

20. Qayum, A., Wei, J., Li, Q., Chen, D., Jiao, X., and Xia, Y. (2019) Efficient decontamination of multi-component wastewater by hydrophilic electrospun PAN/AgBr/Ag fibrous membrane. *Chem. Eng. J.*, **361**, 1255–1263.

21. Formhals, A. (1934) Process and apparatus for preparing artificial threads. US1975504, issued 1934.

22. Zeleny, J. (1914) The Electrical Discharge from Liquid Points, and a Hydrostatic Method of Measuring the Electric Intensity at Their Surfaces. *Phys. Rev.*, **3** (2), 69–91.

23. Stocco, T. D., Bassous, N. J., Zhao, S., Granato, A. E. C., Webster, T. J., and Lobo, A. O. (2018) Nanofibrous scaffolds for biomedical applications. *Nanoscale*, **10**(26), 12228–12255.

24. Ramakrishna, S., Fujihara, K., Teo, W.-E., Lim, T.-C., and Ma, Z. (2005) *An Introduction to Electrospinning and Nanofibers*, World Scientific Publishing Co. Pte. Ltd.

25. Bosworth, L. A. and Downes, S. (2011) *Electrospinning for Tissue Regeneration*, Woodhead Pub.

26. Min Ju, Y., Lee, J., Jin Lee, S., and Xu, W. (2012) Tissue engineering scaffolding using electrospinning, in *Electrospinning for Advanced Biomedical Applications and Therapies* (eds. Neves, N. M.), Smithers Rapra Technology, pp. 89–134.

27. Casper, C. L., Stephens, J. S., Tassi, N. G., Chase, D. B., and Rabolt, J. F. (2004) Controlling surface morphology of electrospun polystyrene fibers: Effect of humidity and molecular weight in the electrospinning process. *Macromolecules*, **37**(2), 573–578.

28. Pai, C. L., Boyce, M. C., and Rutledge, G. C. (2009) Morphology of porous and wrinkled fibers of polystyrene electrospun from dimethylformamide. *Macromolecules*, **42**(6), 2102–2114.

29. Aragón, J., Feoli, S., Irusta, S., and Mendoza, G. (2019) Composite scaffold obtained by electro-hydrodynamic technique for infection prevention and treatment in bone repair. *Int. J. Pharm.*, **557**, 162–169.

30. Soares, R. M. D., Siqueira, N. M., Prabhakaram, M. P., and Ramakrishna, S. (2018) Electrospinning and electrospray of bio-based and natural polymers for biomaterials development. *Mater. Sci. Eng. C*, **92**, 969–982.

31. Bagheri, H., Rezvani, O., Zeinali, S., Asgari, S., Aqda, T. G., and Manshaei, F. (2019) Electrospun nanofibers, in *Solid-Phase Extraction (Handbooks in Separation Science)* (eds. Poole, C. F.), Elsevier, pp. 311–339.

32. Lühmann, T. C., Meinel, L., Groll, J., and Dalton, P. D. (2017) 4.29 Electrospun Fibers for Drug Delivery, in *Comprehensive Biomaterials II*, vol. 4, Elsevier Ltd., pp. 527–548.

33. Tijing, L. D., Woo, Y. C., Yao, M., Ren, J., and Shon, H. K. (2017) Electrospinning for membrane fabrication: Strategies and applications,

in *Comprehensive Membrane Science and Engineering*, vol. 1, Elsevier Ltd., pp. 418–444.

34. Hutmacher, D. W., Woodfield, T. B. F., and Dalton, P. D. (2014) Scaffold design and fabrication, in *Tissue Engineering* (Second Edition) (eds. Blitterswijk, C. A. Van and Boer], J. [De), Academic Press, Oxford, pp. 311–346.
35. Sun, Z., Zussman, E., Yarin, A. L., Wendorff, J. H., and Greiner, A. (2003) Compound core-shell polymer nanofibers by co-electrospinning. *Adv. Mater.*, **15**(22), 1929–1932.
36. Li, D. and Xia, Y. (2004) Direct fabrication of composite and ceramic hollow nanofibers by electrospinning. *Nano Lett.*, **4**(5), 933–938.
37. Qu, H., Wei, S., and Guo, Z. (2013) Coaxial electrospun nanostructures and their applications. *J. Mater. Chem. A*, **1**(38), 11513–11528.
38. Teo, W. E. and Ramakrishna, S. (2006) A review on electrospinning design and nanofibre assemblies. *Nanotechnology*, **17**(14), R89–R106.
39. Deitzel, J., Kleinmeyer, J., Harris, D., and Beck Tan, N. (2001) The effect of processing variables on the morphology of electrospun nanofibers and textiles. *Polymer (Guildf).*, **42**(1), 261–272.
40. Baji, A., Mai, Y. W., Wong, S. C., Abtahi, M., and Chen, P. (2010) Electrospinning of polymer nanofibers: Effects on oriented morphology, structures and tensile properties. *Compos. Sci. Technol.*, **70**(5), 703–718.
41. Baji, A., Mai, Y. W., Wong, S. C., Abtahi, M., and Du, X. (2010) Mechanical behavior of self-assembled carbon nanotube reinforced nylon 6,6 fibers. *Compos. Sci. Technol.*, **70**(9), 1401–1409.
42. Persano, L., Camposeo, A., Tekmen, C., and Pisignano, D. (2013) Industrial upscaling of electrospinning and applications of polymer nanofibers: A review. *Macromol. Mater. Eng.*, **298**(5), 504–520.
43. Li, D., Wang, Y., and Xia, Y. (2004) Electrospinning nanofibers as uniaxially aligned arrays and layer-by-layer stacked films. *Adv. Mater.*, **16**(4), 361–366.
44. Li, D., Wang, Y., and Xia, Y. (2003) Electrospinning of polymeric and ceramic nanofibers as uniaxially aligned arrays. *Nano Lett.*, **3**(8), 1167–1171.
45. Sahay, R., Thavasi, V., and Ramakrishna, S. (2011) Design modifications in electrospinning setup for advanced applications. *J. Nanomater.*, **2011**, 317673.
46. Canejo, J. P. and Godinho, M. H. (2013) Cellulose perversions. *Materials (Basel).*, **6**(4), 1377–1390.

47. Long, Y.-Z., Yan, X., Wang, X.-X., Zhang, J., and Yu, M. (2019) Electrospinning: The setup and procedure, in *Electrospinning: Nanofabrication and Applications* (eds. Ding, B., Wang, X., and Yu, J.), William Andrew Publishing, pp. 21–52.

48. Dosunmu, O. O., Chase, G. G., Kataphinan, W., and Reneker, D. H. (2006) Electrospinning of polymer nanofibres from multiple jets on a porous tubular surface. *Nanotechnology*, **17**(4), 1123–1127.

49. Wei, L., Liu, C., Mao, X., Dong, J., Fan, W., Zhi, C., Qin, X., and Sun, R. (2019) Multiple-jet needleless electrospinning approach via a linear flume spinneret. *Polymers (Basel).*, **11**(12), 2052.

50. Zhan, N., Li, Y., Zhang, C., Song, Y., Wang, H., Sun, L., Yang, Q., and Hong, X. (2010) A novel multinozzle electrospinning process for preparing superhydrophobic PS films with controllable bead-on-string/microfiber morphology. *J. Colloid Interface Sci.*, **345**(2), 491–495.

51. Tijing, L. D., Ruelo, M. T. G., Amarjargal, A., Pant, H. R., Park, C. H., and Kim, C. S. (2012) One-step fabrication of antibacterial (silver nanoparticles/poly(ethylene oxide)): Polyurethane bicomponent hybrid nanofibrous mat by dual-spinneret electrospinning. *Mater. Chem. Phys.*, **134**(2–3), 557–561.

52. Van Kampen, K. A., Scheuring, R. G., Terpstra, M. L., Levato, R., Groll, J., Malda, J., Mota, C., and Moroni, L. (2019) Biofabrication: From additive manufacturing to bioprinting, in *Encyclopedia of Tissue Engineering and Regenerative Medicine*, vol. 1–3, Academic Press, pp. 41–55.

53. Zhang, L. H., Duan, X. P., Yan, X., Yu, M., Ning, X., Zhao, Y., and Long, Y. Z. (2016) Recent advances in melt electrospinning. *RSC Adv.*, **6**(58), 53400–53414.

54. Yan, X., Yu, M., Zhang, L. H., Jia, X. S., Li, J. T., Duan, X. P., Qin, C. C., Dong, R. H., and Long, Y. Z. (2016) A portable electrospinning apparatus based on a small solar cell and a hand generator: Design, performance and application. *Nanoscale*, **8**(1), 209–213.

55. Li, H., Chen, Y., Zhang, X., Zhao, Y., Tao, Y., Li, C., and He, X. (2016) Experimental study on triboelectrostatic beneficiation of wet fly ash using microwave heating. *Physicochem. Probl. Miner. Process.*, **52**(1), 328–341.

56. Brown, T. D., Dalton, P. D., and Hutmacher, D. W. (2016) Melt electrospinning today: An opportune time for an emerging polymer process. *Prog. Polym. Sci.*, **56**, 116–166.

57. Dalton, P. D., Grafahrend, D., Klinkhammer, K., Klee, D., and Möller, M. (2007) Electrospinning of polymer melts: Phenomenological observations. *Polymer (Guildf).*, **48**(23), 6823–6833.

58. Dalton, P. D., Klinkhammer, K., Salber, J., Klee, D., and Möller, M. (2006) Direct in vitro electrospinning with polymer melts. *Biomacromolecules*, **7**(3), 686–690.

59. Zhou, H., Green, T. B., and Joo, Y. L. (2006) The thermal effects on electrospinning of polylactic acid melts. *Polymer (Guildf).*, **47**(21), 7497–7505.

60. Lee, S., and Obendorf, S. K. (2006) Developing protective textile materials as barriers to liquid penetration using melt-electrospinning. *J. Appl. Polym. Sci.*, **102**(4), 3430–3437.

61. Sun, B., Long, Y. Z., Zhang, H. D., Li, M. M., Duvail, J. L., Jiang, X. Y., and Yin, H. L. (2014) Advances in three-dimensional nanofibrous macrostructures via electrospinning. *Prog. Polym. Sci.*, **39**(5), 862–890.

62. Joshi, M. K., Pant, H. R., Tiwari, A. P., Kim, H. J., Park, C. H., and Kim, C. S. (2015) Multi-layered macroporous three-dimensional nanofibrous scaffold via a novel gas foaming technique. *Chem. Eng. J.*, **275**, 79–88.

63. Moreno, I., Navascues, N., Irusta, S., and Santamaria, J. (2015) Electrospun Au/CeO_2 nanofibers: A highly accessible low-pressure drop catalyst for preferential CO oxidation. *J. Catal.*, **329**, 479–489.

64. Zhang, C.-L. and Yu, S.-H. (2014) Nanoparticles meet electrospinning: Recent advances and future prospects. *Chem. Soc. Rev.*, **43**(13), 4423–4448.

65. King, S. G., McCafferty, L., Stolojan, V., and Silva, S. R. P. (2015) Highly aligned arrays of super resilient carbon nanotubes by steam purification. *Carbon N. Y.*, **84**, 130–137.

66. Dai, Y., Liu, W., Formo, E., Sun, Y., and Xia, Y. (2011) Ceramic nanofibers fabricated by electrospinning and their applications in catalysis, environmental science, and energy technology. *Polym. Adv. Technol.*, **22**(3), 326–338.

67. Gu, Y., Jian, F., and Wang, X. (2008) Synthesis and characterization of nanostructured Co_3O_4 fibers used as anode materials for lithium-ion batteries. *Thin Solid Films*, **517**(2), 652–655.

68. Shen, L., Yu, X., Cheng, C., Song, C., Wang, X., Zhu, M., and Hsiao, B. S. (2016) High filtration performance thin film nanofibrous composite membrane prepared by electrospraying technique and hot-pressing treatment. *J. Memb. Sci.*, **499**, 470–479.

69. Kaur, S., Barhate, R., Sundarrajan, S., Matsuura, T., and Ramakrishna, S. (2011) Hot pressing of electrospun membrane composite and its influence on separation performance on thin film composite nanofiltration membrane. *Desalination*, **279**(1–3), 201–209.

70. Pan, J., Ge, L., Lin, X., Wu, L., Wu, B., and Xu, T. (2014) Cation exchange membranes from hot-pressed electrospun sulfonated poly(phenylene oxide) nanofibers for alkali recovery. *J. Memb. Sci.*, **470**, 479–485.

71. He, F., Sarkar, M., Lau, S., Fan, J., and Chan, L. H. (2011) Preparation and characterization of porous poly(vinylidene fluoride-trifluoroethylene) copolymer membranes via electrospinning and further hot pressing. *Polym. Test.*, **30**(4), 436–441.

72. Ali, A. A., Eltabey, M. M., Abdelbary, B. M., and Zoalfakar, S. H. (2015) MWCNTs/carbon nano fibril composite papers for fuel cell and super capacitor applications. *J. Electrostat.*, **73**, 12–18.

73. Yoon, Y. Il, Moon, H. S., Lyoo, W. S., Lee, T. S., and Park, W. H. (2009) Superhydrophobicity of cellulose triacetate fibrous mats produced by electrospinning and plasma treatment. *Carbohydr. Polym.*, **75**(2), 246–250.

74. Michlíček, M., Manakhov, A., Dvořáková, E., and Zajíčková, L. (2019) Homogeneity and penetration depth of atmospheric pressure plasma polymerization onto electrospun nanofibrous mats. *Appl. Surf. Sci.*, **471**, 835–841.

75. Heo, Y., Shin, Y. M., Lee, Y. Bin, Lim, Y. M., and Shin, H. (2015) Effect of immobilized collagen type IV on biological properties of endothelial cells for the enhanced endothelialization of synthetic vascular graft materials. *Colloid. Surface. B*, **134**, 196–203.

76. Liu, Y., Liu, Y., Liao, N., Cui, F., Park, M., and Kim, H. Y. (2015) Fabrication and durable antibacterial properties of electrospun chitosan nanofibers with silver nanoparticles. *Int. J. Biol. Macromol.*, **79**, 638–643.

77. Yu, H., Jia, Y., Yao, C., and Lu, Y. (2014) PCL/PEG core/sheath fibers with controlled drug release rate fabricated on the basis of a novel combined technique. *Int. J. Pharm.*, **469**(1), 17–22.

78. Guo, F., Servi, A., Liu, A., Gleason, K. K., and Rutledge, G. C. (2015) Desalination by membrane distillation using electrospun polyamide fiber membranes with surface fluorination by chemical vapor deposition. *ACS Appl. Mater. Interfaces*, **7**(15), 8225–8232.

79. Liao, Y., Loh, C.-H., Tian, M., Wang, R., and Fane, A. G. (2018) Progress in electrospun polymeric nanofibrous membranes for water treatment: Fabrication, modification and applications. *Prog. Polym. Sci.*, **77**, 69–94.

80. Sohn, J. Y., Im, J. S., Shin, J., and Nho, Y. C. (2012) PVDF-HFP/PMMA-coated PE separator for lithium-ion battery. *J. Solid State Electrochem.*, **16**(2), 551–556.

81. Lee, Y., Ryou, M. H., Seo, M., Choi, J. W., and Lee, Y. M. (2013) Effect of polydopamine surface coating on polyethylene separators as a function of their porosity for high-power Li-ion batteries. *Electrochim. Acta*, **113**, 433–438.

82. Ryou, M.-H., Lee, D. J., Lee, J.-N., Lee, Y. M., Park, J.-K., and Choi, J. W. (2012) Excellent cycle life of lithium-metal anodes in lithium-ion batteries with mussel-inspired polydopamine-coated separators. *Adv. Energy Mater.*, **2**(6), 645–650.

83. Liao, Y., Wang, R., and Fane, A. G. (2013) Engineering superhydrophobic surface on poly(vinylidene fluoride) nanofiber membranes for direct contact membrane distillation. *J. Memb. Sci.*, **440**, 77–87.

84. Fang, L. F., Shi, J. L., Zhu, B. K., and Zhu, L. P. (2013) Facile introduction of polyether chains onto polypropylene separators and its application in lithium ion batteries. *J. Memb. Sci.*, **448**, 143–150.

85. Zhou, F. L., Gong, R. H., and Porat, I. (2009) Mass production of nanofibre assemblies: By electrostatic spinning. *Polym. Int.*, **58**(4), 331–342.

86. Zhou, Z., Lai, C., Zhang, L., Qian, Y., Hou, H., Reneker, D. H., and Fong, H. (2009) Development of carbon nanofibers from aligned electrospun polyacrylonitrile nanofiber bundles and characterization of their microstructural, electrical, and mechanical properties. *Polymer (Guildf).*, **50**(13), 2999–3006.

87. Varesano, A., Carletto, R. A., and Mazzuchetti, G. (2009) Experimental investigations on the multi-jet electrospinning process. *J. Mater. Process. Technol.*, **209**(11), 5178–5185.

88. Fang, D., Chang, C., Hsiao, B. S., and Chu, B. (2006) Development of multiple-jet electrospinning technology. *ACS Symp. Ser.*, **918**, 91–105.

89. Kim, B. S. and Kim, I. S. (2011) Recent nanofiber technologies. *Polym. Rev.*, **51**(3), 235–238.

90. Luo, C. J., Stoyanov, S. D., Stride, E., Pelan, E., and Edirisinghe, M. (2012) Electrospinning versus fibre production methods: From specifics to technological convergence. *Chem. Soc. Rev.*, **41**(13), 4708–4735.

91. Erisken, C., Kalyon, D. M., and Wang, H. (2008) A hybrid twin screw extrusion/electrospinning method to process nanoparticle-incorporated electrospun nanofibres. *Nanotechnology*, **19**(16), 165302.

92. Yarin, A. L. and Zussman, E. (2004) Upward needleless electrospinning of multiple nanofibers. *Polymer (Guildf).*, **45**(9), 2977–2980.

93. Singh, M., Kasper, F. K., and Mikos, A. G. (2013) Tissue engineering scaffolds, in *Biomaterials Science: An Introduction to Materials: Third*

Edition (eds. Ratner, B., Hoffman, A., Schoen, F., and Lemons, J.), Academic Press, pp. 1138–1159.

94. Petrik, S. and Maly, M. (2010) Production nozzle-less electrospinning nanofiber technology. *Mater. Res. Soc. Symp. Proc.*, **1240**, 79–90.

95. Theron, S. A., Yarin, A. L., Zussman, E., and Kroll, E. (2005) Multiple jets in electrospinning: Experiment and modeling. *Polymer (Guildf).*, **46**(9), 2889–2899.

96. Russo, F., Ursino, C., Avruscio, E., Desiderio, G., Perrone, A., Santoro, S., Galiano, F., and Figoli, A. (2020) Innovative poly (vinylidene fluoride) (PVDF) electrospun nanofiber membrane preparation using DMSO as a low toxicity solvent. *Membranes (Basel).*, **10**(3), 36.

97. Tang, Y., Liu, Z., Zhao, K., and Fu, S. (2017) Adsorption and separation properties of positively charged ZrO_2 nanofibrous membranes fabricated by electrospinning. *RSC Adv.*, **7**(67), 42505–42512.

98. Matsumoto, H. and Tanioka, A. (2011) Functionality in electrospun nanofibrous membranes based on fiber's size, surface area, and molecular orientation. *Membranes (Basel).*, **1**(3), 249–264.

99. Agarwal, S., Greiner, A., and Wendorff, J. H. (2013) Functional materials by electrospinning of polymers. *Prog. Polym. Sci.*, **38**(6), 963–991.

100. Ding, B., Wang, M., Wang, X., Yu, J., and Sun, G. (2010) Electrospun nanomaterials for ultrasensitive sensors. *Mater. Today*, **13**(11), 16–27.

101. Zhu, M., Hua, D., Zhong, M., Zhang, L., Wang, F., Gao, B., Xiong, R., and Huang, C. (2018) Antibacterial and effective air filtration membranes by "green" electrospinning and citric acid crosslinking. *Colloid Interface Sci. Commun.*, **23**, 52–58.

102. Wang, Z., Crandall, C., Sahadevan, R., Menkhaus, T. J., and Fong, H. (2017) Microfiltration performance of electrospun nanofiber membranes with varied fiber diameters and different membrane porosities and thicknesses. *Polymer (Guildf).*, **114**, 64–72.

103. Al-Husaini, I. S., Yusoff, A. R. M., Lau, W. J., Ismail, A. F., Al-Abri, M. Z., Al-Ghafri, B. N., and Wirzal, M. D. H. (2019) Fabrication of polyethersulfone electrospun nanofibrous membranes incorporated with hydrous manganese dioxide for enhanced ultrafiltration of oily solution. *Sep. Purif. Technol.*, **212**, 205–214.

104. Shen, K., Cheng, C., Zhang, T., and Wang, X. (2019) High performance polyamide composite nanofiltration membranes via reverse interfacial polymerization with the synergistic interaction of gelatin interlayer and trimesoyl chloride. *J. Memb. Sci.*, **588**, 117192.

105. Shibuya, M., Park, M. J., Lim, S., Phuntsho, S., Matsuyama, H., and Shon, H. K. (2018) Novel CA/PVDF nanofiber supports strategically designed via coaxial electrospinning for high performance thin-film composite forward osmosis membranes for desalination. *Desalination*, **445**, 63–74.

106. Khayet, M., García-Payo, M. C., García-Fernández, L., and Contreras-Martínez, J. (2018) Dual-layered electrospun nanofibrous membranes for membrane distillation. *Desalination*, **426**, 174–184.

107. Santoro, S., Moro, A. J., Portugal, C. A. M., Crespo, J. G., Coelhoso, I. M., and Lima, J. C. (2016) Development of oxygen and temperature sensitive membranes using molecular probes as ratiometric sensor. *J. Memb. Sci.*, **514**, 467–475.

108. Santoro, S., Moro, A. J., Portugal, C., Crespo, J. G., Lima, J. C., and Coelhoso, I. M. (2016) Monitoring oxygen permeation through polymeric packaging films using a ratiometric luminescent sensor. *J. Food Eng.*, **189**, 37–44

109. Zhu, J., Ge, Y., Jasper, S., and Zhang, X. (2017) Physical characterization of electrospun nanofibers, in *Electrospun Nanofibers* (eds. Afshari, M.), Woodhead Publishing, pp. 207–238.

110. Semnani, D. (2017) Geometrical characterization of electrospun nanofibers, in *Electrospun Nanofibers* (eds. Afshari, M.), Woodhead Publishing, pp. 151–180.

111. Hammami, M. A., Croissant, J. G., Francis, L., Alsaiari, S. K., Anjum, D. H., Ghaffour, N., and Khashab, N. M. (2017) Engineering hydrophobic organosilica nanoparticle-doped nanofibers for enhanced and fouling resistant membrane distillation. *ACS Appl. Mater. Interfaces*, **9**(2), 1737–1745.

112. Drioli, E., Santoro, S., Simone, S., Barbieri, G., Brunetti, A., Macedonio, F., and Figoli, A. (2014) ECTFE membrane preparation for recovery of humidified gas streams using membrane condenser. *React. Funct. Polym.*, **79**(1), 1–7.

113. Ursino, C., Simone, S., Donato, L., Santoro, S., De Santo, M. P., Drioli, E., Di Nicolò, E., and Figoli, A. (2016) ECTFE membranes produced by non-toxic diluents for organic solvent filtration separation. *RSC Adv.*, **6**, 81001–81012.

114. Drioli, E., Criscuoli, A., and Curcio, E. (eds.) (2006) Membrane materials, in *Membrane Contactors: Fundamentals, Applications and Potentialities*, vol. 11, Elsevier, pp. 40–104.

115. Homaeigohar, S. S., Mahdavi, H., and Elbahri, M. (2012) Extraordinarily water permeable sol–gel formed nanocomposite nanofibrous membranes. *J. Colloid Interface Sci.*, **366**(1), 51–56.

116. Wang, K., Abdala, A. A., Hilal, N., and Khraisheh, M. K. (2017) Mechanical characterization of membranes, in *Membrane Characterization* (eds. Hilal, N., Ismail, A. F., Matsuura, T., and Oatley-Radcliffe, D.), Elsevier, pp. 259–306.

117. Yoon, K., Hsiao, B. S., and Chu, B. (2009) Formation of functional polyethersulfone electrospun membrane for water purification by mixed solvent and oxidation processes. *Polymer (Guildf).*, **50**(13), 2893–2899.

118. Su, C., Lu, C., Cao, H., Tang, K., Chang, J., Duan, F., Ma, X., and Li, Y. (2018) Fabrication and post-treatment of nanofibers-covered hollow fiber membranes for membrane distillation. *J. Memb. Sci.*, **562**, 38–46.

119. Huhtamäki, T., Tian, X., Korhonen, J. T., and Ras, R. H. A. (2018) Surface-wetting characterization using contact-angle measurements. *Nat. Protoc.*, **13**(7), 1521–1538.

120. Hebbar, R. S., Isloor, A. M., and Ismail, A. F. (2017) Contact angle measurements, in *Membrane Characterization* (eds. Hilal, N., Ismail, A. F., Matsuura, T., and Oatley-Radcliffe, D.), Elsevier, pp. 219–255.

121. Wenzel, R. N. (1936) Resistance of solid surfaces to wetting by water. *Ind. Eng. Chem.*, **28**(8), 988–994.

122. Moradi, R., Karimi-Sabet, J., Shariaty-Niassar, M., and Koochaki, M. A. (2015) Preparation and characterization of polyvinylidene fluoride/graphene superhydrophobic fibrous films. *Polymers (Basel).*, **7**(8), 1444–1463.

123. Arden Pope III, C. and Dockery, D. W. (2006) Health effects of fine particulate air pollution: Lines that connect. *J. Air Waste Manage. Assoc.*, **56**(6), 709–742.

124. Lv, D., Zhu, M., Jiang, Z., Jiang, S., Zhang, Q., Xiong, R., and Huang, C. (2018) Green electrospun nanofibers and their application in air filtration. *Macromol. Mater. Eng.*, **303**(12), 1800336.

125. Canalli Bortolassi, A. C., Guerra, V. G., Aguiar, M. L., Soussan, L., Cornu, D., Miele, P., and Bechelany, M. (2019) Composites based on nanoparticle and pan electrospun nanofiber membranes for air filtration and bacterial removal. *Nanomater. (Basel, Switzerland)*, **9**(12), 1740.

126. Yoon, B. and Lee, S. (2011) Designing waterproof breathable materials based on electrospun nanofibers and assessing the performance characteristics. *Fibers Polym.*, **12**(1), 57–64.

127. Zhang, S., Tang, N., Cao, L., Yin, X., Yu, J., and Ding, B. (2016) Highly integrated polysulfone/polyacrylonitrile/polyamide-6 air filter for multilevel physical sieving airborne particles. *ACS Appl. Mater. Interfaces*, **8**(42), 29062–29072.

128. Ma, H., Burger, C., Hsiao, B. S., and Chu, B. (2011) Ultra-fine cellulose nanofibers: New nano-scale materials for water purification. *J. Mater. Chem.*, **21**(21), 7507–7510.

129. Barhate, R. S. and Ramakrishna, S. (2007) Nanofibrous filtering media: Filtration problems and solutions from tiny materials. *J. Memb. Sci.*, **296**(1), 1–8.

130. Suchecka, T., Biernacka, E., and Piatkiewicz, W. (2003) Microorganism retention on microfiltration membranes. *Filtr. Sep.*, **40**(8), 50–55.

131. Wang, R., Liu, Y., Li, B., Hsiao, B. S., and Chu, B. (2012) Electrospun nanofibrous membranes for high flux microfiltration. *J. Memb. Sci.*, **392–393**, 167–174.

132. Mei, Y., Yao, C., Fan, K., and Li, X. (2012) Surface modification of polyacrylonitrile nanofibrous membranes with superior antibacterial and easy-cleaning properties through hydrophilic flexible spacers. *J. Memb. Sci.*, **417–418**, 20–27.

133. Ma, H., Hsiao, B. S., and Chu, B. (2014) Functionalized electrospun nanofibrous microfiltration membranes for removal of bacteria and viruses. *J. Memb. Sci.*, **452**, 446–452.

134. Parekh, S. A., David, R. N., Bannuru, K. K. R., Krishnaswamy, L., and Baji, A. (2018) Electrospun silver coated polyacrylonitrile membranes for water filtration applications. *Membranes (Basel).*, **8**(3), 59.

135. Ahmed, F. E., Lalia, B. S., and Hashaikeh, R. (2015) A review on electrospinning for membrane fabrication: Challenges and applications. *Desalination*, **356**, 15–30.

136. Santoro, S., Drioli, E., and Figoli, A. (2016) Development of novel ECTFE coated PP composite hollow-fiber membranes. *Coatings*, **6**(3), 40.

137. Arribas, P., Khayet, M., García-Payo, M. C., and Gil, L. (2014) Self-sustained electro-spun polysulfone nano-fibrous membranes and their surface modification by interfacial polymerization for micro- and ultra-filtration. *Sep. Purif. Technol.*, **138**, 118–129.

138. Liao, Y., Wang, R., Tian, M., Qiu, C., and Fane, A. G. (2013) Fabrication of polyvinylidene fluoride (PVDF) nanofiber membranes by electro-spinning for direct contact membrane distillation. *J. Memb. Sci.*, **425–426**, 30–39.

139. Santoro, S., Vidorreta, I. M., Sebastian, V., Moro, A., Coelhoso, I. M., Portugal, C. A. M., Lima, J. C., Desiderio, G., Lombardo, G., Drioli, E., Mallada, R., Crespo, J. G., Criscuoli, A., and Figoli, A. (2017) A non-invasive optical method for mapping temperature polarization in direct contact membrane distillation. *J. Memb. Sci.*, **536**, 156–166.

140. Santoro, S., Vidorreta, I., Coelhoso, I., Lima, J. C., Desiderio, G., Lombardo, G., Drioli, E., Mallada, R., Crespo, J., Criscuoli, A., and Figoli, A. (2019) Experimental evaluation of the thermal polarization in direct contact membrane distillation using electrospun nanofiber membranes doped with molecular probes. *Molecules*, **24**(3), 638.

141. Politano, A., Argurio, P., Di Profio, G., Sanna, V., Cupolillo, A., Chakraborty, S., Arafat, H. A., and Curcio, E. (2017) Photothermal membrane distillation for seawater desalination. *Adv. Mater.*, **29**(2), 1603504.

142. Politano, A., Profio, G. Di, Fontananova, E., Sanna, V., Cupolillo, A., and Curcio, E. (2019) Overcoming temperature polarization in membrane distillation by thermoplasmonic effects activated by Ag nanofillers in polymeric membranes. *Desalination*, **451**, 192–199.

143. Ye, H., Li, X., Deng, L., Li, P., Zhang, T., Wang, X., and Hsiao, B. S. (2019) Silver nanoparticle-enabled photothermal nanofibrous membrane for light-driven membrane distillation. *Ind. Eng. Chem. Res.*, **58**(8), 3269–3281.

Chapter 4

Preparation and Characterization of Inorganic Hollow Fibers

Soudeh Almaie, Mir Saeed Seyed Dorraji, and Hoda Daneshvar
Applied Chemistry Research Laboratory, Department of Chemistry, Faculty of Science, University of Zanjan, Zanjan, Iran
dorraji @znu.ac.ir

Inorganic fiber membranes with high surface-area-to-volume ratio, well-shaped porosity, 3D topography, high permeability and selectivity, and hydrophilic surface have shown great potential in gas separation and water treatment applications. Inorganic membranes can be classified into ceramic materials (metal oxides, nanostructures such as perovskites, zeolites, metal-organic frameworks) and carbon-based materials. There has been considerable progress in the development of inorganic membranes through the incorporation of carbon-based materials to enhance the performance of fibrous membranes. Future research should be focused on the development of the fabrication of fiber membranes with higher permeance, higher mechanical strength, and cost-effective fabrication methods. This chapter reviews the recent progresses in inorganic fiber membranes

Hollow Fibers and Nanofibers in Membrane Science: Preparation, Characterization, and Applications
Edited by Alberto Figoli, Mir Saeed Seyed Dorraji, and Francesco Galiano

ISBN 978-981-4968-03-4 (Hardcover), 978-1-003-25663-2 (eBook)
www.jennystanford.com

with an emphasis on the structure, properties, fabrication approaches, and characterization.

4.1 Introduction

The global demand for clean and safe water supply has been gradually growing over the years. Membrane separation with certain advantages such as low chemical and energy consumption, and excellent separation efficiency has great potential in industrial applications. With advancement in materials science and membrane technology developments, membrane technology has long served as an effective separation process compared to traditional separation techniques [1]. Membrane materials can be classified into two common categories: polymeric membranes and inorganic membranes. Although the polymeric membranes have been widely used in separation technology, their application is often limited due to the fouling problems [2], low physical properties, and the trade-off relationship between permeability and selectivity (upper bound trade-off curve) for various gas separations. This phenomenon, which was first introduced by Robeson, explains the relationship between permeability and selectivity. According to the theory, polymers with high permeability often show low selectivity [3].

Compared to polymeric (organic) membranes, the porous and dense inorganic membranes have been applied for gas/liquid separation extensively due to their excellent thermal stability (in temperatures higher than 400°C), chemical stability, mechanical strength (high stiffness and tensile strength), and good resistance against the degradation caused by the microbial processes [4]. Inorganic membranes can be categorized mainly into ceramic membranes and carbon-based membranes. The ceramic materials such as perovskite oxides, crystalline zeolites, and metal-organic frameworks (MOFs), silica, porous glass, metals [5], and carbon-based materials such as carbon nanofibers (CNFs), carbon nanotube (CNT), and graphene-based or the mixtures of these components are commonly used in inorganic hollow fiber membranes. The structure and morphology of inorganic membranes are divided into porous and non-porous (dense) membranes (as shown in Fig. 4.1) [6, 7].

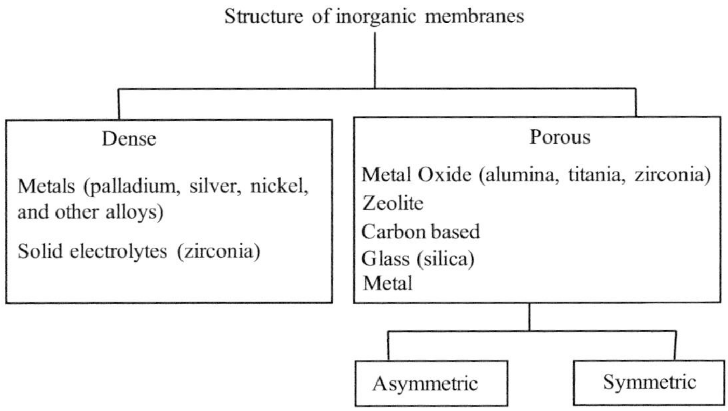

Figure 4.1 Structure of inorganic membranes. Reprinted from Ref. [7] with permission of Springer.

The inorganic asymmetric membranes in the form of flat sheets and hollow fibers are fabricated by multiple steps where the intermediate layer and selective top layer (such as zeolites, MOFs, and carbon-based) are coated on the support layer. Hollow fiber membranes with a high surface-to-volume ratio are preferred over flat-sheet membranes because the fibers are less prone to fouling, which leads to high permeation rates [8]. Recently, the inorganic fiber membranes with well-shaped porosity, 3D topography, high permeability and selectivity, hydrophilic surface, and flexible surface functionalities have attracted much attention in water treatment and gas separation applications. However, despite all the mentioned advantages, inorganic hollow fiber membranes present a series of drawbacks such as fragility and rigid structure, which can hinder their full exploitation. To overcome these limitations, the combination of inorganic and organic materials is often necessary [4]. This chapter summarizes the general features of inorganic materials for inorganic fibers and the fabrication and characterization of inorganic fibers.

4.2 Structure of Inorganic Hollow Fiber Membranes

The structure of hollow fiber membranes has a substantial effect on their performance and applications [6]. Their structure can be classified into asymmetric, symmetric, and sandwich morphology (as shown in Fig. 4.2).

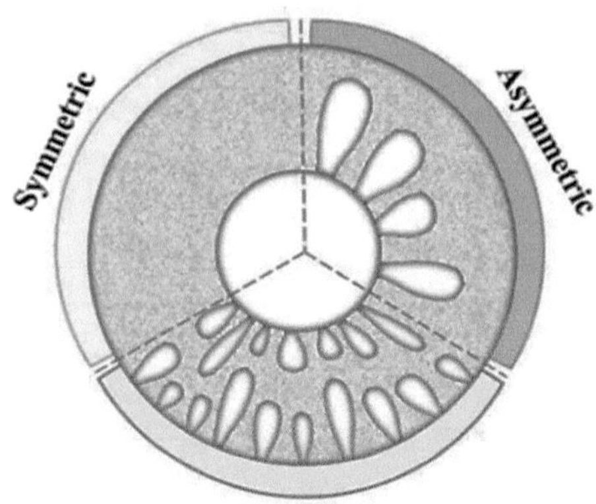

Figure 4.2 Three common structures of the inorganic hollow fiber, including asymmetric (porous), symmetric (dense), and sandwich structures (dense and porous). Reprinted from Ref. [9] with permission of Wiley.

The asymmetric structures, called finger-like structures, include large void channels or large porous structures, which are responsible for reduced mass transport resistance and lower mechanical strength. Figure 4.3 gives an overview of a hierarchical structure of an inorganic membrane. According to the figure, the coarse support layer is first deposited, followed by a microfiltration layer (100 nm), an ultrafiltration layer (10 nm), and then a nanofiltration layer (1 nm), which results in an asymmetric membrane. Finally, the top layer can be porous or non-porous (such as zeolite, MOF, carbon, metal), and it provides the membrane's selectivity [10]. Porous inorganic membranes can be made of porous metals, glasses, metal oxides, perovskites, ceramics, and carbon-based materials [5].

Symmetric membranes are chemically homogenous in composition and structurally uniform. They provide better mechanical strength and higher selectivity. Due to their dense structure, these membranes are impermeable for liquids, but they are ideally selective for gas molecules, including oxygen, hydrogen, and carbon dioxide. Dense or non-porous inorganic fiber membranes are made from either ionic-conducting metal oxides or metallic materials (Pd, Ag, Ni, and alloys). The metallic fiber membranes have attracted increasing attention due to their significant resistance at high temperatures, high permeability, and good selectivity in gas separation. Besides, the combination of the metals with metal oxides such as Al_2O_3, ZrO_2, YSZ, CeO_2, TiO_2, or Fe_2O_3 can improve both the mechanical strength and stability of symmetric membranes [11].

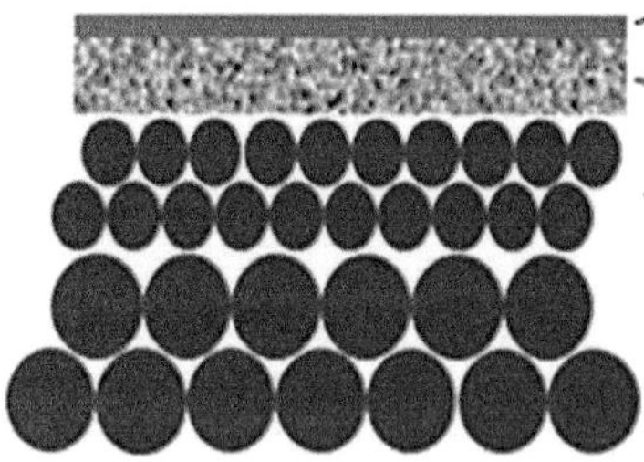

Figure 4.3 Asymmetric supports with hierarchical cross sections are used for minimizing the flow resistance. Reprinted from Ref. [10] with permission of the Royal Society of Chemistry.

Sandwich structure is generally made of a thin spongy layer between two finger-like layers at the inner and outer surfaces. These structures are characterized by high mechanical strength and high permeability and selectivity. For instance, Fig. 4.4 shows the inorganic interlayers with the specific surface and pore sizes below 50 nm, which were used for transition from the microfiltration layer (large pores) to the surface of the thin layer. The intermediate layers are usually fabricated via layer-by-layer deposition for more mechanical strength and chemical stability [12].

Inorganic hollow fiber membranes can be fabricated and modified by ceramics (perovskite oxides, zeolites, and metal-organic frameworks) and carbon-based materials, which are discussed in the following section.

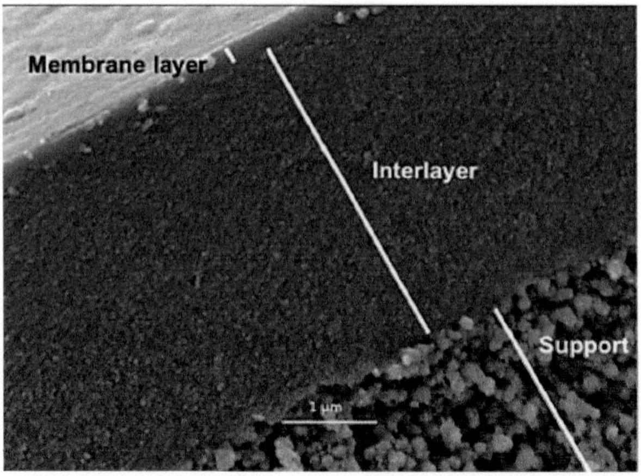

Figure 4.4 A ceramic asymmetric membrane with an interlayer structure. Reprinted from Ref. [12] with permission from the Royal Society of Chemistry.

4.3 Ceramic Hollow Fiber Membranes

Ceramics membranes with a wide range of insoluble oxides are considered excellent materials for membrane fabrication due to their outstanding chemical and thermal stabilities, high water flux, and high temperature resistance [1, 4].

Ceramic hollow fiber membranes can be classified into three main categories: perovskite oxides, zeolites, and metal-organic frameworks, which will be discussed in the next sections.

4.3.1 Perovskite Oxide Hollow Fiber Membranes

Ionic-electronic conducting ceramic membranes such as perovskite oxides have attracted extensive attention recently. Perovskites with ideal cubic structures (mainly ABO_3 based) are synthesized from metallic ions (such as tungstate, molybdate, or vanadate) with various counter ions [13]. These membranes transfer oxygen ions and electrons in different directions; thus, they guarantee high oxygen permeation fluxes. Perovskites hollow fiber membranes with three-dimensional (3D) networks have received attention in fields such as photocatalysis, sensing, solar cells, and optoelectronic devices due to their selectivity, structural stability, the broad light absorption region, and more efficient separation of charges (due to the capability of producing electrons (e^-) and holes (e^-) pairs).

There are different methods to synthesize perovskite materials, including solid-state reaction, coprecipitation, hydrothermal, spray and freeze-drying, and sol-gel process. Among these methods, the solid-state reaction and sol-gel method are conventional and fast for the synthesis of perovskites. Sol-gel techniques such as the Pechini method is widely used for the synthesis of numerous mixed oxides and perovskites due to the feasibility of mixing different compounds in this method.

Perovskite hollow fiber membranes are prepared by three steps: first, perovskite powders, a polymeric binder, and a solvent are mixed in a solution. Subsequently, the prepared solution is spun into hollow fibers. In this process, a hollow fiber geometry is produced, then the hollow fibers are sintered at high temperatures (higher than 1000°C) to burn off the polymeric binder [14].

The main precursors in synthesizing perovskites are metal salts and chelating agents (such as citric acid, ethylenediaminetetraacetic acid (EDTA), and polyvinyl alcohol (PVA)). The ratio between these two precursors is important in the synthesis since it guarantees that metal ions are completely bonded to their structure. Various metal salts with different counter ions such as carbonates, chlorides, hydroxides, and nitrates can be used as precursors for the fabrication of perovskite hollow fibers. Nitrate precursors are commonly used in the synthesis of perovskite oxides; however, they release a large volume of NH_3 gas and also require an additional heating pretreatment before sintering at high temperatures (about 700°C).

To address this issue, Li et al. studied the use of acetate precursor and of citric acid as a chelating agent to prepare $La_{0.6}Sr_{0.4}CoO_{3-\delta}$ (LSC) perovskite oxide powder. LSC powder was used to fabricate hollow fiber membranes via the phase-inversion-sintering technique. In this work, hollow fiber membranes were synthesized by acetate and nitrate precursors and also their oxygen permeation properties were evaluated at temperatures between 700 and 1000°C. The results showed that by increasing the temperature, the oxygen permeation fluxes raised from around 0.055 mL/min cm^2 to 1.2 mL/min cm^2 and 0.83 mL/min cm^2 for LSC hollow fiber membranes with acetate and nitrate precursors, respectively [15].

Currently, the NO_x decomposition process has become more attractive by the integration of perovskite hollow fibers with several catalysts such as metals. For example, Wang et al. investigated the combination of the metallic catalyst (Ni-phyllosilicate) with $BaBi_{0.05}Co_{0.2}Nb_{0.15}O_{3-\delta}$ (BBCN) perovskite to synthesize hollow fiber membranes (as shown in Fig. 4.5a). The catalytic perovskite hollow fiber membranes exhibited better stability in high temperatures (>800°C) and also reduced the negative effects of water and sulfur compounds. Comparing the NO removal capacity between the catalytic BBCN hollow fiber membrane and the BBCN perovskite powders, the NO feeding flow rate increased from 12 to 25 mL/min at low temperatures (675 and 700°C) [16].

In 2019, Ma et al. [17] fabricated perovskite hollow fibers with photocatalytic properties for dye degradation. The catalytic performance of Ag-$La_{0.8}$ $Ca_{0.2}Fe_{0.94}O_{3-\delta}$ (Ag-LCF) was evaluated by the degradation of methylene blue after 75 min. The selectivity of Ag-LCF as perovskite materials in hollow fiber membranes is related to its high efficiency and great performance in removing organic and bacterial pollutants. The results reported that Ag-LCF hollow fibers showed a dye degradation with a removal of up to 90% in the presence of peroxymonosulfate. The reactive oxygen species, such as sulfate and hydroxyl radicals, played a crucial role in dye decolorization [17].

Due to the attractive properties of perovskite hollow fiber membranes, such as larger membrane area, lower membrane mass transfer resistance, mechanical strength, and easy mixture with metallic catalysts, they have a great potential for gas separation applications. However, despite these advantages, one of the challenges of inorganic hollow fiber membranes is their low mechanical strength due to intrinsically low flexibility, fiber wall thickness, and high brittleness. In 2017, An et al. [18] reported that the modification of the porous $La_{0.6}Sr_{0.4}Co_{0.2}Fe_{0.8}O_{3-\delta}$ hollow fibers using a porous $Ba_{0.5}Sr_{0.5}Co_{0.4}Fe_{0.6}O_{3-\delta}$ (BSCF) perovskite binder could significantly improve mechanical properties and the flux oxygen (according to Fig. 4.5b). The porous BSCF perovskite binder not only provided a large surface area and mechanical strength to the hollow fiber bundles but also, as a catalyst, increased the surface oxygen exchange. The LSCF hollow fiber was coated with porous BSCF perovskite layer by using 5 to 10 hollow fiber bundles and the oxygen permeation flux was evaluated in the presence of air or Ar flow and a temperature range from 800 to 1000°C. According to the results, the oxygen permeation flux of the original LSCF hollow fiber was 0.42 mL/min cm^2 at 1000°C and then for 5-LSCF hollow fiber or 10-LSCF hollow fiber, the fluxes reached 0.73 mL/min cm^2 and 0.66 mL/min cm^2, respectively. It is also revealed that with increasing temperature, the oxygen permeation flux raised, and also increasing the Ar flow rate from 40 to 100 mL/min the oxygen flux increased from 0.29 to 0.68 mL/cm^2 min [18].

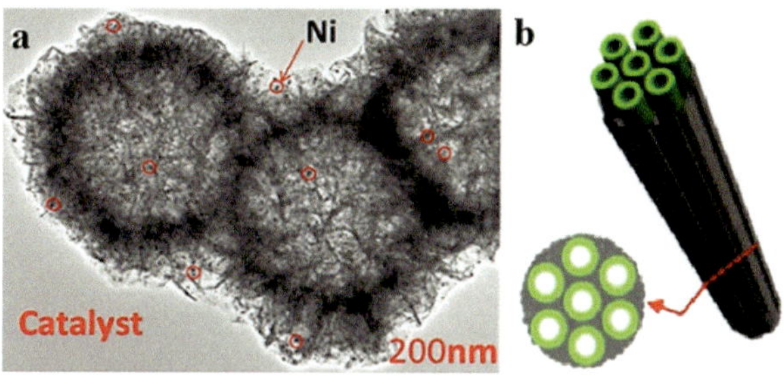

Figure 4.5 (a) Catalyst Ni in perovskite hollow fiber membrane. Reprinted from Ref. [16] with permission of the American Chemical Society. (b) Perovskite hollow fiber with the porous $Ba_{0.5}Sr_{0.5}Co_{0.4}Fe_{0.6}O_{3-\delta}$ (BSCF) binder. Reprinted from Ref. [18] with permission of Elsevier.

In another study, Yang et al. [19] compared oxygen flux of typical perovskite hollow fiber membranes such as $La_{0.6}Sr_{0.4}Co_{0.2}Fe_{0.8}O_3$ (LSCF), $Ba_{0.5}\ Sr_{0.5}Co_{0.8}Fe_{0.2}O_{3-\delta}$ (BSCF), and $(La_{0.8}\ Ca_{0.2})_{1.01}Fe\ O_{3-\delta}$ (LCF). The study indicated that LCF hollow fiber membranes sandwich-structured (outer diameter of 1.75 mm and an inner diameter of 1.25 mm) had better stability, highest mechanical strength, and largest oxygen fluxes in pure CO_2 atmosphere, while the cobalt-containing perovskite hollow fibers (BSCF and LSCF) showed higher oxygen flux values under mild operation conditions. Therefore, to further increase the oxygen flux of LCF, surface modifications were carried out by acid etching and catalyst depositing for surface area enlargement by surface porosity. Compared to original fibers, the oxygen fluxes sharply increased to 0.42 mL/min cm^2 and 0.76 mL/min cm^2 at 900 and 950°C, respectively, after acid treatment. Also, the deposition of porous BSCF catalyst on the outer surface of the LCF hollow fiber showed an increase in the oxygen flux to 0.11 mL/min cm^2. Surface modifications improved the oxygen flux due to the presence of pores with an average size of 2–5 μm. The results showed that the surface structure of LCF membranes played an important role in the oxygen flux.

Overall, the perovskite oxides hollow fibers demonstrated to have good mechanical strength, high selectivity, and chemical stability in the long-term industrial applications [19].

4.3.2 Zeolite Hollow Fiber Membranes

Zeolites are crystalline alumina silicates in the form of AlO_4 and SiO_4 tetrahedra displaying a well-defined channel structure of molecular dimensions. The aluminum-to-silicon ratio plays an important role in the hydrophilicity of the membrane. The high ratio of the aluminum-to-silicon atoms makes membranes more hydrophilic [20].

Zeolite topologies with three-dimensional structures are organized via the connection between metal centers (monovalent and divalent cations (Na^+, K^+, Ca^{2+}, and Mg^{2+})) and organic compounds like ligands or linkers. Zeolites are attractive candidates for numerous industrial applications, such as water treatment, catalysis, ion exchange, adsorption, and gas separation due to the permanent hydrophilicity, surface charge, porosity, antimicrobial properties, high permeability, and solute rejection [21, 22].

The synthesis of crystalline zeolites is performed by hydrothermal treatment when aluminate and silicate solutions are mixed in the temperature range of 30–180°C and at different reaction times (from minutes to days). According to Fig. 4.6a, the precursor materials that are used for the synthesis of zeolites can be classified into four types: (1) framework atoms (Si, Al, Fe, Ti); (2) inorganic materials (Na, K, Rb, Li, metal oxides); (3) organic structure-directing agents (template); (4) H_2O or solvent (as space filler) that can be completely or partially dissolved. The mixing of the initial sources (precursor mixture) can appear in two forms: colloidal (clear) suspensions or dense gel (Fig. 4.6b). The template effect is to stabilize the structure of the zeolite framework and control the size and morphology (Fig. 4.6c) [23].

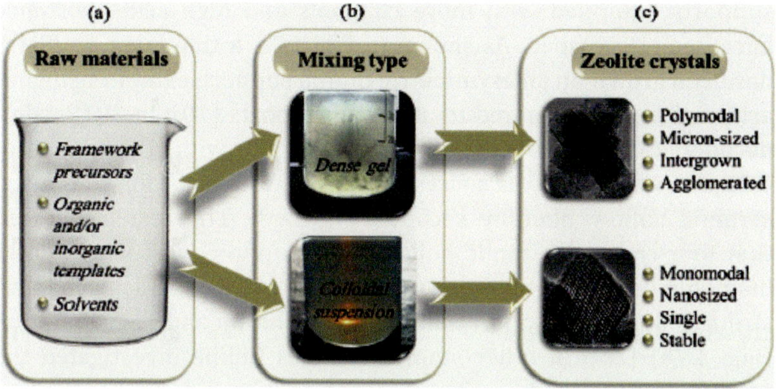

Figure 4.6 (a) Selection of raw materials, (b) preparation of mixing type of dense gel and colloidal suspense, and (c) crystallization of zeolites. Reprinted from Ref. [23] with permission of the Royal Society of Chemistry.

The zeolite hollow fiber membranes are usually prepared by secondary growth approach, which generally involves two steps, including nucleation and crystal growth. The nucleation methods, including vacuum seeding [24], dip-coating, and spin coating technology [1, 5, 25], have been widely used to synthesize zeolite fiber membranes with high quality. Then for crystals growth, commonly used methods are hydrothermal crystallization [15, 26], dry sol-gel, direct gel conversion [20], inner-side, contra-diffusion polymerization [27], and, recently, microwave method. The mechanisms of zeolites growth are categorized into classical and nonclassical methods. In the classical method, the crystal nucleation and growth involves the addition of atoms or molecules, while in the nonclassical method, nucleation and crystal growth are carried out by the aggregation of nanoparticles in solution [23]. In crystal growth, nucleation and layer deposition affect the particle size of fiber support. In the microwave method, the zeolite nanoparticles are prepared with high purity in a short time, and a thin dense layer of zeolite is synthesized by the hydrothermal method.

Recently, hollow-fibers-supported zeolite crystals have gained wide attention due to their superior structural stability and high permeation flux. However, careful selection of support materials is important to ensure stability against acids and mechanical strength of zeolite membranes. For this reason, ceramic hollow fiber supports equipped with more channels and high acid resistance have been developed. As shown in Fig. 4.7a, a thin layer of ZSM-5 has been grown on an asymmetric titania support, exhibiting higher acid resistance compared to alumina supports [10]. In 2018, silica decadodecasil 3R (DD3R) zeolite membrane with a pore size of 0.36 nm × 0.44 nm was successfully prepared on a four-channeled ceramic hollow fiber by secondary growth. The results showed that four-channel ceramic hollow fiber support with larger outer diameter improved both separation performance and mechanical strength during pervaporation in acidic solution and gas separation (Fig. 4.7b) [26]. In other studies, Scholes Coline investigated the growth of crystals of zeolite-silica and cobalt-silica on a γ-alumina and α-alumina support layer, respectively. Both inorganic membranes were synthesized to improve the porous structure of membranes and to optimize membrane selectivity at temperatures below ambient [28]. According to the results, both membranes were able to achieve N_2/He separation at temperatures below ambient; however, the zeolite-silica had more selectivity for N_2 gas. While the cobalt-silica membrane was capable to separate He gas over N_2 at high temperatures. According to Fig. 4.7c, Makhtar et al. reported that the glass hollow fibers could be an excellent support for zeolite growth. In fact, active hydroxyl groups on the glass surface are suitable sites to form strong intermolecular forces between oxygen ions in –OH groups and sodium ions (Na^+) in zeolites, which make them excellent supports for the zeolite growth. The glass hollow fibers are widely used in a range of applications due to chemical inertness, high mechanical, and thermal stabilities [25].

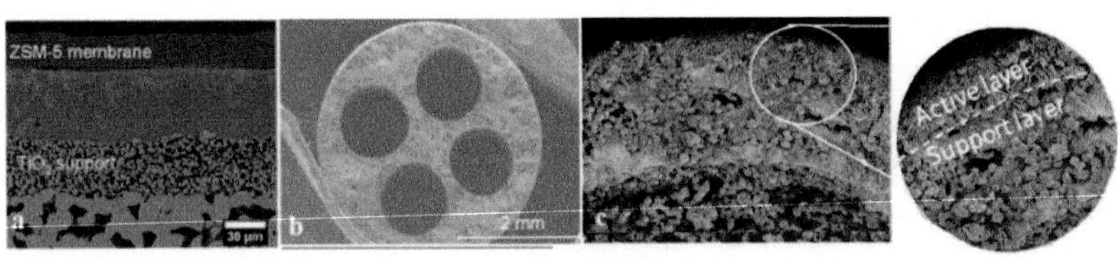

Figure 4.7 SEM image of (a) a supported ZSM-5 membrane on hierarchical TiO_2 support (Reprinted from Ref. [29] with permission of Elsevier), (b) the image of four-channel ceramic hollow fiber as support for DD3R zeolite (Reprinted from Ref. [26] with permission of Elsevier), (c) the zeolite membrane deposited on a glass hollow fiber (Reprinted from Ref. [25] with permission of Elsevier).

Zeolite crystals can randomly grow to the size of 1–2 μm with a zeolite layer thickness of about 5 μm by secondary growth for the fabrication of dense zeolite films, as shown in Fig. 4.8.

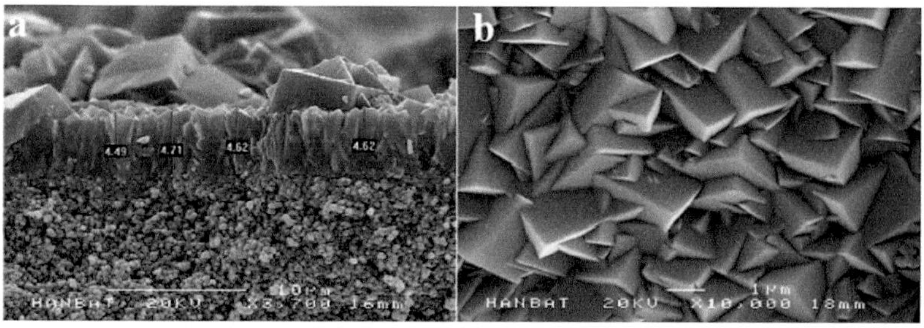

Figure 4.8 SEM images of the NaA zeolite membrane: (a) a cross-sectional view and (b) a top view. Reprinted from Ref. [30] with permission of Elsevier.

Zeolite hollow fiber membranes have demonstrated attractive gas separation properties in industrial applications. Yang et al. [20] investigated pure silica CHA-type (Si-CHA) zeolite hollow fibers with high selectivity showing excellent CO_2/CH_4 separation in both dry and humid conditions. Under a dry CO_2/CH_4 feed (50/50), the membrane showed a high CO_2 permeance up to 3.348×10^{-10} mol/m^{2}sPa at 1 bar pressure. When the pressure was increased from 1 to 7 bar, the flux of CO_2 increased of 3.3 times. With increasing feed gas pressure, the CO_2 performance was improved since CO_2 gas can compete more strongly for adsorption sites in constant water vapor [20]. Also, Yang et al. prepared high-quality zeolites with small pores ((CHA)-type aluminosilicate zeolite (SSZ-13) pore size ~0.38 nm) on α-alumina hollow fiber supports. The H_2 permeance stabilized at about 6.5×10^{-8} mol/m^{2}sPa and excellent H_2/C_3H_8 selectivity was achieved. SSZ-13 zeolite hollow fiber membrane was activated via UV treatment and showed good gas separation for different gas mixtures such as $H_2/(C_3H_6 + C_3H_8)$ and CO_2/CH_4 (Fig. 4.9) [27]. In another report, the performance of an alumina hollow fiber support was investigated when the inner surface was coated with Pd alloy and the outer layer was coated with a zeolite layer. The triple-layered membrane showed good thermal stability and hydrogen separation in the presence of hydrocarbons (C_3H_8 and C_3H_6) at 600°C for 72 h. The results showed that hydrogen permeation decreased to 27%

in the presence of hydrocarbons at 600°C and 1 bar pressure after coating the zeolite layer. This may be ascribed by the mass transfer resistance due to the presence of the outer zeolite layer. However, with increasing temperature, H_2 permeance remained constant in the presence of C_3H_8 and C_3H_6 [31]. The high potential of MFI-type zeolite (Silicalite-1) hollow fiber membranes with pore sizes of 0.56 nm × 0.54 nm have been particularly studied for butane isomer separation at high pressure (9 bar) and also for the removal of higher hydrocarbons (such as n-butane, propane, and ethane) from methane. At 9 bar, the membrane retained a high *n*-butane/methane separation factor of 97 with high *n*-butane permeance of 460 GPU. Also, the results reported a propane/methane separation factor of 48 [32].

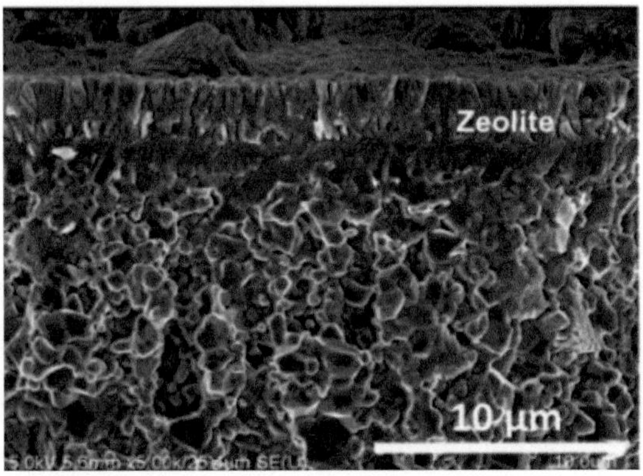

Figure 4.9 Cross-sectional FESEM image of an SSZ-13 zeolite on alumina support. Reprinted from Ref. [27] with permission of Wiley.

4.3.3 Metal-Organic Framework Hollow Fiber Membranes

Metal-organic frameworks (MOFs) are crystalline porous materials obtained through the combination of inorganic metal clusters with organic ligands. The active sites in metals and the functional groups of organic ligands with microporous cavities (<2 nm) make them suitable candidates in gas separation, filtration, and wastewater treatment applications [33].

Compared to polymeric membranes, MOF-based fiber membranes have shown a great potential in the fields of gas separation, nanofiltration, and pervaporation due to their 3D structure, high surface area, uniform porosity, and sharp pore sizes, which make them highly selective and permeable [4]. Compared to zeolites, MOF particles can be fabricated in easy conditions using milder, greener, and cheaper synthesis methods at low or room temperature by dipping a substrate in a linker and metal solution. The substrate surface is first modified with a functional group, such as $-NH_2$ or $-COOH$ to provide available sites for the growth of crystalline MOFs. Then the linker molecule enables MOFs to bind the substrate surface by covalent bonds. After the synthesis of the MOF particles, the solvent molecules are removed at mild temperatures (about 100°C), so MOF layers can be formed on the organic substrate [10].

Common methods for the preparation of MOF thin films are solvothermal [34, 35], hydrothermal, layer-by-layer deposition technique, plasma method, microwave heating, and microfluidics approach. Most of the methods used for the fabrication of zeolite fiber membranes can also be applied for the synthesis of MOF-based fibers. The solvothermal synthesis leads to the formation of micron-size MOF particles, while the sonochemical and spray-drying methods have been used for the preparation of MOF nanocrystals with high surface area [34]. The microfluidics approach can control reaction parameters in one step for the synthesis of MOF-based hollow fiber membranes (particle size and layer thickness). MOFs can be synthesized via the microfluidics method by two main types of organic linker: carboxylates (HKUST-1, MOF-5, IRMOF-3, UiO-66, MIL- 88B, and Cu-BTC) and imidazolates (ZIF-7, ZIF-8, ZIF-67, ZIF-90, ZIF-93, and ZIF-94). Crystallographic structures of MOFs are shown for carboxylates (see Fig. 4.10A–D) and imidazolates (see Fig. 4.10E–F).

The microfluidics method allows addressing the issue of high synthesis temperature and, at the same time, improving the synthesis of MOF-based hollow fiber membranes [36].

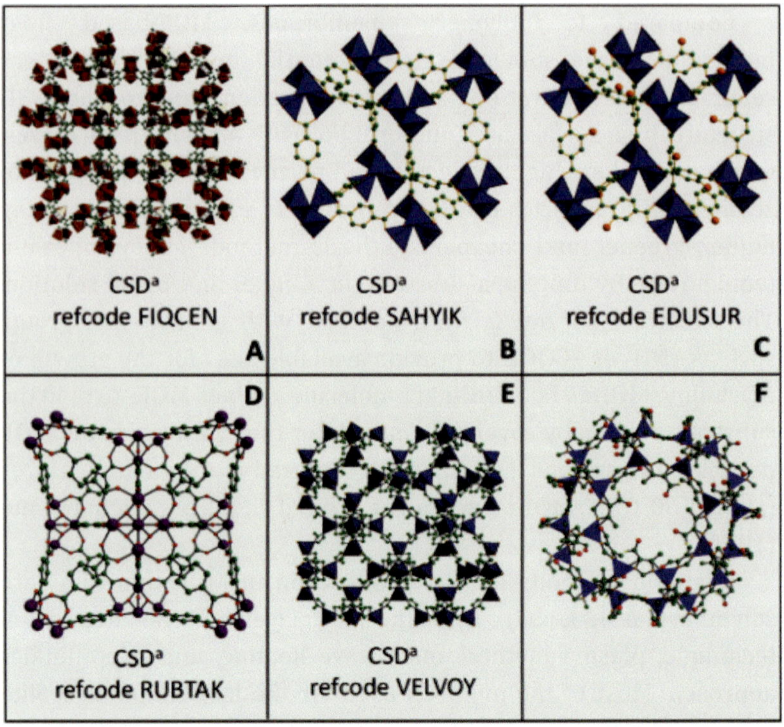

Figure 4.10 The structure of MOFs synthesized by microfluidics method: HKUST-1 (A), MOF-5 (B), IRMOF-3 (C), UiO-66 (Zr) (D), ZIF-8 (E), and ZIF-93 (F). Reprinted from Ref. [36] with permission of the Royal Society of Chemistry.

Due to the difficulties of separating MOFs from water, often MOF particles are coated on a support. Researchers have found that the combination of MOF crystals with organic support (polymers) can have better performance than other conventional inorganic materials [36, 37]. Synthesized MOF–polymer complexes exhibited high porous MOF particles, structural flexibility, high adsorption sites, excellent selectivity, and higher specific surface area. The solvothermal method could limit the application of most of the MOFs in polymeric membrane preparation. However, by using the interfacial contra-diffusion polymerization method, MOF nanoparticles can interact well with the organic support membrane [34]. According to Fig. 4.11, the performance of thin-film nanocomposite (MOF@TFN) with nanoparticle concentration ranging from 0.05 w/w% to 0.1 w/w% was investigated for water vapor/N_2 separation.

The synthesis of small size nanoparticles with hydrophilic groups on the surface of nanoparticles via functionalization reduced their agglomeration on the hollow fiber membranes. The slight agglomeration of MOF nanoparticles in the thin layer decreased the water vapor permeance from 2244 GPU to 1748 GPU; however, the selectivity increased to 542 with increasing thickness of the thin-film layer [34, 35].

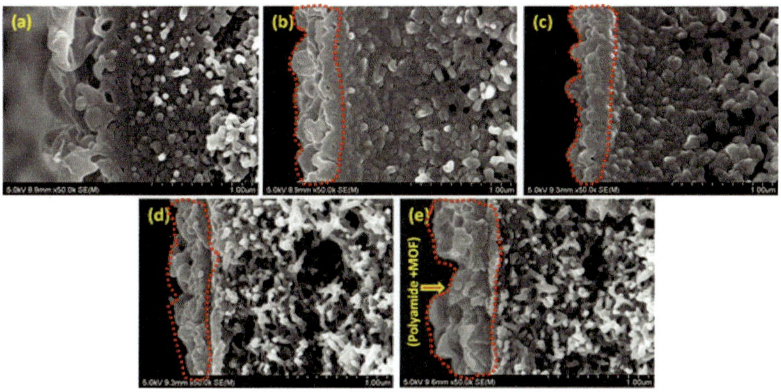

Figure 4.11 FE-SEM cross-sectional images of the TFC and TFN membranes showing the effect of MOF nanoparticle addition in the polyamide membrane matrix. Reprinted from Ref. [38] with permission of Elsevier.

Zeolitic imidazolate frameworks (ZIFs) with three-dimensional structures and zeolite-like topology are a special subclass of MOFs. In the synthesis of ZIFs, Zn^{2+} metals and imidazole organic links meet and react on the interface of the ceramic hollow fiber support by the contra-diffusion method. For example, Huang et al. [40] prepared ZIF-71 hollow fiber membranes with an asymmetric structure (a thin dense layer combined with porous ceramic layer on both sides) with excellent performance in the separation of ethanol–water mixtures (Fig. 4.12a). The performance of hydrophobic ZIF-71 membranes were evaluated for the ethanol recovery from 5 wt% ethanol aqueous solution in the temperature range from 25 to 55°C. The results indicated that the permeate flux was improved significantly to 2.106 g/m^2 h at 25°C [40]. Li et al. successfully synthesized ZIF-8/reduced graphene oxide (ZIF-8/rGO) composite hollow fiber membranes with the thickness of 150 nm, which displayed good reproducibility and long-term stability in the separation of H_2 from

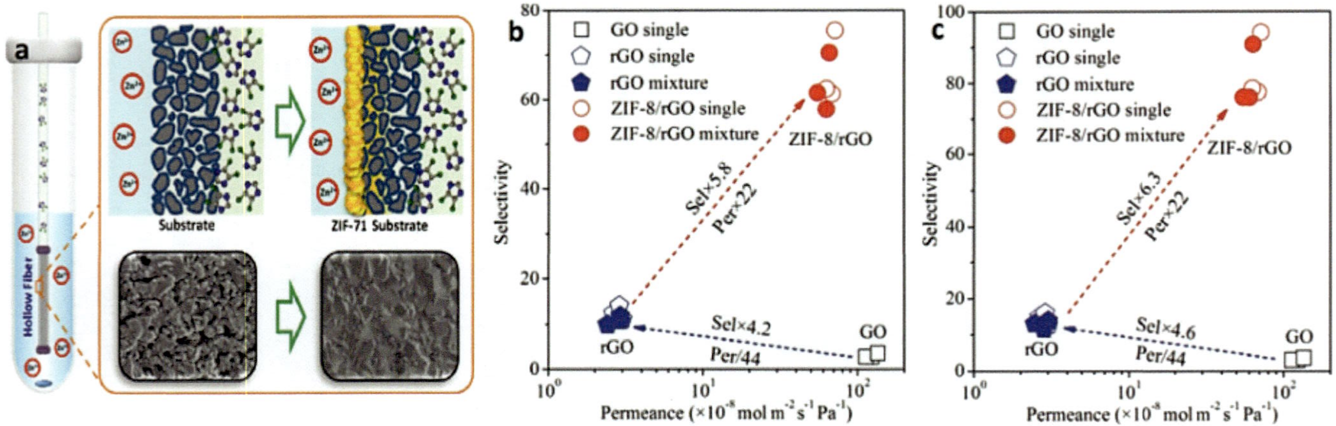

Figure 4.12 (a) Schematic representation of the preparation of ZIF-71 hollow fiber membranes by contra-diffusion. Reprinted from Ref. [40] with permission of the American Chemical Society. Comparison of the GO, rGO, and ZIF-8/rGO hollow fiber membranes for (b) H_2/N_2 and (c) H_2/CH_4 separations. Reprinted from Ref. [39] with permission of Wiley.

CO_2, N_2, and CH_4. During synthesis, the deposition of rGO with carboxyl groups improved by the amino groups on the hollow fiber substrate. Then the residual oxygen-containing groups on the rGO layers coordinated to zinc cation and nucleation sites were grown by interfacial contra-diffusion synthesis. According to Fig. 4.12b,c, the prepared ZIF-8/rGO membranes with GO concentration of 0.2 mg/mL improved both permeance and selectivity. The increase in GO concentration enhanced selectivity but decreased permeance and also ZIF-8 crystalline structure. With respect to the results, the ZIF-8/rGO membranes exhibited excellent separation performance with H_2 permeance over 60×10^{-8} mol/m^{2}sPa, and selectivity up to 25.3 (H_2/CO_2), 70.4 (H_2/N_2), and 90.7 (H_2/CH_4) [39].

4.4 Carbon Family Fiber Membranes

Researchers have attempted to develop efficient inorganic fiber membranes with high flux and selectivity, defect-free support, and reduced cost by using carbon-based materials. Recently, carbon nanotubes (CNTs) and carbon nanofibers (CNFs) have gained great attention due to their unique physicochemical properties, such as excellent chemical, thermal, and mechanical stability, good dimensional stability, and electrical conductivity. Carbon atoms are arranged in a hexagonal lattice structure in 2D graphene and as folded graphene layers in CNTs, which can be assembled as single-wall carbon nanotubes (SWCNTs), double-walled carbon nanotubes (DWCNTs), and multi-walled carbon nanotubes (MWCNTs). CNFs are similar to MWCNTs; however, hollow graphene layers in CNFs are arranged as a stacked cone or cup structure. The performance of carbon-based hollow fiber membranes is better than carbon-based sheet membranes due to their higher surface-area-to-volume ratio.

Gas-phase deposition or vapor deposition techniques such as physical vapor deposition (PVD) and chemical vapor deposition (CVD) are widely used to fabricate structured carbon and metal oxide nanofibers. Usual PVD methods are (1) electron beam evaporation, (2) plasma sputtering, (3) vacuum arc deposition, and (4) pulsed laser deposition. Carbon nanofibers were commonly produced by the CVD method, as shown in Fig. 4.13. Transition metals such as Fe, Ni, Co, Pt, and Cu are put on the substrate, and carbon layers (graphene, CO, or hydrocarbon gas) are grown at temperatures between 500 and 1200°C to make the structure of multilayer carbon nanofibers [41].

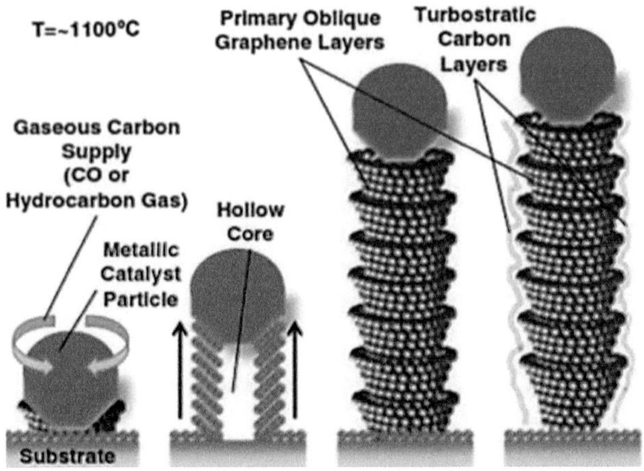

Figure 4.13 Schematic of a single carbon nanofiber with a helical structure formed during the CVD process. Reprinted from Ref. [41] with permission of Elsevier.

Also, electrochemical deposition is one of the most common wet chemical fabrication methods for nanofiber synthesis. Nanofibers can be fabricated using the deposition of materials or nanoparticles on hard/soft templates by vapor deposition on which templates are removed after nanoparticle formation. However, removing templates without damaging the nanofiber structure is difficult. One of the techniques of deposition of nano-fibrous structures is atomic layer deposition (ALD), which is used as layer-by-layer deposition.

Most importantly, hard porous templates such as metallic hollow fiber membranes, ceramic membranes, and aluminum oxide membranes show good resistance to strong acid, anticorrosive capabilities, and stability in high temperature and pressure. They have been applied as porous supports, which are called carbon molecular sieving membranes (CMSMs). The CMSMs with a selective thin carbon layer (such as protected graphene or carbon layers) increase the mechanical strength of hollow membranes. The mechanical stability eliminates the concern of fiber membrane brittleness [42].

For example, Chi et al. fabricated graphene-protected porous nickel hollow fiber (G-Ni-HF) membranes, which had improved mechanical properties against strong acids and also had evaluated CO oxidation after loading TiO_2-Pd catalysts. Chemical resistance tests on G-Ni-HF membranes were evaluated by 0.5 M, 1 M, and 2 M nitric acid. Besides, the graphene layers created the hydrophobic hollow fiber membrane, which enhanced the separation of water-in-oil emulsions. The obtained results showed that catalytic activity of TiO_2-Pd catalysts in the microsphere G-Ni-HF membranes with activation energy 115.7 kJ/mol improved CO oxidation significantly [43]. In another study, the GO layer was deposited on the composite hollow fiber membrane (zeolite-modified alumina (Sil-1-Al_2O_3)) by the vacuum-assisted filtration method (as shown in Fig. 4.14). The GO/2 wt% Sil-1-Al_2O_3 hollow fiber membrane displayed an increase in H_2 and N_2 permeance with the increase in temperature. Also, the performance and stability of GO/2 wt% Sil-1-Al_2O_3 composite hollow fiber membrane were successfully evaluated in two different temperatures: 20°C and 100°C. It revealed that at 20°C, H_2 permeance of 1.22×10^{-7} mol/m^{2}sPa and H_2/N_2 selectivity of 40.7 were achieved, respectively [44].

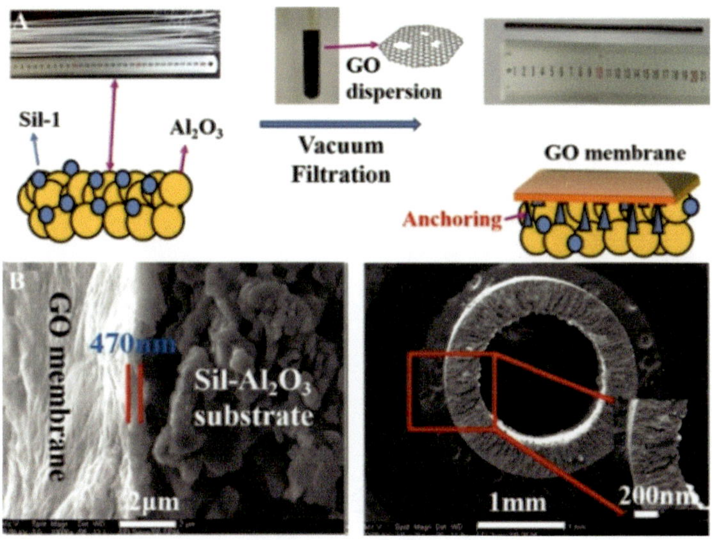

Figure 4.14 (A) Schematic illustration of the structure of the preparation process of GO/Sil-1-Al_2O_3 composite hollow fiber membrane; (B) (SEM) image of the cross section. Reprinted from Ref. [44] with permission of Elsevier.

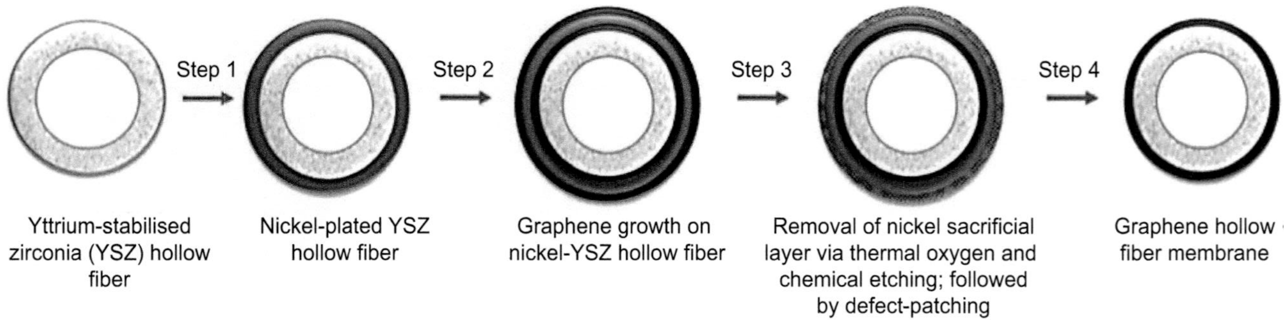

Figure 4.15 The illustration of the preparation procedure for obtaining graphene hollow fiber membranes. Reprinted from Ref. [45] with permission of Elsevier.

In 2020, Chi et al. [43] reported the performance of the graphene layers on supported ceramic hollow fiber to obtain an average methanol permeance of 2.4 L/m^{2}hbar and 98.8% Rose Bengal rejection rate. The results showed that the dye molecules were negatively charged; therefore, they were not adsorbed on the surface of the graphene layer. According to Fig. 4.15, a yttrium-stabilized zirconia (YSZ) hollow fiber was fabricated as support using the phase-inversion method and sintering process. The formation of a dense nickel layer was done using electroless plating. In the next step, the plated nickel–YSZ hollow fiber was annealed at a temperature of 900°C, and two-dimensional graphene synthesis was carried out by methane as a carbon precursor via the CVD process. Finally, using the thermal oxygen etching followed by the chemical leaching method, the nickel layer was removed, obtaining a graphene layer supported on a YSZ hollow fiber [45].

Figure 4.16 Schematic of different module configurations and the setup system. Reprinted from Ref. [42] with permission of Elsevier.

Wey et al. [42] prepared carbon hollow fiber membranes (CHFMs), where thin carbon layers were deposited directly on an alumina hollow fiber substrate. At this point, the coated hollow fiber membrane was cured at 513 K for 6 h and again at 873 K for 2 h during carbonization. The decrease in coating cycles led to thinner fiber membranes. The obtained results showed that the H_2 permeance of the synthesized CHFMs was significantly increased from 1.16 GPU to 11.43 GPU after one or two coating cycles. Besides, the permeance of H_2 enhanced from 11.43 GPU to 116.72 GPU by using dopant concentration of 5 wt% and in the thickness of 1.72 μm.

Figure 4.16 displayed the effects of hollow fiber packing density and packing module on the H_2/alkane separation performance. The packing densities of CHFMs, including 5.54% single, 16.60% triangle, 22.16% square, and 38.78% hexagon, were reported [42].

Fan et al. [46] reported that the rGO/CNT hollow fibers with high porosity and good wettability have excellent chemical and thermal stability in acidic (pH = 2), alkaline, and high-temperature conditions and also displayed improved performance when applied in forward osmosis for water purification and desalination processes (Fig. 4.17a). The thickness of the rGO layer was adjusted by controlling the deposition time or applied voltage in the electrophoretic deposition process (Fig. 4.17b). The optimal rGO/CNT hollow fiber showed high water flux of 22.6 L/m^{2}h [46].

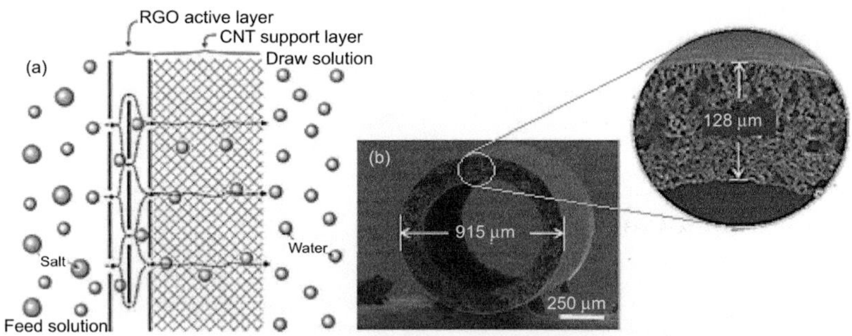

Figure 4.17 (a) Osmosis-based water treatment, (b) SEM image of GO coated on CNT hollow fiber substrate. Reprinted from Ref. [46] with permission of Elsevier.

4.5 Characterization Techniques

The inorganic hollow fiber membranes are commonly investigated via characterization techniques such as field emission scanning electron microscopy (FESEM), X-ray diffraction (XRD), energy dispersive X-ray (EDX), X-ray photoelectron spectroscopy (XPS), and water contact angle (WCA).

The morphology and microstructures of inorganic hollow fiber membranes can be characterized by FESEM analysis. For instance, the microstructure of the GO/Ni hollow fiber showed a homogenous thickness. After the growth of graphene on the porous nickel hollow fiber, wrinkles appeared on the graphene surface, which was due to differences in the thermal expansion coefficients of graphene and nickel [43]. Also, the FESEM analysis can be used to investigate the deposition of nanoparticles with different reactant concentrations on the hollow fibers. In a study, SEM images showed variations in the shape of the zeolite particles on the cross section of the glass hollow fiber [25]. Besides, SEM images of $La_{0.6}Sr_{0.4}CoO_{3-\delta}$ hollow fiber membranes show the dimension of the fibers presenting an outer diameter of 1.53 mm and an inner diameter of 1.08 mm. According to Fig. 4.18, an asymmetric structure is observed where a dense layer was isolated by two porous layers with spongy or finger-like pores. The thicknesses of the inner and outer dense layers were around 6 and 12 μm with a 47 μm thick sandwiched dense layer [15].

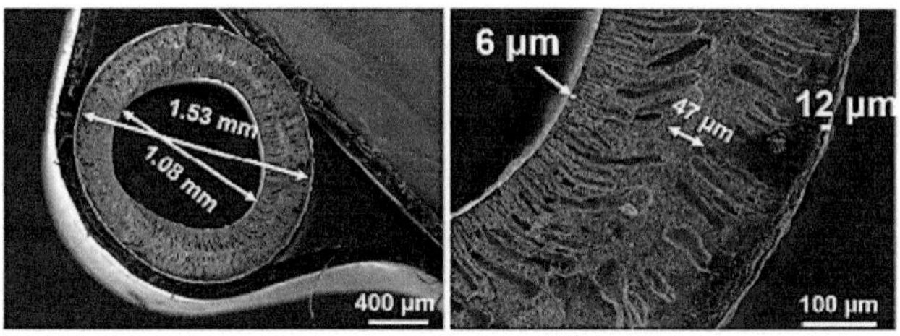

Figure 4.18 LSC hollow fiber membrane. Reprinted from Ref. [15] with permission of Elsevier.

XRD analysis is usually performed to determine the crystal phases of inorganic hollow fibers for both ceramic supports and nanoparticles. The XRD patterns of the inorganic membranes can show how nanoparticles (such as perovskite oxides, zeolites, MOFs, and carbon based) are coated on the surface of hollow fiber membranes and used as a support.

In pure ZIF-8 membranes, three sharp peaks are observed at 2θ values of 7.3°, 12.7°, 18° (Fig. 4.19a). After the deposition of GO layers by hydrothermal treatment, the crystal structures of the ZIF-8 hollow fiber membrane disappeared [39]. As it can be seen, the peak intensity of the ZIF-8 crystal powder is much higher compared to the peak intensity of the ZIF-8 that has grown on porous hollow fiber membranes (Fig. 4.19b) [47].

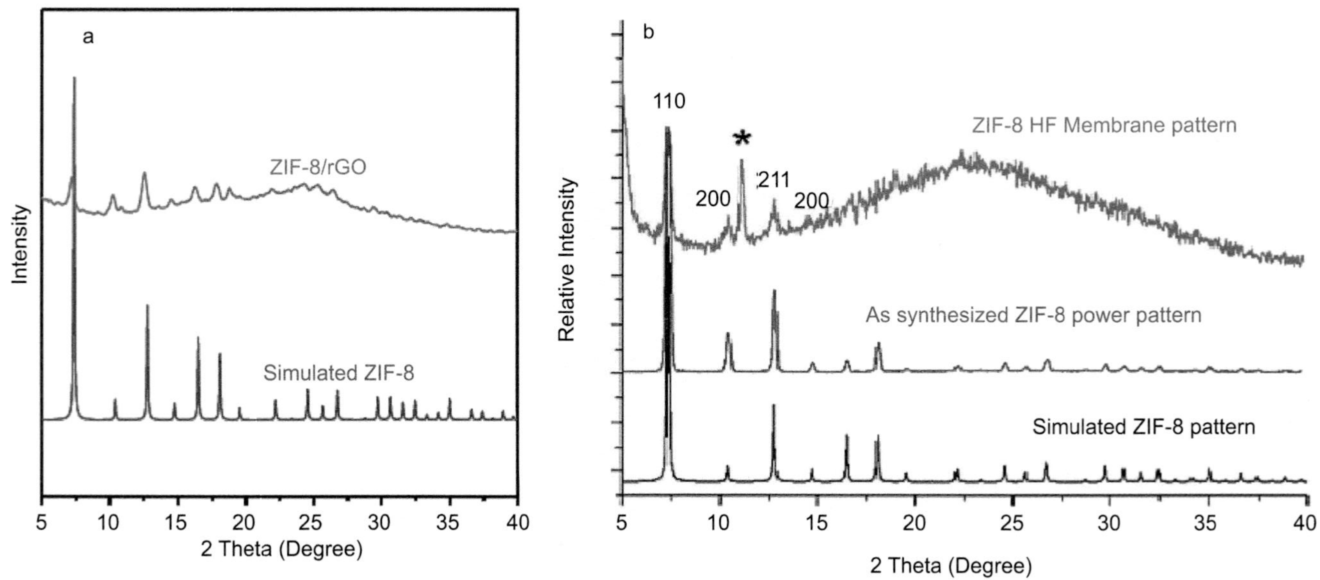

Figure 4.19 The XRD analysis of (a) pure ZIF-8, prepared ZIF-8/GO. Reprinted from Ref. [39] with permission of Wiley. (b) Synthesized ZIF-8 powder, ZIF-8 hollow fiber membrane. Reprinted from Ref. [47] with permission of the American Chemical Society.

Aziz et al. [1] investigated the XRD patterns of raw and calcined ball clay in a ceramic fiber microfiltration membrane. The results showed the presence of kaolinite with sharp peaks at around 2θ = 12.41°, 20.21° (with multiple reflections), and 25.49° in raw ball clay. After calcination at 800°C, the crystalline phase of calcined clay was related to quartz and illite phases. It was observed that after calcination, the kaolinite transformed into metakaolinite, while the quartz phase remained in calcined ball clay [1]. In another study, sodium borosilicate ($Na_2B_4O_7$) in the structure of glass powder was confirmed as amorphous. After sintering and acid etching, the $Na_2B_4O_7$ phase was completely dissolved and the glass membranes contained just silica phase. The increase in sintering temperature appeared as sharp diffraction peaks with different crystal planes [48].

The EDX spectroscopy detects the element distribution on the outer surface, inner surface, or the cross section of the membrane. The EDX mapping shows the molar ratio of elements and the thickness of the top layer on the surface of the fiber membrane or ceramic/metal support. For example, Fig. 4.20 illustrates the elemental mapping of Ni, Co, and Si with their uniform distribution throughout the whole area [49].

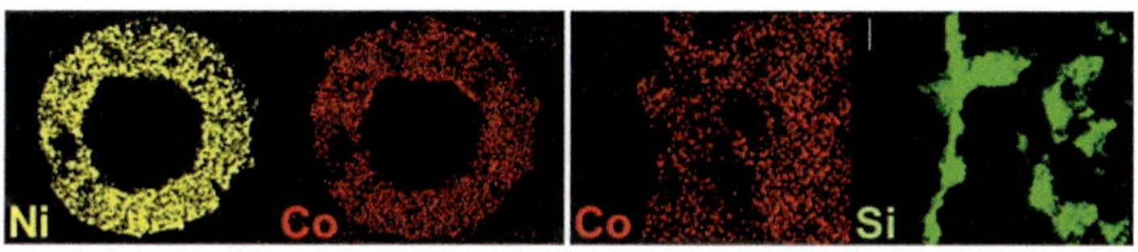

Figure 4.20 EDX mapping of Ni, Co, and Si in a hollow fiber membrane. Reprinted from Ref. [49] with permission of the American Chemical Society.

XPS is used to measure the binding energies of elements in the fresh hollow membranes in comparison with the hollow membranes after permeation. In a study on perovskite hollow membranes, the intensity of XPS peaks for Bi 4f, Co 3p, and Nb 3d of membranes was partially reduced after long-time permeation of steam compared to blank membranes. Also, some weak peaks such as bismuth, cobalt, and niobium disappeared from the inner surface of hollow membrane after long-time permeation swept by steam [50].

The surface hydrophilicity of inorganic hollow membranes is evaluated by the WCA measurement. WCA analysis is usually performed on the outer surface of inorganic hollow fiber membranes.

The surface wettability of carbon/ceramic hollow fiber membranes can differ depending on the final membrane application. Fiz et al. [51] investigated the surface modification of alumina hollow fiber membranes with fluoroalkylsilanes by the grafting method. The WCA analysis demonstrated that the contact angle changed from 32.8° to 111° for the ceramic hollow fiber with hydrophobic tails in their structure. Figure 4.21 shows that the WCA significantly increased after the modification process [51].

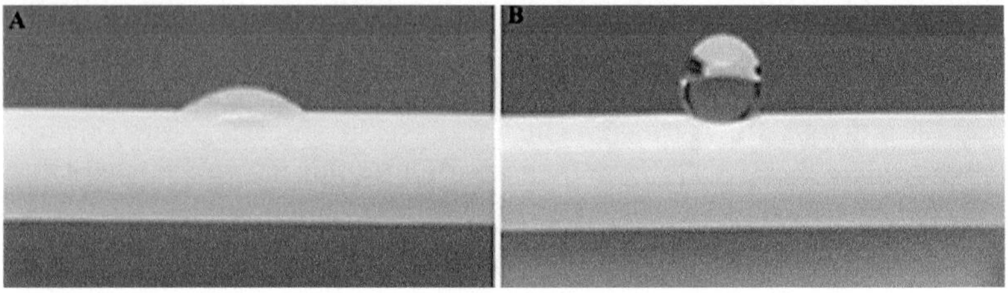

Figure 4.21 Contact angle measure of (A) original fiber and (B) modified fiber membrane. Reprinted from Ref. [51] with permission of the American Chemical Society.

4.6 Conclusion and Future Outlook

The current technology of inorganic hollow fiber membranes is focused on several strategies, including the hierarchical design of individual or hybrid inorganic materials in order to enhance permeation, flux, and selectivity; the fabrication of thin layer and defect-free ceramic supports or interlayer-free hollow fiber membrane; the development of carbon-based layers on hollow membranes for achieving cost-effective fabrication methods.

Inorganic fiber membranes (such as ceramic, perovskite fiber membranes, and zeolite membranes) are widely used due to high surface-area-to-volume ratio, mechanical and thermal long-term stability, hydrophilic surface, and flexible surface functionalities. Inorganic hollow fibers can solve some of the current challenges in polymer membranes, such as structural damages due to the weak chemical resistance in common organic solvents. They can well maintain their performance even after several cycles of filtering and cleaning. Despite the limitation of the expensive raw materials in inorganic membranes, they offer more industrialization and commercial applications, especially for the removal of emerging pollutants from water.

The preparation of inorganic hollow fiber membranes with controlled wall thickness, such as interlayer ceramic membrane and a carbon-based thin layer on the substrate, has opened a window of opportunities to develop gas separation and water treatment membranes. The combination of hollow fibers with nanoparticles, especially graphene/carbon-based materials, has interesting and promising perspectives. Future research needs to develop the fabrication of fiber membranes with higher permeance, higher mechanical strength, and cost-effective fabrication methods. Despite some problems, the inorganic hollow fiber membranes may become the major keys in water treatment, especially for the removal of contaminants under harsh conditions.

References

1. Aziz, M. H. A., Othman, M. H. D., Hashim, N. A., Adam, M. R., and Mustafa, A. (2019) Fabrication and characterization of mullite ceramic hollow fiber membrane from natural occurring ball clay. *Appl. Clay Sci.*, **177**, 51–62.

2. Paiman, S. H., Rahman, M. A., Uchikoshi, T., Nordin, N. A. H. M., Alias, N. H., Abdullah, N., Abas, K. H., Othman, M. H. D., Jaafar, J., and Ismail, A. F. (2020) In situ growth of α-Fe_2O_3 on Al_2O_3/YSZ hollow fiber membrane for oily wastewater. *Sep. Purif. Technol.*, **236**, 116250.

3. Robeson, L. M. (1991) Correlation of separation factor versus permeability for polymeric membranes. *J. Memb. Sci.*, **62**(2), 165–185.

4. Goh, P. S. and Ismail, A. F. (2018) A review on inorganic membranes for desalination and wastewater treatment. *Desalination*, **434**, 60–80.

5. Bhave, R. (2012) *Inorganic Membranes Synthesis, Characteristics and Applications: Synthesis, Characteristics, and Applications*, Springer Science & Business Media.

6. Kayvani Fard, A., McKay, G., Buekenhoudt, A., Al Sulaiti, H., Motmans, F., Khraisheh, M., and Atieh, M. (2018) Inorganic membranes: Preparation and application for water treatment and desalination. *Materials (Basel).*, **11**(1), 74.

7. Ismail, A. F., Rana, D., Matsuura, T., and Foley, H. C. (2011) *Carbon-Based Membranes for Separation Processes*, Springer Science & Business Media.

8. Castro-Muñoz, R., Galiano, F., and Figoli, A. (2020) Recent advances in pervaporation hollow fiber membranes for dehydration of organics. *Chem. Eng. Res. Des.*, **164**, 68–85.

9. Chen, L.-H., Chen, Y.-R., Chou, C.-Y., Chen, C.-H., Ko, C.-C., and Tung, K.-L. (2017) Inorganic membranes in water and wastewater treatment, in *Sustainable Membrane Technology for Water and Wastewater Treatment*, Springer, pp. 121–154.

10. Caro, J. (2016) Hierarchy in inorganic membranes. *Chem. Soc. Rev.*, **45**(12), 3468–3478.

11. Wang, M., Zhou, Y., Tan, X., Gao, J., and Liu, S. (2019) Nickel hollow fiber membranes for hydrogen separation from reformate gases and water gas shift reactions operated at high temperatures. *J. Memb. Sci.*, **575**, 89–97.

12. Smart, S., Lin, C. X. C., Ding, L., Thambimuthu, K., and Da Costa, J. C. D. (2010) Ceramic membranes for gas processing in coal gasification. *Energy Environ. Sci.*, **3**(3), 268–278.

13. Zhang, G., Liu, G., Wang, L., and Irvine, J. T. S. (2016) Inorganic perovskite photocatalysts for solar energy utilization. *Chem. Soc. Rev.*, **45**(21), 5951–5984.

14. Athayde, D. D., Souza, D. F., Silva, A. M. A., Vasconcelos, D., Nunes, E. H. M., da Costa, J. C. D., and Vasconcelos, W. L. (2016) Review of perovskite ceramic synthesis and membrane preparation methods. *Ceram. Int.*, **42**(6), 6555–6571.

15. Li, Y., Li, M., Gao, J., Lun, Y., Li, C., Sunarso, J., and Tan, X. (2020) Characterization of $La_{0.6}Sr_{0.4}CoO_{3-\delta}$ oxygen selective hollow fiber made from acetate precursor-derived powder. *Ceram. Int.*, **46**(3), 3744–3749.

16. Wang, Z., Li, Z., Cui, Y., Chen, T., Hu, J., and Kawi, S. (2019) Highly efficient NO decomposition via dual-functional catalytic perovskite hollow fiber membrane reactor coupled with partial oxidation of methane at medium-low temperature. *Environ. Sci. Technol.*, **53**(16), 9937–9946.

17. Ma, T., Liu, L., Meng, B., Gao, J., Wang, S., and Liu, S. (2019) Heterogeneous activation of peroxymonosulfate via a Ag-$La_{0.8}Ca_{0.2}Fe_{0.94}O_{3-\delta}$ perovskite hollow fiber membrane reactor for dye degradation. *Sep. Purif. Technol.*, **211**, 298–302.

18. An, R., Song, J., Li, Y., Tan, X., Sunarso, J., Zhang, C., Wang, S., and Liu, S. (2017) Bundling strategy to simultaneously improve the mechanical strength and oxygen permeation flux of the individual perovskite hollow fiber membranes. *J. Memb. Sci.*, **527**, 137–142.

19. Yang, D., Yang, N., Meng, B., Tan, X., Zhang, C., Sunarso, J., Zhu, Z., and Liu, S. (2017) A-site excess $(La_{0.8}Ca_{0.2})_{1.01}FeO_{3-\delta}$ (LCF) perovskite hollow fiber membrane for oxygen permeation in CO_2-containing atmosphere. *Energy Fuels*, **31**(4), 4531–4538.

20. Yang, S., Chiang, Y., and Nair, S. (2019) Scalable one-step gel conversion route to high-performance CHA zeolite hollow fiber membranes and modules for CO_2 separation. *Energy Technol.*, **7**(9), 1900494.

21. Tu, M., Wiktor, C., Rösler, C., and Fischer, R. A. (2014) Rapid room temperature syntheses of zeolitic-imidazolate framework (ZIF) nanocrystals. *Chem. Commun.*, **50**(87), 13258–13260.

22. Caro, J. (2016) Supported zeolite and mof molecular sieve membranes: Preparation, characterization, application, in *Zeolites and Zeolite-Like Materials*, Elsevier, pp. 283–307.

23. Grand, J., Awala, H., and Mintova, S. (2016) Mechanism of zeolites crystal growth: New findings and open questions. *CrystEngComm*, **18**(5), 650–664.

24. Min, B., Yang, S., Korde, A., Kwon, Y. H., Jones, C. W., and Nair, S. (2019) Continuous zeolite MFI membranes fabricated from 2D MFI nanosheets on ceramic hollow fibers. *Angew. Chemie Int. Ed.*, **58**(24), 8201–8205.

25. Makhtar, S. N. N. M., Muhammad, N., Rahman, M. A., Abas, K. H., Aziz, A. A., Sokri, M. N. M., Othman, M. H. D., and Jaafar, J. (2020) Zeolite-A deposited on glass hollow fiber for forward osmosis applications. *J. Water Process Eng.*, **33**, 100991.

26. Zhang, Y., Chen, S., Shi, R., Du, P., Qiu, X., and Gu, X. (2018) Pervaporation dehydration of acetic acid through hollow fiber supported DD3R zeolite membrane. *Sep. Purif. Technol.*, **204**, 234–242.

27. Yang, S., Kwon, Y. H., Koh, D., Min, B., Liu, Y., and Nair, S. (2019) Highly selective SSZ-13 zeolite hollow fiber membranes by ultraviolet activation at near-ambient temperature. *ChemNanoMat*, **5**(1), 61–67.

28. Scholes, C. A. (2018) Helium recovery through inorganic membranes incorporated with a nitrogen rejection unit. *Ind. Eng. Chem. Res.*, **57**(10), 3792–3799.

29. Richter, H., Voß, H., Voigt, I., Diefenbacher, A., Schuch, G., Steinbach, F., and Caro, J. (2010) High-flux ZSM-5 membranes with an additional

non-zeolite pore system by alcohol addition to the synthesis batch and their evaluation in the 1-butene/i-butene separation. *Sep. Purif. Technol.*, **72**(3), 388–394.

30. Ahn, H., Lee, H., Lee, S.-B., and Lee, Y. (2007) Dehydration of TFEA/water mixture through hydrophilic zeolite membrane by pervaporation. *J. Memb. Sci.*, **291**(1–2), 46–52.

31. Pati, S., Dewangan, N., Wang, Z., Jangam, A., and Kawi, S. (2020) Nanoporous zeolite-A sheltered Pd-hollow fiber catalytic membrane reactor for propane dehydrogenation. *ACS Appl. Nano Mater.*, **3**(7), 6675–6683.

32. Min, B., Yang, S., Korde, A., Jones, C. W., and Nair, S. (2020) Single-step scalable fabrication of zeolite MFI hollow fiber membranes for hydrocarbon separations. *Adv. Mater. Interfaces*, **7**(19), 2000926.

33. Koo, W.-T., Jang, J.-S., Qiao, S., Hwang, W., Jha, G., Penner, R. M., and Kim, I.-D. (2018) Hierarchical metal–organic framework-assembled membrane filter for efficient removal of particulate matter. *ACS Appl. Mater. Interfaces*, **10**(23), 19957–19963.

34. Allmond, K., Stone, J., Harp, S., and Mujibur, K. (2017) Synthesis and electrospraying of nanoscale MOF (metal organic framework) for high-performance CO_2 adsorption membrane. *Nanoscale Res. Lett.*, **12**(1), 6.

35. Ren, Z., Luo, J., and Wan, Y. (2018) Highly permeable biocatalytic membrane prepared by 3D modification: Metal-organic frameworks ameliorate its stability for micropollutants removal. *Chem. Eng. J.*, **348**, 389–398.

36. Echaide-Górriz, C., Clément, C., Cacho-Bailo, F., Téllez, C., and Coronas, J. (2018) New strategies based on microfluidics for the synthesis of metal–organic frameworks and their membranes. *J. Mater. Chem. A*, **6**(14), 5485–5506.

37. Jin, L., Ye, J., Wang, Y., Qian, X., and Dong, M. (2019) Electrospinning synthesis of ZIF-67/PAN fibrous membrane with high-capacity adsorption for malachite green. *Fibers Polym.*, **20**(10), 2070–2077.

38. Ingole, P. G., Sohail, M., Abou-Elanwar, A. M., Baig, M. I., Jeon, J.-D., Choi, W. K., Kim, H., and Lee, H. K. (2018) Water vapor separation from flue gas using MOF incorporated thin film nanocomposite hollow fiber membranes. *Chem. Eng. J.*, **334**, 2450–2458.

39. Li, W., Shi, J., Li, Z., Wu, W., Xia, Y., Yu, Y., and Zhang, G. (2018) Hydrothermally reduced graphene oxide interfaces for synthesizing high-performance metal–organic framework hollow fiber membranes. *Adv. Mater. Interfaces*, **5**(14), 1800032.

40. Huang, K., Li, Q., Liu, G., Shen, J., Guan, K., and Jin, W. (2015) A ZIF-71 hollow fiber membrane fabricated by contra-diffusion. *ACS Appl. Mater. Interfaces*, **7**(30), 16157–16160.

41. Barhoum, A., Pal, K., Rahier, H., Uludag, H., Kim, I. S., and Bechelany, M. (2019) Nanofibers as new-generation materials: From spinning and nano-spinning fabrication techniques to emerging applications. *Appl. Mater. Today*, **17**, 1–35.

42. Wey, M.-Y., Chen, H.-H., Lin, Y.-T., and Tseng, H.-H. (2020) Thin carbon hollow fiber membrane with Knudsen diffusion for hydrogen/alkane separation: Effects of hollow fiber module design and gas flow mode. *Int. J. Hydrogen Energy*, **45**(12), 7290–7302.

43. Chi, Y., Li, T., Chong, J. Y., Wang, B., and Li, K. (2020) Graphene-protected nickel hollow fibre membrane and its application in the production of high-performance catalysts. *J. Memb. Sci.*, **597**, 117617.

44. Meng, X., Fan, Y., Zhu, J., Jin, Y., Li, C., Yang, N., Zhao, J., Sunarso, J., and Liu, S. (2019) Improving hydrogen permeation and interface property of ceramic-supported graphene oxide membrane via embedding of silicalite-1 zeolite into Al_2O_3 hollow fiber. *Sep. Purif. Technol.*, **227**, 115712.

45. Yunsi, C., Chong, J. Y., Wang, B., and Li, K. (2020) Pristine graphene membranes supported on ceramic hollow fibre prepared via a sacrificial layer assisted CVD approach. *J. Memb. Sci.*, **595**, 117479.

46. Fan, X., Liu, Y., and Quan, X. (2019) A novel reduced graphene oxide/carbon nanotube hollow fiber membrane with high forward osmosis performance. *Desalination*, **451**, 117–124.

47. Marti, A. M., Wickramanayake, W., Dahe, G., Sekizkardes, A., Bank, T. L., Hopkinson, D. P., and Venna, S. R. (2017) Continuous flow processing of ZIF-8 membranes on polymeric porous hollow fiber supports for CO_2 capture. *ACS Appl. Mater. Interfaces*, **9**(7), 5678–5682.

48. Wang, Y., Zhang, Y., and Liu, J. (2018) Preparation of glass hollow fibre nanofiltration membrane and Congo red dye separation. *Phys. Chem. Glas. J. Glas. Sci. Technol. Part B*, **59**(1), 55–59.

49. Cacho-Bailo, F., Etxeberría-Benavides, M., David, O., Téllez, C., and Coronas, J. (2017) Structural contraction of zeolitic imidazolate frameworks: Membrane application on porous metallic hollow fibers for gas separation. *ACS Appl. Mater. Interfaces*, **9**(24), 20787–20796.

50. Wang, Z., Kathiraser, Y., Ang, M. L., and Kawi, S. (2015) High purity oxygen production via BBCN perovskite hollow fiber membrane swept by steam. *Ind. Eng. Chem. Res.*, **54**(24), 6371–6377.

51. Faiz, R., Fallanza, M., Ortiz, I., and Li, K. (2013) Separation of olefin/ paraffin gas mixtures using ceramic hollow fiber membrane contactors. *Ind. Eng. Chem. Res.*, **52**(23), 7918–7929.

Chapter 5

Preparation and Characterization of Mixed Matrix Hollow Fibers

Hoda Daneshvar and Mir Saeed Seyed Dorraji
Applied Chemistry Research Laboratory, Department of Chemistry, Faculty of Science, University of Zanjan, Zanjan, Iran
dorraji@znu.ac.ir

In the recent years, nanomaterials have attracted a great attention worldwide due to their unique properties and their ability to improve membrane properties in terms of hydrophilicity, antifouling properties, membrane permeability, rejection, and adsorption capabilities. Mixed matrix membranes, which were originally designed for gas separation applications, due to the superior separation performance of some nanomaterials, have now emerged as a promising type of membranes in almost all membrane processes. The use of nanomaterials is now considered one of the most feasible ways to alter the chemical and physical properties of the membrane in order to improve its performance. This chapter provides a literature review of the developments of mixed matrix membranes, in hollow fiber configuration, and applied in different membrane

Hollow Fibers and Nanofibers in Membrane Science: Preparation, Characterization, and Applications
Edited by Alberto Figoli, Mir Saeed Seyed Dorraji, and Francesco Galiano

ISBN 978-981-4968-03-4 (Hardcover), 978-1-003-25663-2 (eBook)
www.jennystanford.com

processes. Special attention has been paid to the nanomaterials that have been used in membrane technology, including metal-organic frameworks, zeolites, carbon-based nanomaterials, and metal oxides. The preparation methods and the characterization of hollow fibers mixed matrix membranes have been discussed as well.

5.1 Introduction to Mixed Matrix Membranes

Over the past decades, membrane-based separation technologies have emerged in numerous chemical industrial processes, including CO_2 removal from natural gas, isolation and recovery of hydrogen and O_2/N_2 separation from air, owing to their high energy efficiency, low cost, and ease of operation. The membranes used in gas separation processes can fall into two main categories: polymeric membranes and inorganic membranes. Polymeric membranes are the most common membranes for gas separation because of their excellent processability, easy availability, and adequate mechanical properties. Despite these significant features, some major obstacles, including the trade-off limits between membrane permeability and selectivity (Robeson upper bound limit), poor resistance to plasticization and organic solvents, and aging effects during long-term utilization, restricted the benefits of using polymeric membrane in gas separation processes [1]. The trade-off limit between permeability and selectivity was first demonstrated by Robeson with the upper bound curves in 1991 and then updated in 2008, as shown in Fig. 5.1 [2, 3]. The graph demonstrates that an increase in membrane permeability comes with a decrease in membrane selectivity.

On the other hand, pure inorganic membranes such as metal-organic frameworks (MOFs) or zeolite membranes offer some advantages over organic membranes, such as superior gas separation performance, resistance to plasticization, and higher thermal and chemical stability. Most importantly, inorganic membranes pass the Robeson upper bound limit and approach the desired performance in gas separation. However, the fabrication of continuous defect-free membranes from the single microcrystals remains very challenging and time consuming [5, 6]. Therefore, efforts to solve these problems have led to the fabrication of new hybrid membranes known as mixed matrix membranes (MMMs).

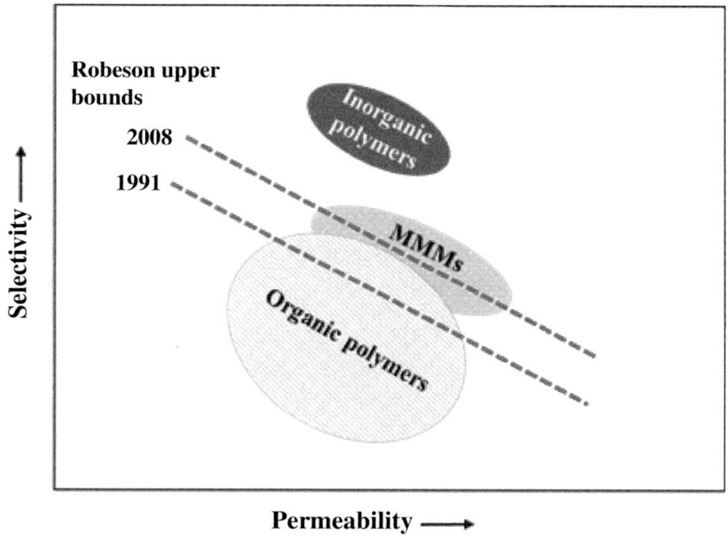

Figure 5.1 Schematic presentation of the trade-off between permeability and selectivity of dense membranes in gas separation with the 1991 and 2008 Robeson upper bounds. Reprinted from Ref. [4] with permission of the Royal Society of Chemistry.

MMMs, which are composed of micro- or nanoscale additives dispersed in a polymeric matrix, can combine excellent flexibility and processability of polymers with the superior gas separation performance of porous inorganic materials (due to their sieving ability) [7–10]. Particularly, additives can significantly enhance the properties and performance of polymeric membranes, while reducing the cost and fabrication difficulties of ceramic membranes.

Although the idea of MMMs was initially developed to surpass the Robeson upper bound limit in gas separation [11], their application was not just confined to this area but has been extended also to other membrane processes, such as water treatment, using adsorptive or photocatalytic membranes. Nanoparticles, in fact, can increase the selectivity of ultrafiltration, nanofiltration, and reverse osmosis membranes by controlling the solute permeation by mechanisms such as adsorption or electrostatic interactions [12]. The use of nanoparticles is also one of the most promising methods to overcome the fouling problem occurring in many membrane processes.

Membrane fouling is a process by which suspended or dissolved particles, colloidal particles, and macromolecules are deposited on the surface or into the pores of the membrane, which results in pore blocking and a severe decrease in flux and selectivity of the membrane. Generally, there are four types of membrane fouling: (1) inorganic fouling, (2) colloid fouling, (3) microbial/biological fouling, and (4) organic fouling. The interaction between membrane and foulant is responsible for membrane fouling and can be affected by different factors, including:

- **Surface hydrophilicity:** It has been well documented that membranes with hydrophilic surfaces are less susceptible to fouling due to the less interaction between the foulant and the membrane surface. In particular, a hydrated layer forms on the surface of highly hydrophilic membranes, which prevents the adsorption and deposition of hydrophobic foulants onto the membrane surface.
- **Surface charge:** Electrostatic interactions between charged solutes and a porous membrane have been an important rejection mechanism. If the membrane surface has a similar charge to the foulant molecule, electrostatic repulsion forces will decrease its adsorption and decrease the membrane fouling.
- **Roughness:** There is a strong correlation between membrane fouling and the surface roughness. It was shown that compared to hydrophobic and rougher membranes, the hydrophilic membranes with smooth membrane surfaces are less prone to colloidal fouling.

Among different methods to modify the physical and chemical characteristics of the membrane to decrease the fouling problem, the addition of nanoparticles is one of the easiest ways to control surface roughness, surface hydrophilicity, and surface charge in order to hinder or mitigate membrane fouling [13, 14].

The wide choice of the polymer matrix and inorganic additives provides vast combinations of these two phases, leading to the rich diversity of MMMs. However, the rational selection of polymer matrix and inorganic material is the critical factor for the preparation of MMMs with favorable properties and performance. In particular, the additive should have an intrinsic molecular sieving ability (for use in gas separation technologies), good interfacial compatibility with the polymeric matrix, and high dispersibility from moderate to high loadings [15, 16].

MMMs are commonly fabricated in either symmetrical or asymmetrical structure. Symmetrical membranes consist of a dense film in which the filler particles are uniformly dispersed. Asymmetrical membranes consist of a dense selective layer coated on a porous support layer (to provide mechanical strength) [17]. Asymmetrical membranes themselves can be prepared as flat films or hollow fibers. Hollow fiber MMMs are optimal for industrial applications due to their high specific area, compact modulus design, flexibility, ease of scalability, and lower final cost [18].

5.2 Filler/Polymer Interface Morphology

Ideally, using porous fillers in the polymer matrix should increase both permeability and selectivity, if the filler is completely getting wet by the polymer solution. However, the formation of interfacial defects is almost inevitable in the fabrication of MMMs. Interface defects can be categorized as follow: (1) interface voids resulting from the detachment of particles from the polymer chains, (2) rigidification of the polymer around the inorganic particles, and (3) partial pore blockage by particles (Fig. 5.2) [19, 20]. All these issues arise from the lack of compatibility between the polymer matrix and the inorganic particle.

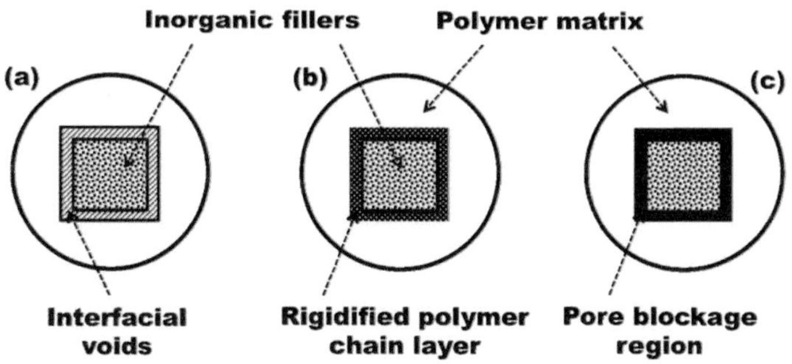

Figure 5.2 Different types of interfacial defects. Reprinted from Ref. [21] with permission of the American Chemical Society.

The most common interfacial defect that can be seen in MMMs is the interfacial void, also known as sieve-in-a-cage morphology, caused by the detachment of particle surface from polymer due to the shrinkage stress during the membrane formation. This type of interfacial defect can severely affect the performance of the membrane and result in a significant loss in selectivity. Rigidification of the polymer is caused by the immobilization of the polymer chains around the filler particles. The most significant effect of rigidification is on membrane permeation, which is decreased with or without a negative impact on selectivity [22]. Pore blockage happens when the polymeric chains penetrate the porous structure of the filler. Blockage always leads to a decrease in membrane permeability, but it is not necessarily considered a problem in the fabrication process. Pore blockage can be used to tailor the molecular sieving properties of porous fillers and increase the selectivity of the membrane [23]. There are several methods to minimize the non-ideal morphology of hollow fiber membranes, such as controlling the particle sizes [24] and surface modification of particles [25].

In general, to obtain homogenous hollow fiber MMMs, utilizing particles with nano or sub-micron size particles is preferred because smaller particles can provide higher polymer/particle interface area, and have a stronger affinity with polymeric matrix and more substantial improvement of membrane performance and morphology [26, 27]. Surface modification is the most effective method to approach enhanced particle/polymer adhesion. The surface functionalization of particles can change the characteristics of the particles and increase the compatibility between particle and polymer, and eliminate the formation of non-selective interfacial voids. And finally, the rational selection of particle/polymer pairs is the first and most crucial step to achieve well-defined hollow fiber MMMs. The polymers that are mostly used in the fabrication of hollow fibers have been thoroughly discussed in Chapter 2. Thus, the next section mainly focuses on inorganic fillers.

5.3 Inorganic Fillers in Hollow Fiber MMMs

The general role of fillers in MMMs is to provide a favored pathway for permeating molecules through the membrane, increasing the selectivity or enhancing specific properties of the membrane (such as antifouling). Concerning the filler, its chemical properties, functional groups on the surface, the distribution of particle sizes, and its shape are important variables that affect the integrity and properties of the final membrane [28]. The fillers that are mainly used in the fabrication of MMMs can be classified into four main categories: MOFs, zeolites, carbon-based materials, and, metal oxides and metal nanoparticles.

5.3.1 Metal-Organic Frameworks

MOFs, also known as novel zeolite-like materials, are a class of 3D porous structures that consist of transition metal cation nodes that are linked together by multidentate organic ligands with a coordination-type connection. Due to the fascinating properties of MOFs, such as tunability in size, shape, chemical microenvironment, and, more importantly, adaptable chemical properties and porosities through the metal–linker combination, these materials are considered a breakthrough in the field of MMMs. Some of the most used MOFs in MMMs are illustrated in Fig. 5.3.

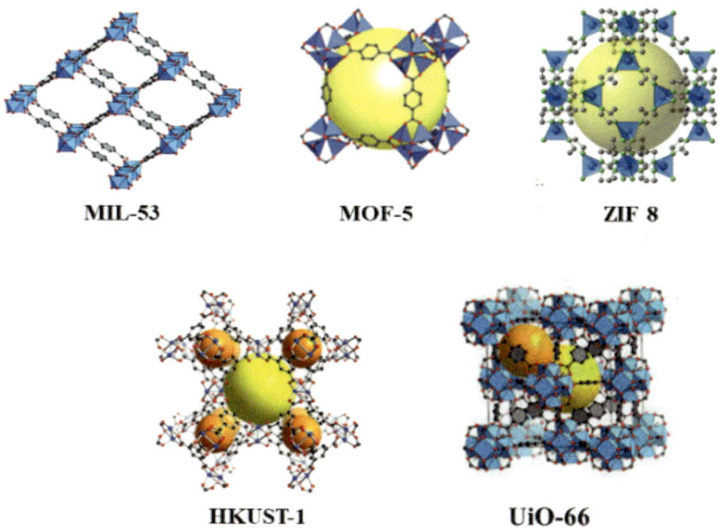

Figure 5.3 Some of the most used MOFs in hollow fiber MMMs. Reprinted from Ref. [29] with permission of the Royal Society of Chemistry.

The diversity in the chemical composition of MOFs allows researchers to design structures with stronger interactions with the polymeric matrix, to reduce the voids in membrane structure and achieve complete dispersion. The interaction between the organic ligand with the surrounding polymer chain prevents interfacial defects and increases the selectivity of the MOF-based MMMs [30]. Besides, by designing MOFs with polar functional groups such as $-NH_2$, -OH and, $-NO_2$ on their ligands, the selectivity of the membrane and compatibility of MOF with polymer chains can be increased. As an example, phenylacetate-functionalized UiO-66-NH_2 shows no defects in membrane structure due to the possible interaction of MOF–NH_2 groups or PSM–amide linkages with imide groups in the Matrimid® polymer through hydrogen bonding, and, π–π stacking interaction between aromatic rings from PhCO-, as shown in Fig. 5.4 [31].

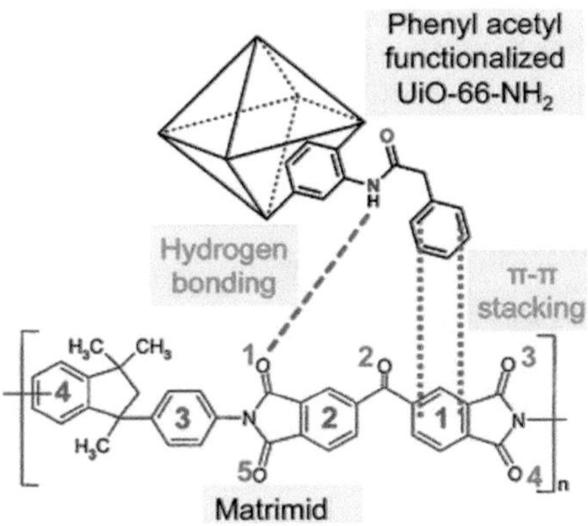

Figure 5.4 The interaction between the Matrimid® polymer and phenyl acetyl modified UiO-66-NH_2. Reprinted from Ref. [31] with permission of the Royal Society of Chemistry.

Zeolitic imidazole framework (ZIF)-8 is composed of Zn^{2+} cations and 2-methylimidazole anions and can be easily synthesized in water. High chemical and thermal stability and suitable pore size of ZIF-8 make this material very interesting in gas separation processes. The pore dimeter of ZIF-8 is 11.6 Å with a six-ring window aperture of 3.4 Å, which is close to the diameter of CO_2 and CH_4 [32]. Benavides et al. [33] prepared BPI-ZIF-8 hollow fiber MMMs for H_2/CO_2 separation. They reported that ZIF-8 contributed to a 65% increase in H_2

permeance at constant ideal selectivity at high temperature (150°C) and pressure (up to 30 bar). Compared to pure polybenzimidazole (PBI) hollow fiber membrane, PBI/ZIF-8 mixed matrix hollow fiber showed a significant improvement in H_2 permeance from 65 GPU to 107 GPU, while the ideal H_2/CO_2 selectivity remained constant at 18 [33].

Recently, Khan et al. [34] fabricated polysulfone (PSf)/ZIF-8 hollow fiber membranes for natural gas purification, with a ZIF-8 concentration ranging from 0.25 to 1.25 wt%. The higher loading percentage of nanoparticles led to the breakage or no fabrication of the membrane at all. Incorporation of ZIF-8 to PSf matrix led to an improvement in thermal stability and glass transition temperature. A significant improvement of 36% in CO_2 permeability and 28% in CO_2/CH_4 selectivity was observed in the membrane containing 0.5 wt% ZIF-8 compared to the pure PSf hollow fiber membrane. The enhancement was due to the uniform distribution of ZIF-8 particles in the polymer matrix and partial pore blockage, which hindered CH_4 permeation (with a kinetic diameter of 3.8 Å) and enhanced CO_2/CH_4 selectivity. However, in slightly higher pressure (8 and 10 bar), the performance of pure PSf membrane was better than that of PSf/ZIF-8, which could be a result of CO_2-induced plasticization at elevated pressure.

Hence, the ZIF-8-based hollow fiber membrane can be a promising candidate for natural gas purification at low feed pressure. Besides, based on this study, the amount of incorporated nanoparticles played an important role in properties and performance determination, as the selectivity of membrane diminished drastically at the higher percentage of ZIF-8 (≥0.75 wt%), as shown in Fig. 5.5.

This observation was possibly due to the agglomeration of the ZIF-8, which limited the contact area and resulted in non-selective voids in membrane structure that allowed both gases to penetrate without discrimination [34].

Mubashir et al. [35] investigated the effect of process parameters on the gas separation performance of CA/NH_2-MIL53-Al hollow fiber membranes. The effect of temperature, CO_2 feed composition, and feed pressure on the separation of CO_2 from CO_2/CH_4 mixture was studied using central composite design coupled with response surface methodology (RSM). According to the results of the experiments, increasing temperature from 30°C to 50°C reduced the CO_2/CH_4 separation factor because of the increased diffusion of CH_4 molecules through the membrane compared to CO_2 molecules. While increasing feed pressure from 3 bar to 15 bar and increment of CO_2 feed composition from 15 to 42.5 vol% increased the separation factor. The results also revealed that the most influential parameter on CH_4 and CO_2 permeation and on CO_2/CH_4 separation was the feed pressure [35].

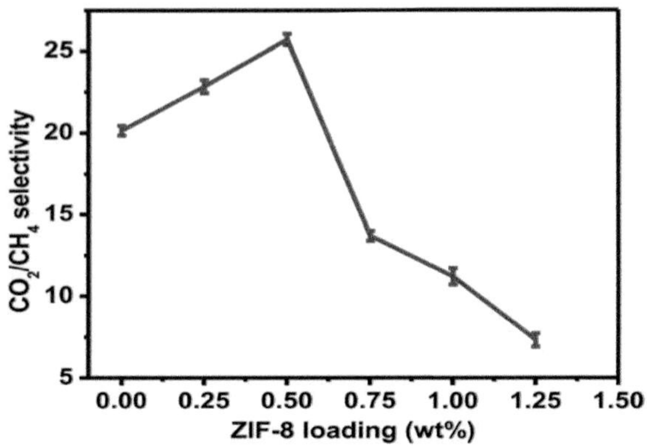

Figure 5.5 The effects of various ZIF-8 loadings on the membrane selectivity. Reprinted from Ref. [34] with permission of Elsevier.

Another study focused on the synthesis of polyimide/$Cu_3(BTC)_2$ hollow fiber MMMs by the dry-/wet-spinning method [36]. The obtained membrane showed smooth and dense layers at the outside and inner surfaces, as well as a finger-like void layer and a sponge-like supporting layer between them. The results of the experiment showed that using 6 wt% of $Cu_3(BTC)_2$ MOF, the H_2 permeance increased by 45% and the selectivity increased two–three times with respect to the pure polyimide membrane [36].

Liu et al. [37] employed MAF-6 (RHO-[Zn(eim)2] (MAF-6, Heim = 2-ethylimidazole)) dispersed in a poly(ether-block-amide) (PEBA) matrix to prepare hollow fiber MMMs on a ceramic tubular substrate to form a composite membrane for the recovery of ethanol from an aqueous solution via pervaporation. The results showed that both flux and separation of the membrane were remarkably enhanced by uniformly incorporating MAF-6 nanoparticles. Particularly, a flux of 4446 g/m^2 and a separation factor of 5.6 were achieved using 7.5 wt% of MAF-6 in the PEBA membrane. The enhanced flux was attributed to the strong interaction of MAF-6 and ethanol molecules as it is evident from the membrane contact angle. The ethanol contact angle of the MAF-6/PEBA membrane was decreased as an indication of the more organophilic nature of the membrane and, therefore, a higher affinity of the nanocomposite hollow fiber for ethanol. On the other hand, because of high hydrophobicity, MAF-6 showed a very weak interaction with water molecules. Thus, the diffusion resistance for water molecules was higher than that of ethanol, which also increased the separation factor [37].

5.3.2 Zeolites

Zeolites are a group of microporous crystalline aluminosilicates that are constructed from corner-sharing tetrahedra SiO_4 and AlO_4 as building units, with the presence of cations such as Ca^{2+}, Na^+, K^+, and Mg^{2+}. The hydrophilicity and the adsorption properties of the zeolite are dramatically dependent on the Si/Al ratio. The molecular sieving properties of zeolites make them suitable candidates to be used as fillers in MMMs, and for many years they have been successfully employed as adsorbents for CO_2/CH_4 separation.

The selectivity of MMMs is greatly affected by the pore size of the inorganic filler according to the diffusion model. Therefore, it is important to consider the pore size of zeolite as a criterion for filler selection, as well as the size of the particle itself. Generally, smaller particles are preferred, due to their higher surface area and better interaction with neighboring polymer chains. The pore size of zeolites lies typically in the 0.4–1.2 nm range, which is ideal, especially for gas separation applications.

In 2008, Ismail et al. synthesized silane-modified zeolite 4A/polyethersulfone (PES) hollow fiber MMMs for gas separation application, and the results proved that Dynasylan Ameo silane successfully improved the adhesion between the polymer and zeolite surface. Dynasylan Ameo is a silicon-based coupling agent with the general formula of $(RO)_3$-Si-$(CH_2)_3$-X, where RO is a hydrolyzable group, and X is an organofunctional group of amino. Consequently, this modification yielded good compatibility between PES–zeolite 4A and improved the gas separation performance of hollow fiber MMMs. PES-modified zeolite 20 wt% silane yielded a significant selectivity of 28.75 and 4.78 for O_2/N_2 and CO_2/CH_4, respectively [38].

Rezaei-Dasht Arzhandi et al. prepared ZSM5/polyvinylidine fluoride (PVDF) hollow fiber MMMs for CO_2 removal via membrane contactor. The water contact angle of the membrane increased from 84° to 104°, after the addition of 5 wt% ZSM5 nanoparticles, indicating an increased hydrophobicity. The increased hydrophobicity alleviated the wetting of the membrane and enhanced the diffusion of the gas phase through the membrane. Therefore, CO_2 absorption was increased by 177% [39].

Despite the numerous publications on using zeolites as fillers in MMMs, this family of fillers has not gained commercial acceptance, mostly due to their poor interfacial compatibility with polymer chains [40].

5.3.3 Carbon-Based Fillers

Carbon-based materials are a unique platform of materials for small molecule separations based on transport through controlled pore sizes (0.3 nm to 0.7 nm). These materials have received worldwide attention due to their large surface area, excellent mechanical strength, and high thermal stability. But most importantly, they can be easily functionalized with various functional groups in order to introduce new properties, which ultimately enhance the membrane filtration performance. Therefore, carbon-based materials are great candidates to be used as fillers in MMMs for gas separation, water treatment, and preparation of antifouling membranes [41].

In a study conducted by Favvas et al. [42], phenol-functionalized multi-walled carbon nanotubes (MWCNTs)/co-polyimide-based hollow fiber membranes, with enhanced gas permeation properties, were prepared. The SEM images revealed a good dispersion of carbon nanotube in the sample with the highest concentration of filler (4% w/w). The results showed that the permeance increased with the increase in MWCNT concentration into the membrane matrix. The permeance of CH_4 increased from 0.82 GPU to 13.9 GPU at 100°C with the addition of 4% w/w MWCNT. Hence, the membrane exhibits high permeance coefficients and can successfully be used for gas separation applications (up to 100°C) in industrial processes [42].

Another study focused on the role of pretreatment of MWCNT (dry air oxidation and surfactant dispersion) on the gas separation performance of MWCNT/polyetherimide (PEI) hollow fiber membranes [43]. It was found that after pretreatment of MWCNT, the membrane permeability increased by 60%, while the selectivity of O_2/N_2 gas pair remained the same. The authors stated that dry air oxidation can open the caps at the end of the nanotubes and allows the hollow structure to be more accessible to the passage of gas molecules. On the other hand, surfactant dispersion using Triton X100 avoided the agglomeration of MWCNT. Consequently, the better interaction between the filler and polymer matrix could reduce the formation of voids, guaranteeing a high selectivity of the membrane [43].

Zahri et al. prepared graphene oxide (GO)/PVDF hollow fiber membranes for gas separation. The authors reported that the addition of GO could create new small passes for CO_2 molecule, restricting N_2 and CH_4 permeation and resulting in a 158% increase in CO_2/N_2 selectivity [44].

Mukherjee et al. [45] prepared carbon nanoparticle/PSf hollow fiber mixed matrix ultrafiltration membranes for the absorption of benzene, toluene, and phenol. The rejection values increased with increasing nanoparticle concentration. As can be seen in Fig. 5.6, the rejection values for benzene increased with the addition of nanoparticles. But as more benzene molecules were adsorbed onto the membrane surface, the adsorption resistance increased, resulting in a reduction in permeate flux. The membrane containing 2 wt% of carbon nanoparticles was considered the optimum, showing a permeate flux of 7.32×10^7 m^3/m^2s and a rejection of 97.7% [45].

In 2019, Park et al. [46] reported the fabrication of GO-incorporated hollow fiber PES membranes as pressure retarded osmosis (PRO) thin-film composite membranes. Their results indicated that the incorporation of 0.2 wt% GO nanosheets in the PES hollow fiber support layer enhanced the initial PRO flux from 38.62 L/m^2h to 43.74 L/m^2h, without deteriorating the mechanical properties. The enhancement in initial flux was due to the increased hydrophilicity and the decreased internal concentration polarization of the membrane.

In a research by Modi et al. [47], PES hollow fibers with embedded Fe_3O_4 nanoparticles, decorated with carboxylated GO nanosheets (Fe_3O_4–cGO), were prepared as high flux, antifouling ultrafiltration membranes for the separation of proteins. The water contact angle of PES hollow fibers was reduced from 79.3±1.4° to 50.6 ± 0.5° upon addition of 0.1 wt% Fe_3O_4–cGO nanohybrid, which was due to the presence of oxygen-containing surface functional groups on these modified membranes. Moreover, the higher pure water flux (110.0 ± 3.8 L/m^2·h) and the high flux recovery ratio of 97.8% were considered a great improvement in membranes performance.

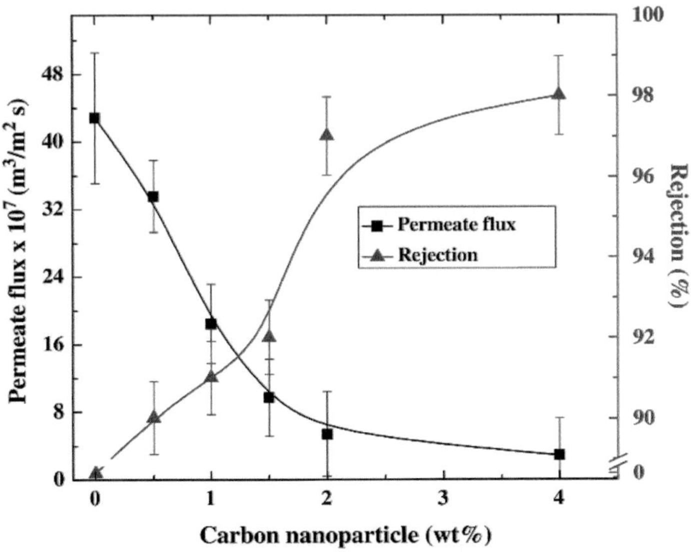

Figure 5.6 The effect of carbon nanoparticle concentration on permeate flux and rejection of benzene. Reprinted from Ref. [45] with the permission of Elsevier.

In 2020, Kumari et al. [48] prepared carboxylated MWCNT (MWCNTs-COOH) and TPGS (a vitamin E derivative) modified PES mixed matrix hollow fiber membranes with enhanced permeation flux and fouling resistance. The prepared membranes were characterized in terms of surface roughness, hydrophilicity, and separation properties. The AFM results showed that the roughness parameter (R_a) of the membrane decreased from 8.9 nm to 7.4 nm after the blending of 0.025 wt% MWCNTs-COOH in mixed matrix hollow fiber membrane. The functional groups on the surface of MWCNTs, on one hand, and the low concentration of MWCNTs-COOH, on the other hand, led to the uniform distribution of nanoparticles in the membrane matrix. Moreover, the water contact angle significantly decreased for the membrane containing 0.025 wt% MWCNTs-COOH. It is worth mentioning that both surface roughness and water contact angle increased when the nanoparticle concentration increased to 0.05 wt%, probably due to the agglomeration of MWCNTs-COOH nanoparticles at higher concentrations. The high hydrophilicity and low surface roughness of modified mixed matrix hollow fiber membrane, containing 1 wt% of TPGS and 0.025 wt% of MWCNTs-COOH, resulted in ~187% higher permeation as compared to that of the pristine PES hollow fiber membrane, and with a flux recovery of 89.3 ± 2.1% toward proteins [48].

5.3.4 Metal Oxide and Metal Nanoparticles

Nanosized metal oxides have been widely applied in different water treatment processes due to their large surface area, a large number of active sites for adsorption, and unique physical and chemical properties.

Isloor et al. [49] fabricated PPSU/SnO_2 ultrafiltration hollow fiber membranes and investigated their ability to remove the toxic reactive dyes from the aqueous media. The results demonstrated that the membrane containing 0.4 wt% of SnO_2 had a lower contact angle (63.7°), pure water flux of 362.9 L/m^2h, and a high potential for dyes rejection (94% rejection for Reactive Black-5, and 73% rejection for Reactive Orange-16). They also discussed that the Donnan exclusion effect was the responsible mechanism of rejection rather than the size exclusion effect (steric hindrance). The molecular weights of Reactive Black-5 and Reactive Orange-16 dyes are 991.82 g/mol and 617.54 g/mol, respectively. The fabricated hollow fiber membranes were in the range of ultrafiltration, so the size exclusion could not be the main reason for the high rejection values measured. On the other hand, the prepared membranes were negatively charged due to the presence of SnO_2, the same charge borne by the two dyes investigated. Thus, the electrostatic repulsion force by the Donnan exclusion effect was the more predominant mechanism for the removal of dyes [49].

Ahmad et al. [50] investigated the antifouling behavior of PES/PVP/TiO_2 mixed matrix hollow fiber membrane for the treatment of aqueous solutions containing humic acid. Based on the analyses, the presence of 2 wt% of TiO_2 led to the lowest fouling resistance, high permeate flux of 23.851 kg/m^{2}h, and high humic acid rejection of 97.89%. The improved antifouling properties can be justified by the improved hydrophilicity of the membrane, which prevented humic acid to be deposited onto the membrane surface. This explanation was supported by the water contact angle, which increased from 72.93° to 69.47° after the addition of TiO_2 nanoparticles. It was observed that a further increase in nanoparticle percentage slightly increased the overall fouling resistance due to the TiO_2 agglomeration. Therefore, well distribution and dispersion of TiO_2 are important for enhancing the overall membrane surface hydrophilicity and antifouling properties [50].

Simone et al. [51] modified PES hollow fibers membrane by introducing TiO_2 nanoparticles (with a concentration ranging from 0.3 to 1 wt%) in the polymeric dope and evaluated the photocatalytic performance of the resultant mixed matrix hollow fiber membranes. The SEM images revealed that the hollow fibers with a high concentration of TiO_2 showed an irregular inner contour, which could be due to the impaired coagulation of the fiber inner surface. The best morphology and highest pure water permeability (75 L/h m^2 bar) were achieved with the addition of 0.3 wt% of TiO_2. Moreover, the photocatalytic activity of the membranes was tested by following the degradation of methylene blue under UV light irradiation. The prepared MMMs were photoactive toward methylene blue degradation, even if a very fast increase in the permeability value revealed that the selective layer was damaged during UV-A irradiation, demonstrating that the prepared membranes were not stable under UV light [51].

As mentioned before, the agglomeration of nanoparticles is the main obstacle of MMMs. In order to address this issue, Galiano et al. [52] used two different approaches in order to improve the dispersion of TiO_2 nanoparticles in the dope solution evaluating (1) a chemical and (2) a chemical/mechanical modification of the nanoparticles. The simple chemical modification was carried out by functionalizing TiO_2 nanoparticles with the surfactant sodium dodecylsulfate (SDS). In the second method, the SDS modification was combined with a mechanical dispersion of nanoparticles using ball milling and ultrasonication. The results of experiments showed that the chemical/mechanical modified TiO_2 exhibited the best performance in terms of long-term stability. Moreover, the photocatalytic activity of the resulting membranes was evaluated by following the photodegradation of methylene blue. The results showed that the membrane containing 0.5 wt% TiO_2 can effectively remove methylene blue up to 97% in 300 min [52].

5.4 Fabrication of MMMs

In the preparation of MMMs, commonly, the inorganic particles should have selectivity superior to the pure polymer. The main procedure for fabricating hollow fiber MMMs is through conventional hollow fiber preparation methods described in Chapter 2. Instead of the dope polymer solution, dope particle/polymer suspension is used. There are three different methods (Fig. 5.7) to prepare the particle/polymer dope suspension, prior to the spinning process: (1) dispersion of the filler in a solvent and, subsequently, the addition of polymer into the particle dispersion; (2) addition of a predetermined amount of particles to the prepared polymer solution; and (3) preparation of separate solutions of polymer and particle suspension and the following addition of the particle suspension into the polymer solution [53].

The first and the third methods are generally preferred since they permit a suitable dispersion of the inorganic particles due to the low viscosity of the suspension.

However, it should be noted that the fabrication of hollow fiber MMMs with minimum defects remains challenging. As mentioned before, the performance of hollow fiber MMMs can be significantly affected by loading percentage and agglomeration of particles, particle size, and spinning process parameters such as dope composition and its extrusion rate, take-up speed, and air-gap distance [54].

For example, Mubashir et al. reported that an increase in selectivity and a decrease in permeability and non-selective voids in membrane structure were observed with the increment of take-up speed. They also concluded that the optimization of spinning parameters could be considered a feasible and efficient method to enhance the performance of hollow fiber MMMs [55]. Additionally, it should be mentioned that in the spinning of MMMs, a high particle loading percentage (>20 wt%) usually leads to a significant suspension viscosity responsible for particle agglomeration, which may cause severe issues for practical fiber spinning.

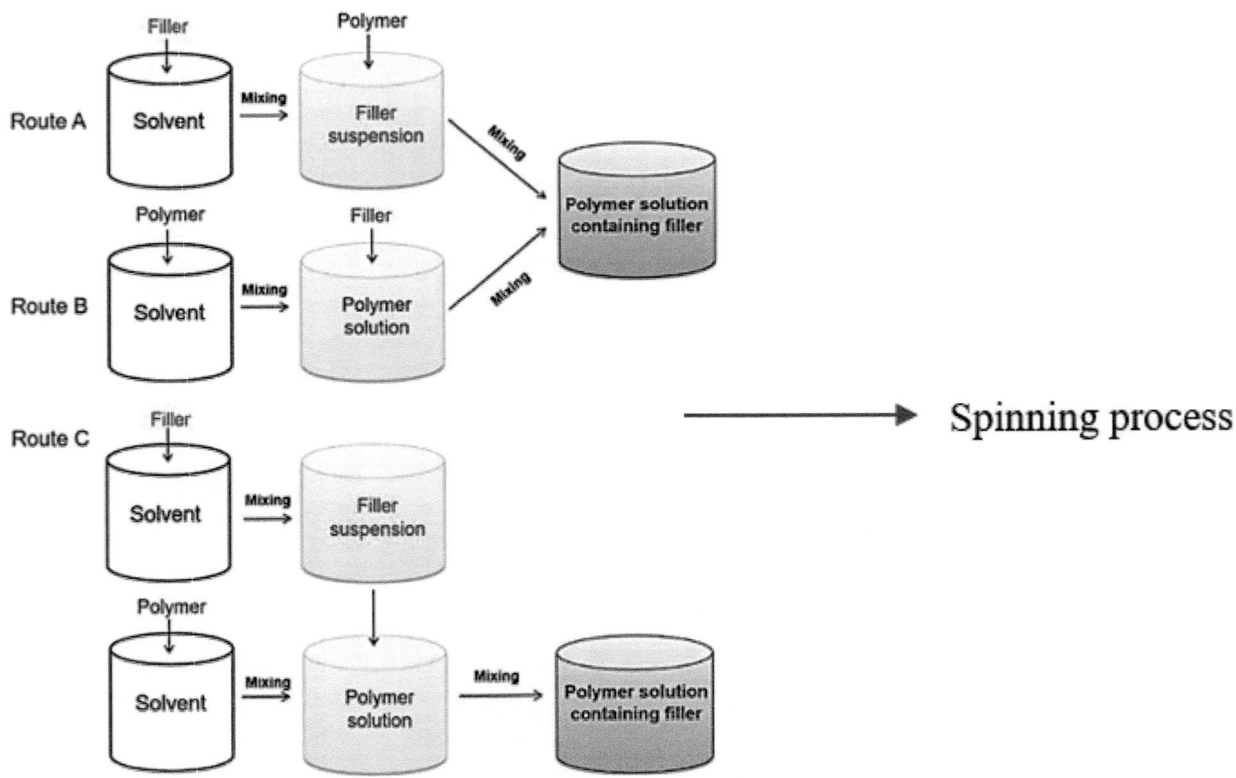

Figure 5.7 Different methods used for the preparation of the dope suspension. Reprinted from Ref. [30] with permission from the Royal Society of Chemistry.

5.5 Characterization of Hollow Fiber MMMs

The main characterization techniques employed for hollow fiber MMMs are SEM, EDX, FTIR, DSC, TGA, XRD, and, AFM.

SEM is used to study the surface and cross-sectional morphology of the prepared MMMs, as shown in Fig. 5.8.

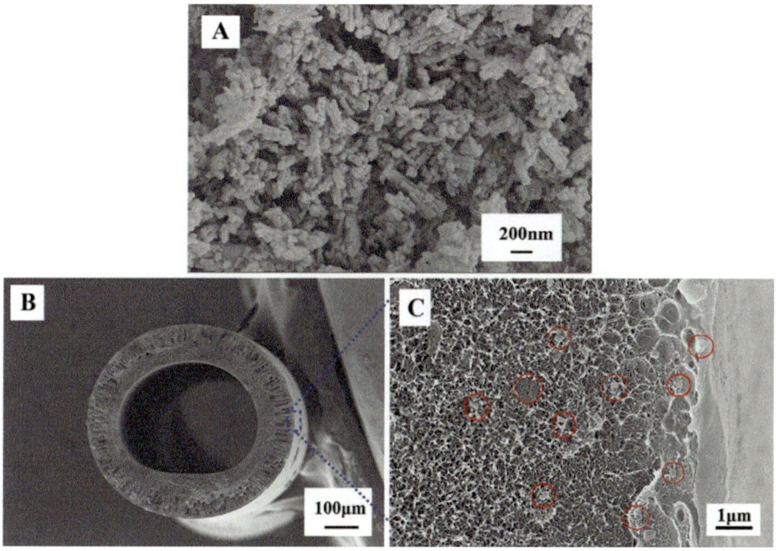

Figure 5.8 SEM images of (A) MIL-53 and (B and C) hollow fiber MMMs cross sections (red circles represent the presence of MIL-53 particles). Reprinted from Ref. [54] with permission of the Royal Society of Chemistry.

Additionally, any change in membrane structure can be thoroughly evaluated using SEM analysis. For example, Fig. 5.9 illustrates the cross-sectional image of PSf hollow fiber membranes before and after the addition of GO. The SEM reveals that some finger-like structures beneath the outer surface layer were diminished after the addition of GO. SEM observation allowed the authors to conclude that the addition of GO to the membrane has retarded the phase-inversion process, thus suppressing the development of a finger-like structure near the outer surface of the membrane. EDX analysis provides a map in which the distribution of inorganic filler elements into the polymeric matrix can be observed [53].

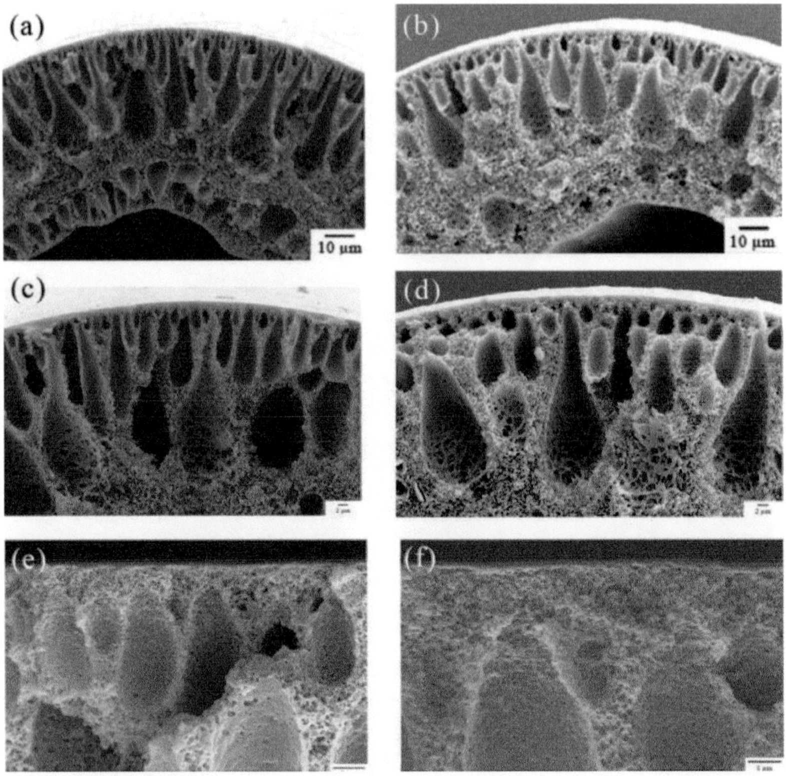

Figure 5.9 SEM cross-sectional images of (a, c, e) pristine PSf and (b, d, f) GO/PSf hollow fiber MMMs at different magnifications. Reprinted from Ref. [44] with permission of the Royal Society of Chemistry.

FTIR analysis is used to study the interactions between the inorganic particle and polymer chains, as well as the functional groups of the particles. Usually, the FTIR spectra of inorganic particles, pure polymer membrane, and MMMs are compared to investigate the presence of fillers in the polymeric matrix. The FTIR spectra of MIL-53 (as the filler), neat cellulose acetate hollow fiber membrane, and MIL-53/CA hollow fiber MMMs are shown in Fig. 5.10. The FTIR spectra of hollow fiber MMMs showed the peaks of both filler and polymer phases and thus confirmed the presence of MIL-53 in membrane structure.

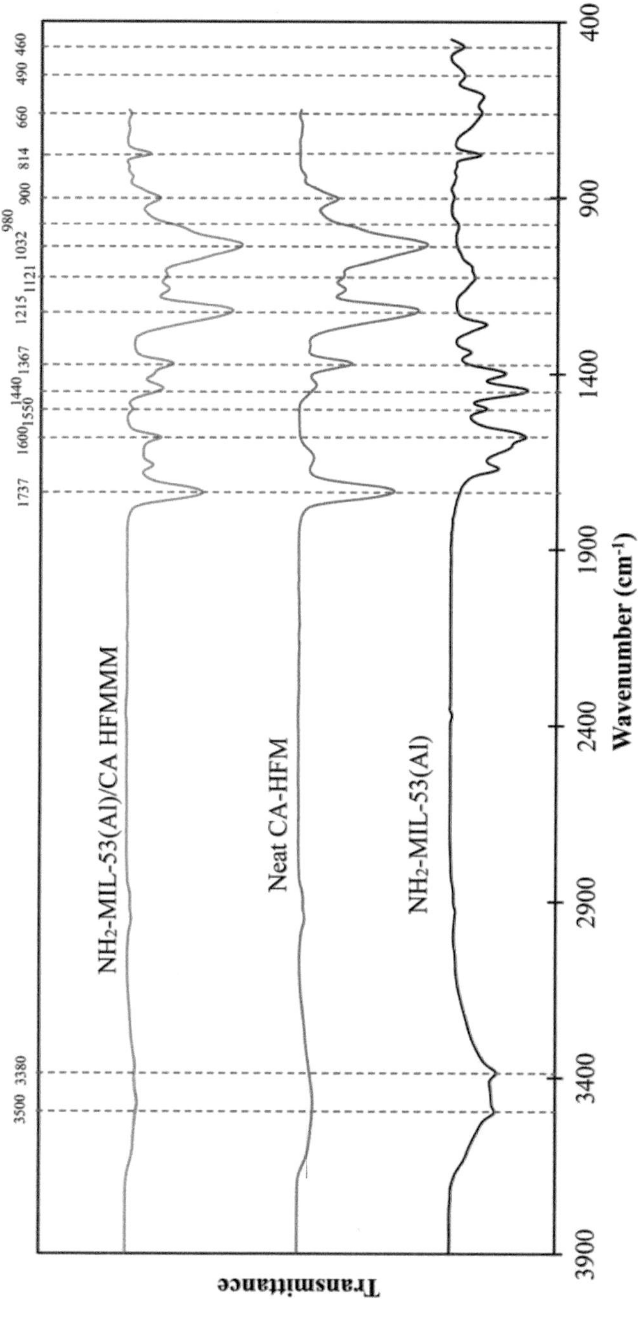

Figure 5.10 The FTIR spectra of MIL-53 MOF, neat cellulose acetate hollow fiber membrane, and MIL-53/CA hollow fiber MMMs. Reprinted from Ref. [56] with permission of the American Chemical Society.

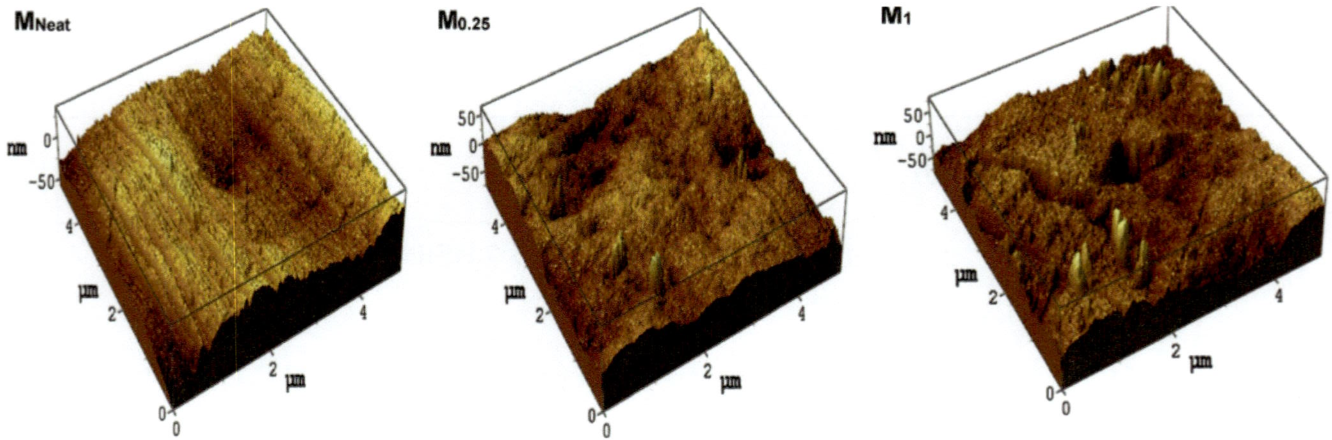

Figure 5.11 AFM images of the surfaces of neat PSf and PSf/ZIF-8 hollow fiber membranes fabricated at various ZIF-8 loading: M_{Neat}—0 wt%, $M_{0.25}$—0.25 wt%, and M_1—1 wt%. Reprinted from Ref. [34] with permission of Elsevier.

DSC and TGA measurements are used, respectively, to determine the T_g of the MMMs, at different filler loadings, and the thermal stability. The DSC measurements are of great importance because when the T_g of the MMMs is much higher than the value of the bare polymer, the rigidification of the chains can happen. Also, the specific interaction between both phases of filler and polymer can be predicted from the T_g value. XRD analysis is used to study the crystallinity and confirm the synthesis of the nanoparticles. In addition, it indicates whether there has been a modification of the particles into the polymeric matrix [53].

The AFM images are basically used to evaluate the roughness of the membrane surface before and after the addition of the nanoparticles. As the membrane morphology, hydrophilicity, and performance are greatly dependent on the surface roughness, AFM is one of the most important characterization techniques for MMMs. Generally, an increased amount of nanoparticles causes agglomeration, which leads to a high roughness of membrane surface and adversely affects its performance. For example, Fig. 5.11 shows the AFM images of neat and mixed matrix hollow fiber membrane with 0.25 wt% and 1 wt% ZIF-8 nanoparticles. The mean surface roughness (R_a) of the membranes increased from 9.70 nm (for the neat membrane) to 11.30 and 18.70 for the membranes with 0.25 and 1.0 wt% ZIF-8, respectively. It is clear that the roughness parameters increased slightly at the low concentration of nanoparticles, which indicates the better compatibility between ZIF-8 and the polymer matrix. However, a higher concentration of ZIF-8 significantly increased the membrane roughness due to a higher accumulation and less dispersion of the nanoparticles [34].

5.6 Conclusion

The use of nanomaterials has opened a new window in membrane technology specially to address the issues related to the trade-off effect between permeability and selectivity or rejection. In fact, the use of nanomaterials in the polymer matrix is a very promising and easy way to improve the chemical and physical properties of the membranes. MMMs can combine the excellent flexibility and processability of polymers with the special and unique properties of nanomaterials. Depending on the nature of the used nanomaterial, new features, including antifouling properties, adsorptive capacities, and photocatalytic activity, can be achieved. The key factors in fabricating a MMM are the compatibility between nanoparticles and polymer matrix and the uniform distribution of nanoparticles into the polymer matrix. Additionally, there are still some other issues present in the fabrication of MMMs. Among them, the agglomeration of the nanoparticles remains one of the main obstacles that hamper the preparation of MMMs with a uniform nanoparticle distribution. The future of the mixed matrix hollow fiber membranes is greatly related to the development of new methods to minimize the agglomeration of nanoparticles in the polymer matrix in order to fully exploit all the advantages stemming from the various nanomaterials.

References

1. Basu, S., Khan, A. L., Cano-Odena, A., Liu, C., and Vankelecom, I. F. J. (2010) Membrane-based technologies for biogas separations. *Chem. Soc. Rev.*, **39**(2), 750–768.
2. Robeson, L. M. (2008) The upper bound revisited. *J. Memb. Sci.*, **320**(1–2), 390–400.
3. Robeson, L. M. (1991) Correlation of separation factor versus permeability for polymeric membranes. *J. Memb. Sci.*, **62**(2), 165–185.
4. Jeazet, H. B. T., Staudt, C., and Janiak, C. (2012) Metal-organic frameworks in mixed-matrix membranes for gas separation. *Dalt. Trans.*, **41**(46), 14003–14027.
5. Rangnekar, N., Mittal, N., Elyassi, B., Caro, J., and Tsapatsis, M. (2015) Zeolite membranes: A review and comparison with MOFs. *Chem. Soc. Rev.*, **44**(20), 7128–7154.
6. Bux, H., Liang, F., Li, Y., Cravillon, J., Wiebcke, M., and Caro, J. (2009) Zeolitic imidazolate framework membrane with molecular sieving properties by microwave-assisted solvothermal synthesis. *J. Am. Chem. Soc.*, **131**(44), 16000–16001.
7. Dechnik, J., Sumby, C. J., and Janiak, C. (2017) Enhancing mixed-matrix membrane performance with metal–organic framework additives. *Cryst. Growth Des.*, **17**(8), 4467–4488.
8. Sorribas, S., Zornoza, B., Téllez, C., and Coronas, J. (2014) Mixed matrix membranes comprising silica-(ZIF-8) core–shell spheres with ordered meso–microporosity for natural-and bio-gas upgrading. *J. Memb. Sci.*, **452**, 184–192.

9. Tanh Jeazet, H. B., Sorribas, S., Román-Marín, J. M., Zornoza, B., Téllez, C., Coronas, J., and Janiak, C. (2016) Increased selectivity in CO_2/CH_4 separation with mixed-matrix membranes of polysulfone and mixed-MOFs MIL-101 (Cr) and ZIF-8. *Eur. J. Inorg. Chem.*, **2016**(27), 4363–4367.

10. Zornoza, B., Tellez, C., Coronas, J., Gascon, J., and Kapteijn, F. (2013) Metal organic framework based mixed matrix membranes: An increasingly important field of research with a large application potential. *Microporous Mesoporous Mater.*, **166**, 67–78.

11. Kang, Z., Peng, Y., Qian, Y., Yuan, D., Addicoat, M. A., Heine, T., Hu, Z., Tee, L., Guo, Z., and Zhao, D. (2016) Mixed matrix membranes (MMMs) comprising exfoliated 2D covalent organic frameworks (COFs) for efficient CO2 separation. *Chem. Mater.*, **28**(5), 1277–1285.

12. Mukherjee, R., Bhunia, P., and De, S. (2019) Nanofiltration range desalination by high flux graphene oxide impregnated ultrafiltration hollow fiber mixed matrix membrane. *J. Clean. Prod.*, **213**, 393–405.

13. Ahmad, A. L., Abdulkarim, A. A., Ismail, S., and Ooi, B. S. (2015) Preparation and characterisation of PES-ZnO mixed matrix membranes for humic acid removal. *Desalin. Water Treat.*, **54**(12), 3257–3268.

14. Li, X., Fang, X., Pang, R., Li, J., Sun, X., Shen, J., Han, W., and Wang, L. (2014) Self-assembly of TiO_2 nanoparticles around the pores of PES ultrafiltration membrane for mitigating organic fouling. *J. Memb. Sci.*, **467**, 226–235.

15. Denny, M. S., Moreton, J. C., Benz, L., and Cohen, S. M. (2016) Metal–organic frameworks for membrane-based separations. *Nat. Rev. Mater.*, **1**(12), 1–17.

16. Dechnik, J., Gascon, J., Doonan, C. J., Janiak, C., and Sumby, C. J. (2017) Mixed-matrix membranes. *Angew. Chemie Int. Ed.*, **56**(32), 9292–9310.

17. Baker, R. W. and Lokhandwala, K. (2008) Natural gas processing with membranes: An overview. *Ind. Eng. Chem. Res.*, **47**(7), 2109–2121.

18. Monsalve-Bravo, G. M. and Bhatia, S. K. (2018) Comparison of hollow fiber and flat mixed-matrix membranes: Theory and simulation. *Chem. Eng. Sci.*, **187**, 174–188.

19. Nik, O. G., Chen, X. Y., and Kaliaguine, S. (2012) Functionalized metal organic framework-polyimide mixed matrix membranes for CO_2/CH_4 separation. *J. Memb. Sci.*, **413**, 48–61.

20. Basu, S., Cano-Odena, A., and Vankelecom, I. F. J. (2010) Asymmetric Matrimid®/[$Cu_3(BTC)_2$] mixed-matrix membranes for gas separations. *J. Memb. Sci.*, **362**(1–2), 478–487.

21. Vinh-Thang, H. and Kaliaguine, S. (2013) Predictive models for mixed-matrix membrane performance: a review. *Chem. Rev.*, **113**(7), 4980–5028.

22. Moore, T. T. and Koros, W. J. (2005) Non-ideal effects in organic–inorganic materials for gas separation membranes. *J. Mol. Struct.*, **739**(1–3), 87–98.

23. Klaysom, C. and Shahid, S. (2019) Zeolite-based mixed matrix membranes for hazardous gas removal, in *Advanced Nanomaterials for Membrane Synthesis and its Applications*, Elsevier, pp. 127–157.

24. Bae, T., Lee, J. S., Qiu, W., Koros, W. J., Jones, C. W., and Nair, S. (2010) A high-performance gas-separation membrane containing submicrometer-sized metal–organic framework crystals. *Angew. Chemie*, **122**(51), 10059–10062.

25. Tien-Binh, N., Vinh-Thang, H., Chen, X. Y., Rodrigue, D., and Kaliaguine, S. (2015) Polymer functionalization to enhance interface quality of mixed matrix membranes for high CO_2/CH_4 gas separation. *J. Mater. Chem. A*, **3**(29), 15202–15213.

26. Perez, E. V, Balkus Jr, K. J., Ferraris, J. P., and Musselman, I. H. (2009) Mixed-matrix membranes containing MOF-5 for gas separations. *J. Memb. Sci.*, **328**(1–2), 165–173.

27. Zhang, C., Dai, Y., Johnson, J. R., Karvan, O., and Koros, W. J. (2012) High performance ZIF-8/6FDA-DAM mixed matrix membrane for propylene/propane separations. *J. Memb. Sci.*, **389**, 34–42.

28. Lan, T., Kaviratna, P. D., and Pinnavaia, T. J. (1994) On the nature of polyimide-clay hybrid composites. *Chem. Mater.*, **6**(5), 573–575.

29. Silva, P., Vilela, S. M. F., Tomé, J. P. C., and Paz, F. A. A. (2015) Multifunctional metal–organic frameworks: From academia to industrial applications. *Chem. Soc. Rev.*, **44**(19), 6774–6803.

30. Lin, R., Hernandez, B. V., Ge, L., and Zhu, Z. (2018) Metal organic framework based mixed matrix membranes: An overview on filler/polymer interfaces. *J. Mater. Chem. A*, **6**(2), 293–312.

31. Venna, S. R., Lartey, M., Li, T., Spore, A., Kumar, S., Nulwala, H. B., Luebke, D. R., Rosi, N. L., and Albenze, E. (2015) Fabrication of MMMs with improved gas separation properties using externally-functionalized MOF particles. *J. Mater. Chem. A*, **3**(9), 5014–5022.

32. Tan, J. C., Bennett, T. D., and Cheetham, A. K. (2010) Chemical structure, network topology, and porosity effects on the mechanical properties of Zeolitic Imidazolate Frameworks. *Proc. Natl. Acad. Sci.*, **107**(22), 9938–9943.

33. Etxeberria-Benavides, M., Johnson, T., Cao, S., Zornoza, B., Coronas, J., Sanchez-Lainez, J., Sabetghadam, A., Liu, X., Andres-Garcia, E., and Kapteijn, F. (2020) PBI mixed matrix hollow fiber membrane: Influence of ZIF-8 filler over H_2/CO_2 separation performance at high temperature and pressure. *Sep. Purif. Technol.*, **237**, 116347.

34. Khan, I. U., Othman, M. H. D., Jilani, A., Ismail, A. F., Hashim, H., Jaafar, J., Zulhairun, A. K., Rahman, M. A., and Rehman, G. U. (2020) ZIF-8 based polysulfone hollow fiber membranes for natural gas purification. *Polym. Test.*, **84**, 106415.

35. Mubashir, M., Leng, C. T., Keong, L. K., and Jusoh, N. (2020) Study on the effect of process parameters on CO_2/CH_4 binary gas separation performance over NH_2-MIL-53 (Al)/cellulose acetate hollow fiber mixed matrix membrane. *Polym. Test.*, **81**, 106223.

36. Hu, J., Cai, H., Ren, H., Wei, Y., Xu, Z., Liu, H., and Hu, Y. (2010) Mixed-matrix membrane hollow fibers of $Cu_3(BTC)_2$ MOF and polyimide for gas separation and adsorption. *Ind. Eng. Chem. Res.*, **49**(24), 12605–12612.

37. Liu, Q., Li, Y., Li, Q., Liu, G., Liu, G., and Jin, W. (2019) Mixed-matrix hollow fiber composite membranes comprising of PEBA and MOF for pervaporation separation of ethanol/water mixtures. *Sep. Purif. Technol.*, **214**, 2–10.

38. Ismail, A. F., Kusworo, T. D., and Mustafa, A. (2008) Enhanced gas permeation performance of polyethersulfone mixed matrix hollow fiber membranes using novel Dynasylan Ameo silane agent. *J. Memb. Sci.*, **319**(1–2), 306–312.

39. Rezaei-DashtArzhandi, M., Ismail, A. F., Goh, P. S., Wan Azelee, I., Abbasgholipourghadim, M., Ur Rehman, G., and Matsuura, T. (2016) Zeolite ZSM5-filled PVDF hollow fiber mixed matrix membranes for efficient carbon dioxide removal via membrane contactor. *Ind. Eng. Chem. Res.*, **55**(49), 12632–12643.

40. Galizia, M., Chi, W. S., Smith, Z. P., Merkel, T. C., Baker, R. W., and Freeman, B. D. (2017) 50th anniversary perspective: polymers and mixed matrix membranes for gas and vapor separation: A review and prospective opportunities. *Macromolecules*, **50**(20), 7809–7843.

41. Weitkamp, J. (2000) Zeolites and catalysis. *Solid State Ionics*, **131**(1–2), 175–188.

42. Favvas, E. P., Stefanopoulos, K. L., Nolan, J. W., Papageorgiou, S. K., Mitropoulos, A. C., and Lairez, D. (2014) Mixed matrix hollow fiber membranes with enhanced gas permeation properties. *Sep. Purif. Technol.*, **132**, 336–345.

43. Goh, P. S., Ng, B. C., Ismail, A. F., Aziz, M., and Hayashi, Y. (2012) Pre-treatment of multi-walled carbon nanotubes for polyetherimide mixed matrix hollow fiber membranes. *J. Colloid Interface Sci.*, **386**(1), 80–87.
44. Zahri, K., Wong, K. C., Goh, P. S., and Ismail, A. F. (2016) Graphene oxide/polysulfone hollow fiber mixed matrix membranes for gas separation. *RSC Adv.*, **6**(92), 89130–89139.
45. Mukherjee, R. and De, S. (2016) Novel carbon-nanoparticle polysulfone hollow fiber mixed matrix ultrafiltration membrane: Adsorptive removal of benzene, phenol and toluene from aqueous solution. *Sep. Purif. Technol.*, **157**, 229–240.
46. Park, M. J., Lim, S., Gonzales, R. R., Phuntsho, S., Han, D. S., Abdel-Wahab, A., Adham, S., and Shon, H. K. (2019) Thin-film composite hollow fiber membranes incorporated with graphene oxide in polyethersulfone support layers for enhanced osmotic power density. *Desalination*, **464**, 63–75.
47. Modi, A. and Bellare, J. (2019) Efficient separation of biological macromolecular proteins by polyethersulfone hollow fiber ultrafiltration membranes modified with Fe_3O_4 nanoparticles-decorated carboxylated graphene oxide nanosheets. *Int. J. Biol. Macromol.*, **135**, 798–807.
48. Kumari, P., Modi, A., and Bellare, J. (2020) Enhanced flux and antifouling property on municipal wastewater of polyethersulfone hollow fiber membranes by embedding carboxylated multi-walled carbon nanotubes and a vitamin E derivative. *Sep. Purif. Technol.*, **235**, 116199.
49. Isloor, A. M., Nayak, M. C., Prabhu, B., Ismail, N., Ismail, A. F., and Asiri, A. M. (2019) Novel polyphenylsulfone (PPSU)/nano tin oxide (SnO_2) mixed matrix ultrafiltration hollow fiber membranes: Fabrication, characterization and toxic dyes removal from aqueous solutions. *React. Funct. Polym.*, **139**, 170–180.
50. Ahmad, A. L., Pang, W. Y., Shafie, Z. M. H. M., and Zaulkiflee, N. D. (2019) PES/PVP/TiO_2 mixed matrix hollow fiber membrane with antifouling properties for humic acid removal. *J. Water Process Eng.*, **31**, 100827.
51. Simone, S., Galiano, F., Faccini, M., Boerrigter, M. E., Chaumette, C., Drioli, E., and Figoli, A. (2017) Preparation and characterization of polymeric-hybrid PES/TiO_2 hollow fiber membranes for potential applications in water treatment. *Fibers*, **5**(2), 14.
52. Galiano, F., Song, X., Marino, T., Boerrigter, M., Saoncella, O., Simone, S., Faccini, M., Chaumette, C., Drioli, E., and Figoli, A. (2018) Novel photocatalytic PVDF/Nano-TiO_2 hollow fibers for environmental remediation. *Polymers (Basel).*, **10**(10), 1134.

53. Donato, L., Garofalo, A., and Algieri, C. (2016) Mixed-matrix membranes: Preparation and characterization for biorefining, in *Membrane Technologies for Biorefining*, Elsevier, pp. 61–84.

54. Zhu, H., Jie, X., Wang, L., Kang, G., Liu, D., and Cao, Y. (2016) Effect of MIL-53 on phase inversion and gas separation performance of mixed matrix hollow fiber membranes. *RSC Adv.*, **6**(73), 69124–69134.

55. Mubashir, M., Yeong, Y. F., Chew, T. L., and Lau, K. K. (2019) Optimization of spinning parameters on the fabrication of NH_2-MIL-53 (Al)/cellulose acetate (CA) hollow fiber mixed matrix membrane for CO_2 separation. *Sep. Purif. Technol.*, **215**, 32–43.

56. Mubashir, M., Yeong, Y. F., Chew, T. L., and Lau, K. K. (2019) Comparison of post-treatment methods on the performance of hollow fiber membranes containing metal organic framework in gases separation. *Ind. Eng. Chem. Res.*, **58**(17), 7120–7130.

Chapter 6

Hollow Fibers in Micro- and Ultrafiltration Processes of the Agrofood Production

C. Conidi and A. Cassano
Institute on Membrane Technology, National Research Council of Italy (ITM-CNR), via P. Bucci, Cubo 17/C Rende (CS), Italy
c.conidi@itm.cnr.it; a.cassano@itm.cnr.it

Microfiltration (MF) and ultrafiltration (UF) processes represent well-established operation units in agrofood applications. Among MF and UF membranes, hollow fiber (HF) membranes have had a significant commercial impact on these processes, thanks to their capacity to combine a large membrane area in a restricted volume, low transmembrane drop, and flexibility.

This chapter gives an overview of the typical applications of MF and UF membranes in HF configuration in specific areas of the agrofood production, such as fruit juice clarification, wine and beer processing, milk and dairy industry. The chapter also discusses their combination with other membrane unit operations aimed at the recovery of valuable compounds from agrofood processing wastewater.

Hollow Fibers and Nanofibers in Membrane Science: Preparation, Characterization, and Applications
Edited by Alberto Figoli, Mir Saeed Seyed Dorraji, and Francesco Galiano

ISBN 978-981-4968-03-4 (Hardcover), 978-1-003-25663-2 (eBook)
www.jennystanford.com

6.1 Introduction

Microfiltration (MF) and ultrafiltration (UF) are well-known membrane processes based on the use of a selective barrier through which fluids and solutes are selectively transported under a hydrostatic pressure as driving force. These processes have found a large application in different areas of food production, including dairy, fruit juices and pulps production, beer and wine, beet and cane sugar, meat, water, and wastewaters, thanks to their advantages over conventional separation methods [1].

MF membranes typically contain pores with diameters in the range of 0.1–10 μm and are widely used in separating suspended particles, colloids, and macromolecules from solution. The separation mechanism is based on a sieving effect, and particles are separated according to their dimensions, although the separation is influenced by interactions between the membrane itself and particles being filtered.

UF membranes are typically asymmetric in structure with pore sizes in the skin layer of 2–10 nm. They are capable of separating larger materials such as colloids, fats, proteins, and bacteria, while allowing sugars and low-molecular-weight compounds to pass through the membrane. These membranes are typically characterized by their molecular weight cutoff (MWCO), defined as the equivalent molecular weight of the smallest species that exhibit 90% rejection.

Most commercial MF and UF membranes are formed from organic polymers or from inorganic materials (ceramic, metallic, carbon, etc.). According to their geometry, membranes can be subdivided in tubular, HF, spiral wound, and flat sheet [2].

As a preferable configuration for industrial applications, HF membrane modules are widely used in different applications of the agrofood industry due to the larger membrane area per unit volume and enhanced productivity. These advantages can be attributed to the typical configuration of HF membrane modules, which include a large number of fibers straightly packed in the shell housing, and not requiring spacers between the membranes.

This chapter provides an overview of the typical applications of HF membranes in specific areas of the agrofood production such as fruit juice clarification, wine and beer processing, milk and dairy industry. The treatment of agrofood processing wastewaters is also presented and discussed in the light of the increased attention toward the valorization of agrofood residues through the separation and recovery of high-added value compounds.

6.2 Clarification of Fruit Juices

The clarification of fruit and vegetable juices by MF and UF processes is the most developed application of membrane technology in food processing [3]. Conventional clarification processes involve many steps such as enzymatic treatment, centrifugation, use of fining agents (gelatin, silica sol, bentonite), filtration with diatomaceous earth, centrifugation, and sterilization for removal of microorganisms. These processes are characterized by different drawbacks in terms of prolonged processing time, high energy consumption, and significant juice loss. Additionally, the use of fining agents is characterized by the risk of dust inhalation with consequent health problems due to handling and disposal, significant high costs, and environmental problems due to their disposal [4].

Clarification procedures based on MF and UF processes have emerged as a valid alternative and attractive substitute to the traditional juice clarification and have become a commercial success. Typical advantages over conventional methodologies are in terms of increased juice yields, possibility of operating in a single step, reduction in working times, elimination of gelatin, adsorbents, and other filtration aids, reduction in waste products, elimination of needs for pasteurization, and reduction in enzyme utilization. In addition, MF and UF operations are very efficient in preserving the freshness, aroma, and nutritional value of juice, as the separation process requires no heat application or the use of chemical agents [5].

Fruit juices are a colloidal mixture of soluble and suspended solids containing both microsolutes such as sugars, organic acids, pigments, vitamins, mineral salts, and macrosolutes such as proteins, enzymes, pectic substances, polyphenols, and microorganisms. MF and UF membranes retain large species such as microorganisms, lipids, proteins and colloids, while small solutes such as vitamins, salts, and sugars flow together with water. Therefore, in these processes, the juice is separated into a clarified fraction free of spoilage microorganisms (permeate) and a fibrous concentrated pulp (retentate) [6]. Among the most used configurations for the clarification of fruit juices, HF membrane modules are usually preferred on industrial scale since they are self-supported, more flexible, and easy to handle, ensuring high surface-area-to-volume ratio, in comparison with tubular and plate-and-frame membrane modules. Consequently, they result in consistent space saving, improved productivity, and reduction in operating costs also connected to their maintenance, since these modules can be back flushed [7, 8]. HF membranes used in fruit juice clarification have been prepared with different polymeric materials. In particular, polysulfone (PS), polyethersulfone (PES), polyacrylonitrile (PAN), modified poly(ether ether ketone) (PEEK-WC), polypropylene (PP), and polyvinylidene fluoride (PVDF) are common polymers used to prepare HF membrane modules for juice clarification. Ceramic HF membranes are preferred due to their chemical and mechanical stabilities, but material and production costs are generally greater for ceramic membranes than for polymeric membranes [9]. Furthermore, HF membranes can be prepared with different pore diameters making them suitable for feeds with different characteristics of viscosity, suspended solid concentration, etc.

The feed solution is commonly forwarded in the lumen side, while the permeate stream is removed from the shell side.

Operating conditions, including cross-flow velocity (CFV), temperature, transmembrane pressure (TMP), and volume reduction factor (VRF), as well as membrane properties, such as MWCO or pore size and membrane material, have a significant effect on permeate flux and physicochemical properties of the clarified juice. In particular, high permeate fluxes are requested for filtration to be practical and economic, and product quality should at least meet those obtained by the other standard clarification methods [10]. Additionally, productivity is strongly affected by concentration polarization and fouling phenomena. Concentration polarization is a reversible phenomenon because of the accumulation of rejected solutes such as macromolecules and other larger solutes (such as pectins and proteins) on the membrane surface, leading to an additional hydraulic resistance to the permeate flux with time during the filtration process. On the other hand, membrane fouling leads to a long-term flux decline caused by the deposition of retained particles (colloids, macromolecules, dissolved organics, etc.) onto the membrane surface or in the membrane pores [11, 12].

Membrane fouling not only decreases permeate flux but also affects membrane longevity due to a more frequent membrane cleaning and replacement; therefore, it is a key factor affecting the economic and commercial viability of a membrane system.

The optimization of operating conditions, the identification of proper pretreatment systems, and the selection of membranes with specific properties in terms of molecular weight cutoff, morphology, and hydrophobicity/hydrophilicity are typical approaches to minimize membrane fouling phenomena. In particular, the adsorption of colloidal material is strongly affected by the membrane material. Besides, the surface structure and flux resistance of the fouling layer is affected by the membrane morphology. Dense fouling layers are detected on smooth surfaced membranes. On the contrary, rough membranes produce a looser surface fouling layer with a lower flow resistance per unit thickness of foulant.

A limiting flux is generally reached when the transmembrane pressure is increased up to a specific value; after that, any further pressure increase determines no significant increase in the permeate flux. The existence of a limiting flux can be related to the formation of a viscous and gelatinous-type layer, which is responsible for an additional resistance to the permeate flux in addition to that of the membrane. An increase in cross-flow velocity enhances wall shear stress on the membrane surface reducing concentration polarization and reversible fouling on the membrane surface; consequently, the mass transfer coefficient is enhanced, leading to an increase in the permeation flux [2].

The optimization of operating conditions and membrane performance (permeate flux and quality of clarified juice) in the production of fruit juices with MF and UF membranes in HF configuration has been investigated by different authors. For instance, Tasselli et al. [8] studied the effect of different operating and fluid dynamics conditions, such as transmembrane pressure (TMP), feed flowrate, temperature and suspended solids concentration on the permeate flux during the UF of kiwifruit juice using PEEK-WC HF membranes. The results indicated an increase in membrane productivity by increasing the feed flowrate and the temperature. On the other hand, an increase in pressure was not efficient in terms of flux improvement due to the thickening or compression of the deposited layer onto the membrane surface and adsorption in the membrane pores. A limiting flux was reached when the operating pressure was increased up to a specific value; after that, any further increase determined no significant increase in the permeate flux (Fig. 6.1). The highest permeate flux values were obtained at a temperature of 25°C, a TMP of 0.75 bar, and a feed flowrate of 40 L/h. In these conditions, the UF process allowed a remarkable improvement in juice clarity and a complete removal of suspended solids without significant loss of other physicochemical characteristics of the original juice.

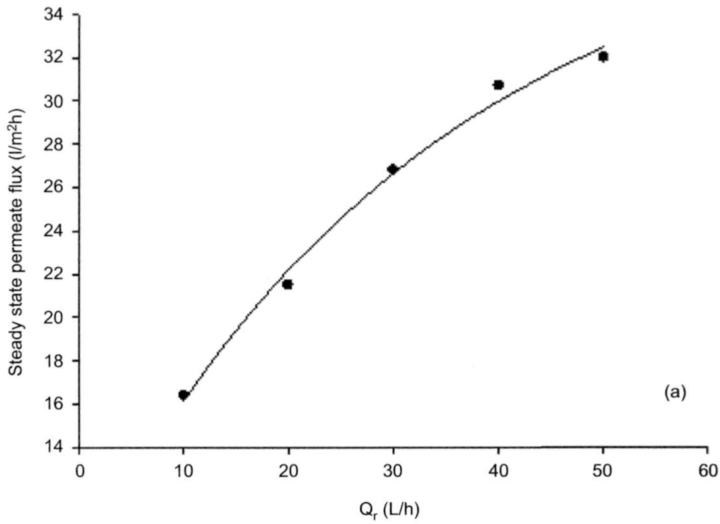

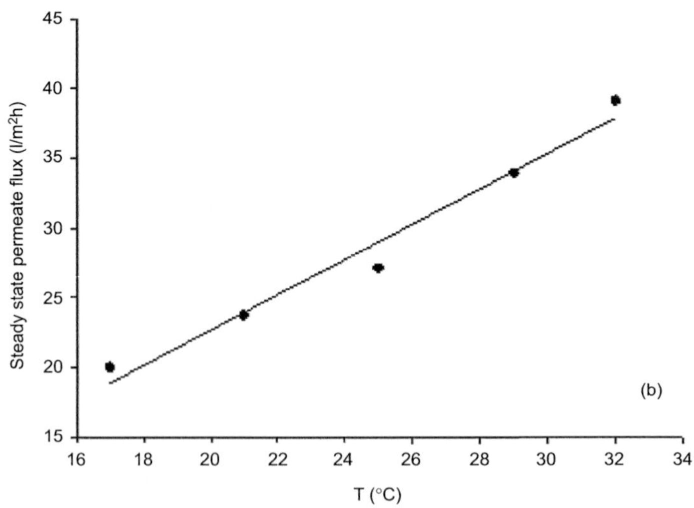

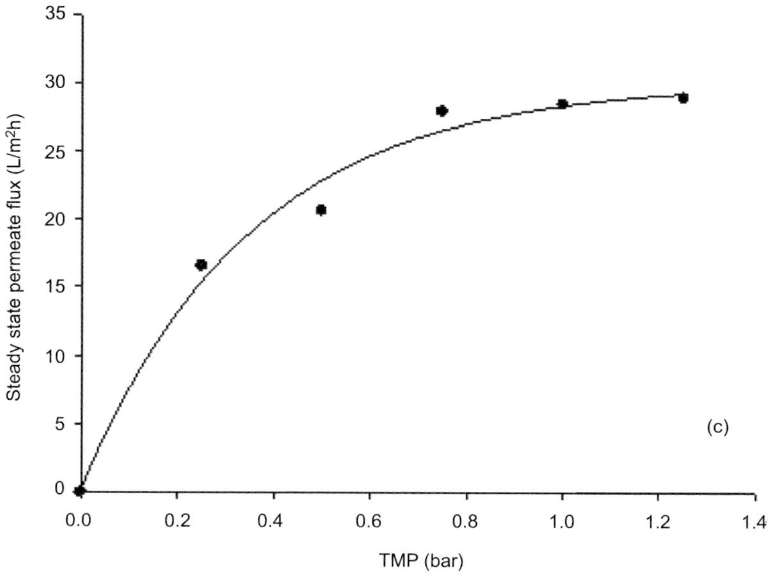

Figure 6.1 Clarification of kiwifruit juice with PEEK-WC HF membranes. Effect of (a) feed flowrate (Q_f), (b) temperature (T), and (c) transmembrane pressure (TMP) on the steady-state permeate flux. Reprinted from Ref. [8] with permission of Elsevier.

In a similar approach, Cassano et al. [13] evaluated the effect of TMP and CFV on permeate flux, membrane fouling, and quality of clarified juice in the clarification of pomegranate juice by using PEEK-WC and PS HF membranes prepared in laboratory through the phase-inversion process.

For both membranes, an increase in axial velocities reduced concentration polarization, enhanced the mass transfer coefficient, and increased the steady-state permeate flux. However, an increase in pressure was not efficient in terms of flux improvement. PS membranes exhibited higher productivity, lower indexing of fouling, and lower rejection toward biologically active compounds (flavonoids and total phenols) in comparison to PEEK-WC membranes. Overall, experimental results indicated that the membrane material plays a key role in the separation of bioactive compounds from the juice.

In a previous work, Laorko et al. [14] investigated the effect of operating conditions, pore size, and MWCO on permeate flux and quality of pineapple juice clarified by MF and UF HF membranes. For this purpose, PS membrane modules with a pore size ranging from 0.1 to 0.2 μm and MWCO from 30 to 100 kDa were tested. According to the obtained results, HF membranes with a pore size of 0.2 μm were considered the most suitable for juice clarification since they exhibited the highest permeate flux (37 L/m^2h) and the highest recovery of polyphenols and other antioxidant compounds in the selected operating conditions (CFV, 3.4 m/s; TMP, 0.7 bar).

Sagu et al. [15] investigated the effect of TMP and CFV on permeate flux, quality, and taste of banana juice ultrafiltered with PS HF membranes with MWCO of 10, 27, and 44 kDa, respectively. The 27 kDa membrane resulted as the most suitable membrane for the juice clarification due to its highest productivity and lowest rejection toward soluble solids. For this membrane, the increase in feed flowrate from 10 to 20 L/h did not affect the permeate fluxes; on the other hand, a decline in permeate flux by increasing the operating pressure was observed. The variation in operating conditions during the clarification process did not influence the juice quality. Most of the analyzed compounds, including total soluble sugars (TSS), total dissolved solids (TDS), total polyphenols, and proteins, were recovered in the UF permeate fraction, while the alcohol insoluble solids were completely retained by the UF membrane. Results of storage studies indicated that the clarified juice can be stored for 1 month under refrigerated conditions, maintaining the nutritional qualities and taste without adding any preservative substance.

Galiano et al. [16] evaluated the effect of membrane material on the functional properties of pomegranate juice clarified by PVDF and PS membranes in HF configuration prepared in laboratory according to the phase-inversion methodology. PVDF membranes showed lower retention toward healthy compounds, including flavonoids and anthocyanins, in comparison to PS membranes. Therefore, the PVDF clarified juice exhibited greater antioxidant effects than the juice clarified with PS membranes. In addition, the treatment with PVDF membranes enriched the juice in α-amylase and α-glucosidase inhibitors. The different retention of bioactive compounds of PVDF and PS membranes was attributed to the chemical nature of both polymers. PVDF membranes consist of alternating CH_2 and CF_2 units, conferring a hydrophobic nature to the material. These membranes have less tendency to form hydrogen bonds and van der Waals interactions with the hydroxyl groups of polyphenols and, consequently result highly permeable to phenolic compounds when compared with PS membranes. These results indicated that the clarification of pomegranate juice with PVDF HF membranes is an efficient method to produce a high-quality clarified juice suitable for the production of functional foods and beverages.

More recently, Morittu et al. [17] analyzed the in vitro antioxidant and hypoglycemic activities and the in vivo effect of pomegranate juice clarified by PVDF HF membranes. The whole results showed that the clarified juice presented higher antioxidant activity as well as higher α-glucosidase and α-amylase inhibitory activity in comparison to the natural juice (Table 6.1). Therefore, the use of PVDF HF membranes is a useful tool to obtain significant improvements in the juice quality.

The influence of membrane material, MWCO, and VRF on the quality, content of antioxidant compounds, and permeate flux in the clarification of blood orange juice has been studied by Conidi et al. [18]. Three HF UF membranes with different MWCO (50 and 100 kDa) and membrane materials (PS and PAN) were used. All selected membranes retained suspended solids, producing a clear juice with a brilliant red color and a physicochemical composition similar to that of the natural fresh juice. In selected operating conditions (TMP, 0.5 bar; feed flowrate, 130 L/h; temperature, 20°C; VRF, 3), a PS 100 kDa membrane provided the best results in terms of permeate flux (steady-state permeate flux of 7.12 L/m^2h) and lower rejection toward total polyphenols, total antioxidant activity, and total anthocyanins (17.5%, 15%, and 7.3%, respectively) in comparison with other investigated membranes (Fig. 6.2).

An improved color and clarity of mandarin juice through the removal of suspended solids was also achieved by using PEEK-WC and PS HF membranes prepared in laboratory [19]. PEEK-WC membranes showed a lower rejection toward TSS, TAA, and total polyphenols in comparison with the PS membranes in agreement with their lower rejection toward dextrans with specific molecular weight.

PS HF membranes with MWCO of 100 kDa (Amersham Biosciences Corp. Model UFP-100-E-4A) produced a clear juice from the juice of Xoconostle (a native Mexican fruit with limitations for its commercialization) with high antioxidant activity (21.40 ± 1.1 TEAC), high phenolic (27.50 ± 0.2 mg gallic acid/L), and betalain (18.80 ± 0.6 mg/L) contents [20].

Table 6.1 Antioxidant and hypoglycemic activities of natural juice and clarified juice with PVDF HF membranes [17]

Sample	Antioxidant Activity				Hypoglycemic Activity	
	DPPH Test (IC_{50} μg/mL)	FRAP Test[#] (μM Fe (II)/g)	β-Carotene Bleaching Test (IC_{50} μg/mL)		α-Amylase Inhibitory Assay (IC_{50} μg/mL)	α-Glucosidase Inhibitory Assay (IC_{50} μg/mL)
			t 30 min	t 60 min		
Natural juice	782.6 ± 3.7^{A}	5.1 ± 1.0^{C}	51.5 ± 1.9^{A}	55.7 ± 1.7^{A}	67.1 ± 2.3^{B}	80.1 ± 2.8^{A}
Clarified juice	734.2 ± 2.8^{B}	14.1 ± 0.6^{B}	19.7 ± 1.1^{B}	44.1± 1.0^{B}	76.6 ± 2.2^{A}	68.6 ± 2.0^{B}
Positive control[*]	5.0 ± 0.8^{C}	63.2 ± 4.5^{A}	1.0 ± 0.04^{C}	1.0 ± 0.03^{C}	50.0 ± 0.9^{C}	35.5 ± 1.2^{C}
p	<0.0001	<0.0001	<0.0001	<0.0001	<0.0001	<0.0001
R squared	1.000	0.9927	0.9975	0.9984	0.9803	0.9918

Data are expressed as means ± SD (n = 3). [#] At concentration of 2.5 mg/mL. [*]Positive control: ascorbic acid for DPPH test, BHT for FRAP test, propyl gallate for β-carotene bleaching test, acarbose for α-amylase, and α-glucosidase inhibitory assays. A–C: different letters along the column indicate significant differences $p < 0.01$. NJ: natural juices, CJ: clarified juices, and BHT: butylated hydroxytoluene

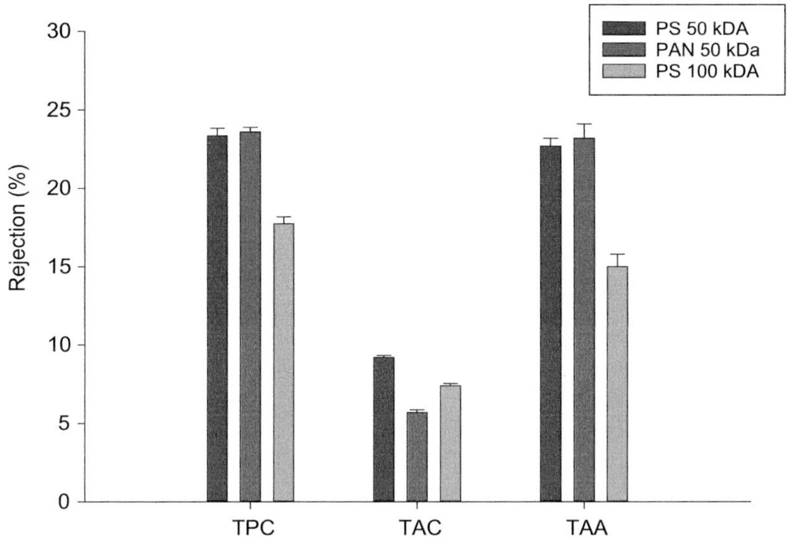

Figure 6.2 Effect of MWCO and membrane material on the rejection of HF UF membranes toward total polyphenols (TPC), total anthocyanin content (TAC), and total antioxidant activity (TAA) (PS, polysulphone; PAN, polyacrylonitrile) (adapted from Ref. [18]).

The pretreatment of fruit juices is another important aspect that affects the performance of MF and UF HF membranes in terms of juice quality and permeate flux. In particular, in order to enhance filtration performance and reduce membrane fouling, fruit juices are usually submitted to a pretreatment method before the clarification process, with the aim to decrease the content of macromolecules responsible for the high viscosity of the juice and that build upon the membrane surface or in the membrane pores. The pretreatments include mainly the application of fining agents, pectolic enzymes, chitosan, or centrifugation. Domingues et al. [21] evaluated the effect of different pretreatments, such as centrifugation, clarification with chitosan, and enzymatic treatment, on the physical–chemical characteristics of passion juice clarified with HF membranes. The investigated pretreatments presented a positive effect on the permeate flux; however, the highest permeate flux values were obtained through the addition of chitosan. The MF process reduced both color and turbidity of the juice, and the quality of the clarified juice was not affected by the pretreatment. On the other hand, in

Table 6.2 Hollow fiber membranes used in fruit juices clarification

Fruit Juice	Process	Membrane Type	Operating Conditions	References
Banana	UF	PS, 10, 27, and 0.44 kDa, 0.026 m^2	TMP, 3.5, 5.2, 6.9, 10.4 bar; Q_f, 10, 15, 20 L/h, T, 25°C	[15]
Blood orange	UF	PAN, 50 KDa, 0.16 m^2, (DCQ II-006C-PAN50, Blue Star Membrane Technology)	TMP, 0.5 bar; Q_f, 140 L/h; T, 20°C	[18]
		PS, 50 kDa, 0.16 m^2 (DCQ II-006C-PS50, Blue Star Membrane Technology)	TMP, 0.5 bar; Q_f, 140 L/h; T, 20°C	
		PS, 100 kDa, 0.16 m^2 (DCQ II-006C-PS100, Blue Star Membrane Technology)	TMP, 0.5 bar; Q_f, 140 L/h; T, 20°C	
Clementine mandarin	UF	PEEK-WC (prepared in laboratory), 37 cm^2	TMP, 0.3 bar; T, 25°C	[19]
		PS (prepared in laboratory), 32 cm^2	TMP, 0.3 bar; T, 25°C	
Jamun	UF	PS, 10, 50, 100 kDa; 0.005 m^2 (Explorer 12)	TMP, 10, 15, 20 psi, Q_f, 10 L/h, T, 30°C	[24]
Kiwifruit	UF	PEEW-WC (prepared in laboratory), 20–30 cm^2	TMP, 75 kPa; Q_f, 40 L/h; T, 25°C	[8]
Kiwifruit	UF	PEEW-WC (prepared in laboratory), 26.9 cm^2	TMP, 1.6 bar; Q_f, 70 L/h; T, 25°C	[25]
Passion fruit	MF	PA, 0.3 μm, 0.0158 m^2 (PAM Membranas Seletivas LTDA, Brazil)	TMP 1.0 bar; Q_f, 325 L/h; T, 25°C	[26]
Passion fruit	MF	Polieterimide, 0.40 μm, 0.056 m^2 (PAM Membranas Seletivas LTDA, Brazil)	TMP, 1.0 bar; T, 25°C	[21]
Pineapple	UF	PS, 100 kDa, 0.12 m^2 (AIG Technology Corporation)	TMP, 0.2, 0.8, 2 bar; T, 20, 30, 40°C; CFV, 1.19 m/s	[27]

Fruit Juice	Process	Membrane Type	Operating Conditions	References
Pineapple	MF	PS, 0.2 μm, 0.011 m^2 (Amersham Biosciences, Little Chalfont, UK)	TMP, 0.7 bar; CFV, 3.4 m/s, T, 20°C	[14]
Pineapple	MF	PS, 0.2 μm, 0.011 m^2 (Amersham Biosciences, Little Chalfont, UK)	TMP, 1.0 bar; CFV, 1.2 m/s; T, 20°C	[28]
Pomegranate	UF	PEEK-WC (prepared in laboratory), 46 cm^2	TMP, 0.96 bar; Q_f, 1166 mL/min, T, 25°C)	[29]
Pomegranate		PEEK-WC (prepared in laboratory), 26.9 cm^2	TMP, 115 kPa; CFV, 4 m/s; T, 25°C)	[30]
		PS (prepared in laboratory), 32.3 cm^2	TMP, 115 kPa; CFV, 4 m/s; T, 25°C)	
Pomegranate	MF	PVDF, PS (prepared in laboratory), 0.13 μm	TMP, 0.6 bar; Q_f, 30 L/h; T, 25°C	[16]
Pomegranate	UF	Cellulose triacetate, 150 kDa, 0.26 m^2 (FUC 1582, Microdyn-Nadir, Wiesbaden, Germany)	TMP, 0.6 bar, Q_f, 400 L/h; T, 25°C	[31]
Pomegranate		PVDF (prepared in laboratory), 0.13 μm, 27.6 cm^2	TMP, 0.6 bar, Q_f, 30 L/h; T, 25°C	[17]
Sugarcane	MF	PAN, 0.1 μm, 0.0027 m^2	TMP, 104 kPa, CFV, 0.369 m/s	[22]
Sugarcane	UF	PS, 10, 30, 44, 70 kDa; 0.0022 m^2	TMP, 69 kPa; Q_f, 30 L/h	[32]
Rasberry	UF	PS, 30 kDa, 1.8 m^2 (Ultraflux®AV 1000S, Fresenius Medical Care, Germany)	TMP, 1 bar; Q_f, 1 L/min; T, 22°C	[33]

Legend: MF, microfiltration; UF, ultrafiltration; PS, polysulphone; PVDF, polyvinylidenefluoride; PAN, polyacrylonitrile; PEEK-WC, modified poly(ether ether ketone); TMP, transmembrane pressure; Q_f, feed flowrate; T, temperature; CFV, cross-flow velocity.

centrifuged and enzyme-treated samples, cake formation was found to be the major fouling factor, while internal pore blocking occurred during the filtration of the chitosan-pretreated sample. These results indicated a key role of the applied pretreatment on the predominant fouling mechanism in the MF of passion fruit with HF membranes.

The MF process of centrifuged sugarcane juice with HF membranes was studied by Panigrahi et al. [22] in order to achieve the maximum flux, minimum retention of sucrose, and maximum retention of total solids. The MF process produced a removal of turbidity and total solids of about 98% and 26%, respectively, with a remarkable improvement in clarity (of about three times). The nutritional quality in terms of vitamin C, total soluble solids, and sucrose of centrifuged juice was preserved in the MF permeate. Microbiological analysis confirmed that MF successfully improved the stability of the juice for a long time if stored at low temperature, due to declined rate of fermentation owing to a large reduction in population of yeasts and molds.

Carrin et al. [23] investigated the pectinase immobilization on HF PS membranes used for the clarification of apple juice analyzing the influence of the enzyme layer on the resistance to permeate flux and the capability of the physically immobilized enzyme to hydrolyze the gel layer. Results showed that the immobilization of the enzyme on HF membranes retarded the rate of formation of the gel layer during the juice clarification and greatly extended the membrane operation, by keeping permeate flux constant during prolonged periods at a reasonable yield.

Typical applications of MF and UF HF membranes in fruit juice clarification are summarized in Table 6.2.

6.3 Wine and Beer Processing

Wine is a fermented alcoholic beverage of high commercial importance, produced from the fermentation of grape juice through several operations (pressing, decanting, filtration, bottling) and processes (alcoholic and malolactic fermentations). Crude wine is a complex solution with different macromolecules, colloids, microorganisms, yeasts, numerous solutes (organic acids, salts, and polyphenols), and large particles as potassium hydrogen tartrate. In addition, it presents a turbid aspect that is not very well accepted by consumers, and usually it needs to be clarified. Conventional separation techniques such as centrifugation, static defecation, and depth filtration are often unsuccessful and difficult to realize. These processes, being often discontinuous, require high energy consumption for the temperature control and high quantities of additives (bentonite, activated carbons, silica sol, and diatomaceous earth) with consequent problems of disposal and environmental impact. MF and UF processes offer several advantages over traditional wine clarification in terms of reduction in energy consumption; low temperatures; elimination of earth use and related associated environmental problems; removal of bacteria, cells, and spores; reduction in pectin content, proteins, and colloids; and combination of clarification, stabilization, and sterile filtration in one single continuous operation and economic and operational benefits [34, 35].

In particular, wine clarification by MF removes bacteria, virus, and yeast, reducing the turbidity and producing a clear, sterile wine without affecting the organoleptic characteristics [36]. On the other hand, UF can be used either before the fermentation (for clarifying the must) or after (for treating the finished wine). UF of the must is used for the removal of colloids, high-molecular-weight tannins, polysaccharides, proteins, and undesirable microorganisms. The main limiting factor of these processes is the permeate flux decay over time, which causes membrane fouling and concentration polarization and gel formation with consequent reduction in the membrane productivity as well as the potential retention of some value compounds with a loss of organoleptic characteristics [37]. As reported previously, the optimization of operating conditions and the selection of the most suitable membrane material and configuration are key factors to reduce these phenomena and to improve membrane performance.

Cassano et al. [38] investigated the effect of operating conditions on membrane fouling and permeate fluxes of PS HF membranes (100 kDa, China Blue Star Membrane Technology Co., Ltd., Beijing, China) used for the clarification of grape must destined to the production of white wine.

Experimental results showed that the steady-state permeate fluxes increased by increasing the applied TMP until reaching the maximum value of about 17 L/m^{2}h at about 1 bar; further increases in pressure did not produce a significant improvement in the membrane performance. On the other hand, the steady-state permeate flux increased by increasing both temperature and the feed flowrate in the range of investigated values. In terms of must quality, a decrease in the content of soluble solids and polyphenols was observed by increasing the applied pressure due to an increase in the rejections of the UF membrane toward these parameters at higher pressures (Fig. 6.3). This behavior can be attributed to the reduction in membrane pore size due to the increased membrane fouling at higher operating pressures. The best conditions for processing the grape must in order to achieve the maximum permeation flux were fixed at 40°C, 1 bar as operating pressure, and 750 L/h as feed flowrate. On the other hand, the best conditions to reduce membrane fouling and the loss of total polyphenols and soluble solids were fixed at 30°C, 0.2 bar, and 750 L/h.

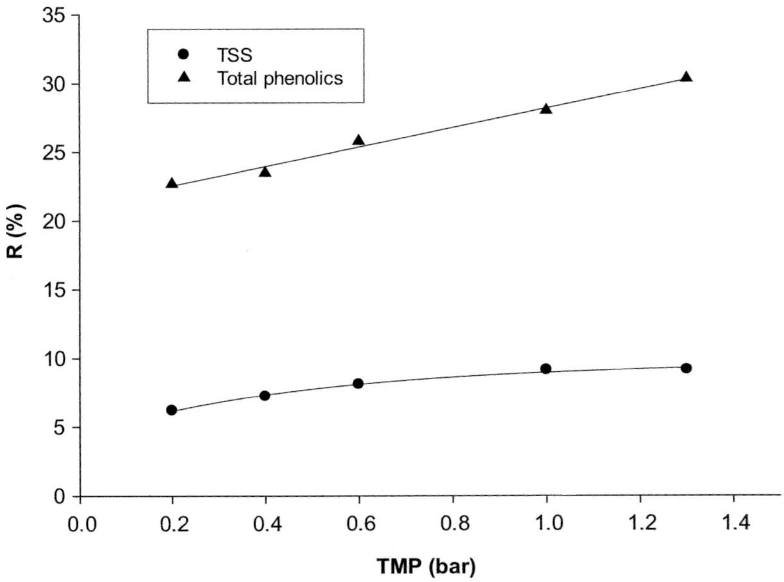

Figure 6.3 Clarification of grape must with HF UF membranes. Observed rejection for total soluble solids (TSS) and polyphenols at different transmembrane pressures (Q_f, 440 L/h; T, 15°C). Reprinted from Ref. [38] with permission of Elsevier.

In the brewing process, MF and UF operations have been successfully used for the clarification of rough beer and pasteurization producing a clear and bright beer and ensuring, at the same time, microbiological stability. Conventional beer clarification employs filter press or pressure vessel filters, which are commonly precoated with porous particles mainly of diatomaceous earth (but also with perlites, cellulose, or active carbon) as the filter aids, which play an important role not only in acting as a second filtration barrier, but also in absorbing haze components [39].

Diatomaceous earth has been the historic foundation of beer filtration and clarification, as standard industrial practice for more than 100 years. However, diatomaceous earth is associated with handling and disposal of the powder, larger amounts of effluents, and production of a not completely clear filtrate [40].

Cross-flow MF and UF processes offer the advantages of elimination of using filter aids and associated problems, with reduction in beer losses, high solids handling capacity, and the possibility to substitute heat pasteurization with additional cost saving [41]. Loss reduction concerns the recovery of extract during wort clarification and beer recovery from the fermentation and maturation tank bottoms. MF and UF membranes allow the production of a permeate stream of acceptable quality from this brewing by-product, including flavor and haze, with minimal loss of original gravity, color, and bitterness. The beer recovered can be recycled to the maturation vessel or sent to the final clarification [36].

In this context, membranes in HF configuration have been employed, not only at lab scale but also at industrial level, due to their advantages in terms of high packing density, low power consumption, and capacity to withstand back-flushing procedures. For instance, Heineken and Norit Membrane Technology patented the first industrial plant used for the clarification of rough beer equipped with 10 polymeric HF membrane modules (PES; pore size, 0.45 μm; length, 1 m; inner diameter, 1.5 mm; filtration area, 9.3 m^2) [42]. The HF membrane modules showed average permeate fluxes of about 100 L/m^2h, in selected operating conditions (temperature, 0°C; TMP 1.6 bar; CFV, 2 m/s), producing a clarified beer, which respected the European Brewery Convention (EBC) standards in terms of turbidity (close to 0.6 EBC unit), bitterness, total extract, color, and protein content. In order to control membrane fouling, 7 min of black flushing was applied each hour to remove the reversible fouling that has built up. At the end of each experimental run, a cleaning procedure combining an enzymatic step, an alkaline step, and a strongly oxidative step has been successful in maintaining run times between 7 and 10 h for about 50 runs. More recently, in order to further increase the permeate fluxes and to have membranes with a higher life span to reduce the contribution of the annual membrane replacement, ceramic HF membrane modules have been tested for beer clarification. A ceramic membrane module equipped with 40 HF membranes in Al_2O_3 (InoCep® membrane module, Hyflux Ltd, Singapore) and nominal pore size of 0.8 μm was tested by Cimini and Moresi [43] for the clarification and stabilization of lager beer. Experimental results showed that working in total recycle mode, steady-state permeation fluxes tended to a limiting flux and increased by increasing the CFV in the range of 0.5–0.6 m/s. Average permeate fluxes of 128 ± 33 L/m^{2}h at an operating transmembrane pressure of 2.4 bar, temperature of 10°C, and CFV of 2.5 m/s, with a periodic CO_2 back flushing (2 min, applied every 50–60 min), were obtained. In similar operating conditions, HF ceramic membranes with pore size of 1.4 μm exhibited higher permeate fluxes (average permeate fluxes of 196.4 ± 38.0 L/m^{2}h) during rough beer clarification in comparison with 0.2, 0.5, and 0.8 μm membranes (permeate fluxes in the range of 32.3 ± 2.8 to 49.4 ± 0.6 L/m^{2}h) [44]. In terms of physicochemical composition, 0.2 and 0.5 pore size membranes showed a turbidity in the permeate fractions lower than 1.5 EBC unit (the initial feed value was 4.8 ± 2.2), and a significant reduction in density, viscosity, and color, with respect to the original feed (Table 6.3). The 1.4 μm HF ceramic membrane was proposed as a real valid alternative to conventional powder filters due to its high productivity and no practical impact on the permeate quality. A finishing step with microbially rated membrane cartridges of 0.45–0.65 μm to remove bacteria and spores from clarified beer allows to produce crude beer, thus avoiding a pasteurization step.

Table 6.3 Effect of membrane pore size on average steady-state permeation flux and main chemicophysical characteristics of beer samples collected from the brewery maturation tank and centrifuged (RB), permeate (P) and retentate (R). Operating conditions: TMP, 2.4 bar; CFV, 2.5 m/s; *T*, 10°C (adapted from Ref. [44])

Pore Size (μm)	Type of Sample	pH	$\rho_{20°C}$* (g/L)	$\eta_{20°C}$* (mPa s)	$H_{20°C}$ (EBC)	$H_{0°C}$* (EBC)	*A* [% (v/v)]	*C* (EBC)	Steady-State Permeate Flux (L/m²h)
	RB	4.46 ± 0.05	1006.9 ± 0.2	1.53 ± 0.01	4.8 ± 2.2	5.33 ± 0.05	5.33 ± 0.05	7.1 ± 0.4	
0.2	P	4.45 ± 0.02	1004.7 ± 0.1	1.42 ± 0.00	1.17 ± 0.05	5.34 ± 0.08	5.34 ± 0.08	4.9 ± 0.6	37.1 ± 4.8
	R	4.46 ± 0.03	1006.9 ± 0.1	1.52 ± 0.01	4.2 ± 2.2	5.34 ± 0.08	5.34 ± 0.08	7.3 ± 0.2	
0.5	P	4.45 ± 0.03	1005.2 ± 0.1	1.47 ± 0.01	1.4 ± 0.2	5.11 ± 0.16	5.11 ± 0.16	5.3 ± 0.5	32.3 ± 2.8
	R	4.46 ± 0.05	1006.6 ± 0.1	1.54 ± 0.01	5.0 ± 0.1	5.33 ± 0.16	5.33 ± 0.16	7.1 ± 0.0	
0.8	P	4.45 ± 0.05	1006.7 ± 0.3	1.51 ± 0.01	2.7 ± 0.1	5.33 ± 0.02	5.33 ± 0.02	6.1 ± 0.0	49.4 ± 0.6
	R	4.46 ± 0.03	1007.0 ± 0.1	1.53 ± 0.01	4.8 ± 0.2	5.38 ± 0.02	5.38 ± 0.02	6.8 ± 0.0	
1.4	P	4.45 ± 0.02	1006.7 ± 0.1	1.53 ± 0.01	2.85 ± 0.03	5.25 ± 0.10	5.25 ± 0.10	6.7 ± 0.0	196.4 ± 38.0
	R	4.46 ± 0.03	1007.0 ± 0.3	1.55 ± 0.01	5.2 ± 0.1	5.25 ± 0.10	5.25 ± 0.10	7.0 ± 0.0	

Legend: ρ, density; η, viscosity; *H*, turbidity; *A*, alcohol content, *C*, color.

6.4 Milk and Whey Processing

Pressure-driven membrane operations are widely developed in the dairy industry for both milk and whey processing with improved separation capabilities when compared with traditional technologies. The main applications of cross-flow MF in the dairy industry include the separation of bacteria and somatic cells from milk and whey, the defatting of whey and globular milk fat removal, and the fractionation of milk proteins. UF is widely used for the fractionation of cheese whey and preconcentration of milk for cheese manufacturing.

UF is a viable approach to separate lactose and protein from whey with high yield and purity. HF UF membranes with MWCO of 8 kDa have been used in a diafiltration mode to improve the yield of protein in the retentate. The UF permeate was then treated by NF to produce a concentrated lactose in the retentate [45]. The proposed approach allowed to recover 90% of lactose and 80% of proteins in the form of freeze-dried products. The recovered whey proteins and lactose can further be converted to high-value end products such as whey protein concentrate (WPC) and bioethanol, respectively.

A high degree of lactose removal from goat's milk was obtained by Sofuwani et al. [46] by using 10 kDa UF membrane in a cross-flow HF system. TMP values of 0.55 bar and feed flowrate of 0.18 L/min produced the best flux value without reaching the limiting flux region.

An innovative HF process to produce lactose-reduced skim milk was developed by Novalin et al. [47]. Skim milk was pumped in the lumen of an HF membrane module (consisting of a bundle of capillaries made of PS membranes of 10 kDa) and was ultrafiltered through the HF membranes into the shell side where a β-galactosidase solution is recirculated. In this way, continuous, enzymatic lactose hydrolysis is possible without inherent problems of immobilized enzymes. Molecules with a lower size than the MWCO value such as small proteins, salts, and lactose pass through into the shell side. There, lactose is enzymatically converted into glucose, galactose, and a small amount of oligosaccharide by β-galactosidase. The conversion rate at a skim milk flowrate of 9.9 L/h and at a temperature of 23 ± 2°C was of 78.11%.

The following parameters were estimated in order to achieve a cost-effective process with a productivity of 360 g/L·h: 15 ± 2°C, 240 U/mL of β-galactosidase, an enzyme solution flowrate of 25 L/h, and a skim milk flowrate of about 9 L/h [48].

MF membranes can be used for the pretreatment of whey before whey protein concentration by UF. PES and PVDF MF membranes in HF configuration were found to remove fat and microorganisms from whey, enhancing the permeation flux for whey UF process. Cleaning protocols with an order of 0.1% NaOH, 200 ppm NaClO, and 0.5% NaOH resulted efficient for both PES and PVDF membranes [49].

An integrated membrane process based on the use of MF, UF, NF, and RO membranes was investigated by Rektor and Vatai [50] for Mozzarella whey processing. MF membranes in HF configuration retained most of the fat content (98.7%) due to their small pore size (0.2 μm) in relation to the sizes of the fat balls (between 0.1 and 20 μm). Proteins were also retained for 67%, while the loss of lactose was only 19.5%.

Virgin whey, free of fat, casein, spores, bacteria, and cheesemaking foulants can be obtained as permeate from the MF of slightly acidified skim milk prior to cheesemaking [51]. Liquid virgin whey protein isolates (LVWPI) were obtained by concentrating and fractionating virgin whey through a two-stage UF system consisting of 10 kDa spiral-wound and HF PS membrane modules [52]. The produced LVWPI contained 26.1% w/w total solids, about 91% of which were whey proteins.

The use of high membrane shear rates is one of the most efficient factors to increase the permeate flux of HF membranes through the reduction in concentration polarization in UF and cake build-up MF [53]. This approach has been also applied to HF membrane modules by attaching them to a sliding rod connected to a rotating disk.

Vibrating PVDF HF membranes in submerged configuration have been used for the separation and concentration of milk proteins [54]. The system allowed high transmission of whey proteins (α-lactalbumin and β-lactoglobulin) with full rejection of casein micelles. It had significantly better performance and separation efficiency when compared with a conventional cross-flow system.

The use of ultrasound and gas bubbling treatment by direct injection of N_2 was found to enhance the performance and cleaning processes of HF membranes used in the UF of skimmed milk powder solution. Under this treatment, the permeation flux was enhanced up to 288.57% and the fouling percentage was decreased below 15.55% after cleaning the fouled membrane [55].

Recently Schopf et al. [56] used the in situ magnetic resonance imaging (MRI) methodology to characterize the milk protein deposits on ceramic HF membranes (α-Al_2O_3, 0.1 μm, from MANN + HUMMEL GmbH, Ludwigsburg, Germany) applied for milk protein fractionation under processing conditions relevant for practical situations. Milk proteins resulted highly concentrated on the surface with an amount of casein 11 times higher than whey proteins. This behavior was attributed to the larger diameter of the caseins (50–400 nm) in comparison to the pore size of the membrane. The deposited caseins contributed also to retain whey proteins, which are significantly smaller than caseins (mean diameter of 8 nm). The protein concentration and the height of the deposit resulted in strict correlation. Moreover, pressure affected the filtration behavior on a microscopic length scale. A dynamic MF system consisting of a vibrating HF membrane module was tested by Beier et al. [57] for the filtration of bakers yeast suspensions. The shear rate induced by vibrations allowed to obtain high permeation fluxes although operating at low feed cross-flow velocity and TMPs.

6.5 Agrofood Wastewaters Processing

Nowadays, the food industry produces high quantities of solid and liquid wastes that mainly result from production, preparation, consumption, and disposal processes. The characteristics of these wastes depend mainly on the product being processed (e.g., fruit, vegetable, oils, dairy, meat, and fish) and the processing methods. On average they contain large amounts of macropollutants such as suspended solids (SS), chemical oxygen demand (COD), biochemical oxygen demand (BOD) in addition to nitrogen, phosphorous, pesticides, herbicides, and cleaning chemicals [58].

Several techniques have been proposed to reduce the environmental pollution generated by agrofood wastewaters, including physical and physicochemical processes (screening, sedimentation, centrifugation, adsorption, etc.), biological treatment processes (aerobic or anaerobic, activated sludge, aerated lagoon treatment, etc.), and advanced treatments, including membrane operations. Combinations of treatment technologies are often used to develop the most cost-effective and environmentally acceptable solutions for waste management. On the other hand, residues and wastes generated from industrial and agrofood processes are considered renewable feedstock enriched by bio-based compounds that can be reused and recycled into the different food chains satisfying environmental EU policy within the logic of the circular economy [59, 60].

The identification of adequate methods for the recovery of these compounds from their original sources is crucial and can be accomplished through several steps (i.e., coagulation and precipitation of impurities, adsorption on resins, etc.) that require the use of chemicals, solvents, and high temperatures with high processing cost and low efficiency.

Membrane processes offer several advantages over traditional technologies, including mild operating conditions of temperature and pressure, therefore preserving the functional properties of food products, non-use of chemical agents or solvents, and, consequently, product contamination. In addition, they are characterized by high selectivity toward specific compounds, simple equipment, relatively easy scale-up, reduced number of processing steps, and low energy consumption [61].

In the light of these properties, pressure-driven membrane processes have been largely investigated in the last years for the recovery of valuable compounds from different agrofood processing wastes [62]. Among these, wastewaters arising from the olive oil extraction process, named as olive mill wastewaters (OMWs), have been widely studied.

OMWs are effluents with an intensive violet-dark brown color and strong specific olive oil smell constituted on average by 83.4% of water, 1.8% of inorganic salts, and 14.8% of organic compounds. Organic substances determine the high polluting load of these wastewaters: BOD and COD may be as high as 100 and 200 g/L, respectively. The organic fraction contains sugars, tannins, polyphenols, polyalcohols, pectins, lipids, proteins, and organic acids [63].

MF and UF processes have been largely investigated as a pretreatment step in integrated membrane systems to reduce the organic load and other undesired compounds (i.e., suspended solids, sediments, colloidal particles, microorganisms) from OMWs, in order to produce phenolic fractions, which can be further purified and concentrated [64, 65]. PVDF UF membranes in HF configuration (HFS, 0.02 μm, 0.4 m^2, Toray) removed all suspended solids and 33% of total organic carbon (TOC) from raw OMWs; these membranes showed a low rejection toward low-molecular-weight polyphenols such as hydroxy-tyrosol, protocatechuic acid, catechol, tyrosol, caffeic acid, p-coumaric acid (rejections values of about 2.1%), and a rejection toward total antioxidant activity (TAA) of about 8.3%. Therefore, most of phenolic compounds were recovered in the UF permeate and purified from suspended solids that were completely retained in the retentate fraction [66]. However, the pretreatment of OMWs by MF or UF is considered a critical step of the process for the selective fractionation of OMWs due to the rapid decline in permeate flux due to severe fouling phenomena [64]. Acidification, enzymatic treatment, centrifugation, and classical pre-filtration represent useful approaches to reduce the formation of fouling layers in the pretreatment of OMWs with membranes. Gebreyohannes et al. [67] evaluated the performance of a biocatalytic membrane reactor in which the separation capability of MF membranes was combined with the enzymatic hydrolysis of pectines considered the main responsible of fouling phenomena in the MF treatment of OMWs. In particular, pectinases were immobilized on MF polyethylene HF membranes with 0.4 μm pore size and internal/external diameter of 0.41/0.65 mm. Flux performance as a function of time was monitored and compared with inert membranes. The whole results showed that the steady-state flux through the enzyme immobilized on the membrane surface was 50% higher than that observed for the inert membrane. The degradation of deposited pectins and the continuous removal of product inhibition from the reaction site were considered the main factors responsible for the better performance of the biocatalytically active membrane. Analyses of membrane fouling index indicated that surface-biofunctionalized membranes exhibited the least fouling tendency and significantly improved flux.

More recently, a photoactive gel has been introduced on the surface of polyethylene HF membranes for enhancing the fouling resistance during the MF of OMWs [68]. Titanium oxide (TiO_2) nanoparticles (NPs) were dispersed in polyacrylic acid (PAA) and patterned with poly (diallyldimethyl ammonium chloride) (PDDA) in a hybrid complex that was deposited onto the shell side of HF membranes. These membranes were assembled in tubular modules and tested in a submerged configuration under irradiation at two different pH conditions. The obtained results showed that the decline in the permeate flux has been limited over 6 h of process without the need of using further chemical reagents or expensive back-flushing procedures to further clean the membranes. A fast and reversible pH switch was able to remove residues, thereby preserving the integrity of the layer-by-layer (LBL) hybrid complex onto the membrane surface.

MF and UF processes have also been used in integrated systems for the recovery of valuable compounds from wine by-products such as grape seeds, lees, sludge, and liquors. Giacobbo et al. [69] investigated the use of MF polyimide (PI) HF membrane (0.4 μm, PAM Membranas Seletivas, Brazil) for the recovery of polyphenols from wine lees generated during the second racking of Merlot grape winemaking. At the optimal conditions, the selected membranes allowed removal of suspended colloidal particles producing a clear permeate from a solution diluted 10 v/v with a recovery rate of 21% of the total content of polyphenols. Similarly, an aqueous extraction combined with an MF process was used as a pretreatment for the recovery of antioxidant compounds from the first racking effluent, which is one of the most polluting effluent of the wine industry [70]. According to the experimental results, the MF of a solution diluted 50 V/V with HF membranes (0.4 μm, PAM Membranas Seletivas, Brazil) allowed achievement of a limpid permeate with a recovery of 1 g of polyphenols and 1 g of polysaccharides per liter of raw effluent.

In a similar approach, PVDF HF membranes prepared in laboratory according to the phase-inversion methodology were characterized and tested to clarify an aqueous extract from red wine lees [71]. The prepared membranes presented a thickness of 0.49 ± 0.9 mm, an outer diameter of 2.29 ± 0.54 mm, and an inner diameter of 1.30 ± 0.65 mm. The average pore diameter was 0.13 μm; a very high porosity (81.0 ± 4.5%) was measured due to a typical finger-like morphology. The prepared membranes removed all suspended solids from the wine lees extract, producing a clarified solution enriched in sugars and antioxidant compounds, including polyphenols, anthocyanins, and resveratrol (Table 6.4). Steady-state permeate flux values were about 53 kg/m^{2}h in the selected conditions of TMP (0.8 bar), feed flowrate (580 mL/min), and temperature (24 ± 2°C).

Table 6.4 General composition of aqueous extracts from red wine lees before and after clarification with PVDF HF membranes (adapted from Ref. [71])

Parameter	Raw Extract	Clarified Extract	Retentate
Total suspended solids (%)	6.4 ± 0.2	n.d.	19.3 ± 1.4
TSS (°Brix)	3.5 ± 0.2	2.5 ± 0.1	4.5 ± 0.4
TAA (mM Trolox)	3.8 ± 0.9	3.2 ± 0.1	2.9 ± 0.6
Total polyphenols (mg/L galllic acid)	1160.7 ± 5.6	982.1 ± 1.7	1331.6 ± 3.6
Sugars (mg/L glucose)	23.8 ± 0.3	16.8 ± 1.0	28.6 ± 0.1
Anthocyanins (mg/L cyanidin 3-glucoside)	369.1 ± 17.8	328.9 ± 6.1	434.1 ± 0.7
Resveratrol (mg/L)	0.29 ± 0.06	0.24 ± 0.02	0.30 ± 0.04

Polymeric MF and UF HF membranes have also been successfully employed for the clarification and recovery of valuable compounds from the by-products of the citrus industry. In general, the citrus-processing industry generates large amounts of waste material such as peels and seed residues, which may account for up to 50% of the total fruit weight. Most of the waste residue from commercial juice extractors is shredded, limed, cured, and pressed into press liquors and press cakes, which are then processed independently. Press liquor is a complex mixture containing soluble sugars (sucrose, glucose, and fructose), insoluble carbohydrates, fibers, essential oil, carotenoids, polyphenols (mainly flavonoids), and organic acids whose recovery against waste-destruction technologies is very attractive for new business opportunities.

Simone et al. [72] investigated the clarification and recovery of phenolic compounds from orange press liquor with MF PVDF membranes in HF configuration prepared by the dry/wet spinning techniques. The HF membranes with a pore size of 0.22 μm allowed to remove all suspended solids from the feed solution, preserving phenolic compounds in the clarified liquor. The permeate stream enriched in antioxidant compounds can be further processed to produce formulations of nutraceutical or pharmacological interest. Similarly, suspended solids were completely removed from orange press liquor by using PS HF membranes with MWCO of 100 kDa;

flavonoids and anthocyanins were recovered in the clarified liquor according to the low rejection of the membranes toward these compounds (lower than 1%) [30].

Bergamot (*Citrus Bergamia Risso*) is a natural citrus fruit widely used for the extraction of essential oil from peel and employed in the pharmaceutical, cosmetic, or food industry. The juice has not found a real use in the food industry due to its bitter taste, and it is considered a waste of the essential oil extraction; however, it has been recognized as a source of flavonoids, in particular, neohesperidosides of hesperidin and naringenin such as brutieridin and melitidin, which exhibit statin-like properties [73]. Therefore, their recovery and purification are considered of great interest to develop natural formulations, which can be used as functional ingredients for fortified foods and beverages. PS HF membranes with MWCO of 100 kDa (from China Blue Star Membrane Technologies, Beijing, China) have been used for the clarification of bergamot juice as a pretreatment step devoted to remove suspended solids and macromolecules [74, 75]. Particularly, the raw juice was depectinized in order to hydrolyze both high and low esterified pectins and then ultrafiltered in selected operating conditions. This step allowed the production of a clear juice with a sugar content unchanged in comparison with the fresh juice. Most of flavonoids, polyphenols, and organic acids were recovered in the UF permeate as showed by the total antioxidant activity of the clarified juice, which was only 9% lower than that of the depectinized juice.

Polypropylene-grafted 3-hydroxy-2,2-bis(hydroxymethyl)propyl acrylate HF membranes (PP-g-PEA) have been successfully prepared by Wang et al. [76] through the continuous UV irradiation method. These membranes exhibited high water flux (about 4.5 times of the original PP) and can be used in the field of MF and UF processes for water treatment and food processing.

Recently, Li et al. [77] investigated an integrated membrane process for the recovery and purification of proteins from potato starch wastewater based on the use of self-made HF UF and NF membranes. In particular, PS HF membranes prepared through the conventional dry-wet spinning technique retained 85.62% of potato proteins with high molecular weight and 57.3% of COD; most of the potato proteins with low molecular weight were rejected by the NF membrane.

6.6 Future Outlook

Membrane fouling and complex hydrodynamics are major challenges that limit the large-scale application of HF membranes also in agrofood productions. In situ monitoring techniques for HF membrane fouling, including scanning electron microscopy (SEM), energy dispersive X-ray (EDX), Fourier transform infrared (FTIR), provide useful information on the properties of fouling deposition on the membrane surface and within the membrane pores. However, these destructive methods provide little information on membrane fouling deposition and distribution in HF systems during practical applications. The use of non-invasive techniques (i.e., electrical impedance spectroscopy, streaming potential measurements, etc.) and in situ monitoring techniques for local filtration characterization (i.e., experimental segmentation, doppler ultrasonography, and scintigraphic methods) offers a better comprehension about the formation and transport properties of foulants deposited on the membrane surface. Unfortunately, most of these methodologies are still limited to lab-scale applications. Future R&D investigations should be addressed to industrial applications where severe fouling phenomena occur due to the variability of feed properties and seasonal environmental conditions. A combination of filtration hydrodynamic models with in situ monitoring techniques will also be a promising approach for the design and optimization of HF membrane modules for practical applications.

Membrane surface modification techniques, including membrane surface grafting and coating, as well as polymer blending with hydrophilic polymers are other areas of remarkable investigation interest to enhance the antifouling characteristics of HF membranes for MF and UF applications.

References

1. Cuperus, F. P. (1998). Membrane processes in agro-food. State-of-the-art and new opportunities. *Sep. Purif. Technol.*, **14**, pp. 233–239.
2. Cheryan, M. (1998) *Ultrafiltration and Microfiltration Handbook* (Technomic Publishing Company, Lancaster).
3. Cassano, A., Conidi, C., and Drioli, E. (2010). Physico-chemical parameters of cactus pear (*Opuntia ficus-indica*) juice clarified by

microfiltration and ultrafiltration processes. *Desalination*, **250**, pp. 1101–1104.

4. Eykamp, W. (1995) *Membrane Separation Technology: Principles and Applications*, eds. Noble, R. D. and Stern S. A., Chapter "Microfiltration and Ultrafiltration" (Elsevier, Amsterdam), p. 37.
5. Echavarria, A. P., Torras, C., Pagan J., and Ibarz, A. (2011). Fruit juice processing and membrane technology application. *Food Eng. Rev.*, **3**, pp. 136–158.
6. Cassano, A., Donato, L., and Drioli, E. (2007). Ultrafiltration of kiwifruit juice: Operating parameters, juice quality and membrane fouling. *J. Food Eng.*, **79**, pp. 613–621.
7. Chung, T. S., Qin, J. J., and Gu, J. (2000). Effect of shear rate within the spinneret on morphology, separation performance and mechanical properties of ultrafiltration polyethersulfone hollow fiber membranes. *Chem. Eng. Sci.*, **55**, pp. 1077–1091.
8. Tasselli, F., Cassano, A., and Drioli, E. (2007). Ultrafiltration of kiwifruit juice using modified poly(ether ether ketone) hollow fibre membranes. *Sep. Purif. Technol.*, **57**, pp. 94–102.
9. Terra, M. N., Madrona, G. S., Ferreira, F. B., Cardoso, V. L., and Reis, M. H. M. (2019). High performance of asymmetric alumina hollow fiber membranes for the clarification of Genipap (*Genipa americana* L.) fruit extract. *Food Bioprocess Technol.*, **12**, pp. 27–38.
10. Qaid, S., Zait, M., El Kacemi, K., El Midaoui, A., El Hajjil, H., and Taky, M. (2017). Ultrafiltration for clarification of Valencia orange juice: Comparison of two flat sheet membranes on quality of juice production. *J. Mater. Environ. Sci.*, **8**, pp. 1186–1194.
11. Yazdanshenas, M., Tabatabaee-Nezhad, S. A. R., Soltanieh, M., Roostaazad, R., and Khoshfetrat, A. B. (2010). Contribution of fouling and gel polarization during ultrafiltration of raw apple juice at industrial scale. *Desalination*, **258**, pp. 194–200.
12. Hojjatpanah, G., Emam-Djomeh, Z., Kalbasi Ashtari, A., Mirsaeedghazi, H., and Omid, M. (2011). Evaluation of the fouling phenomenon in the membrane clarification of black mulberry juice. *Int. J. Food Sci. Technol.* **46**, pp. 1538–1544.
13. Cassano, A., Conidi, C., and Tasselli, F. (2015). Clarification of pomegranate juice (*Punica granatum* L.) by hollow fibre membranes: Analyses of membrane fouling and performance. *J. Chem. Technol. Biotechnol.*, **90**, pp. 859–866.

14. Laorko, A., Zhenyu, L., Tongchitpakdee, S., Chantachum, S., and Youravong, W. (2010). Effect of membrane property and operating conditions on phytochemical properties and permeate flux during clarification of pineapple juice. *J. Food Eng.*, **100**, pp. 514–521.

15. Sagu, S. T., Karmakar, S., Nso, E. J., Kapseu, C., and De, S. (2014). Ultrafiltration of banana (*Musa acuminata*) juice using hollow fibers for enhanced shelf life. *Food Bioprocess Technol.*, **7**, pp. 2711–2722.

16. Galiano, F., Figoli, A., Conidi, C., Menichini, F., Bonesi, M., Loizzo, M. R., Cassano, A., and Tundis, R. (2016). Functional properties of *Punica granatum* L. juice clarified by hollow fiber membranes. *Processes*, **4**, 21.

17. Morittu, V. M., Mastellone, V., Tundis, R., Loizzo, M. R., Tudisco, R., Figoli, A., Cassano, A., Musco, N., Britti, D., Infascelli, F., and Lombardi, P. (2020). Antioxidant, biochemical, and in-life effects of *Punica granatum L.* natural juice vs. clarified juice by polyvinylidene fluoride membrane. *Foods*, **9**, 242.

18. Conidi, C., Destani, F., and Cassano, A. (2015). Performance of hollow fiber ultrafiltration membranes in the clarification of blood orange juice. *Beverages*, **1**, pp. 341–353.

19. Cassano, A., Tasselli, F., Conidi, C., and Drioli, E. (2009). Ultrafiltration of Clementine mandarin juice by hollow fibre membranes, *Desalination*, **241**, pp. 302–308.

20. Castro-Muñoz, R., Fíla, V., Barragán-Huerta, B. E., Yáñez-Fernández, J., Piña-Rosas, J. A., and Arboleda-Mejía, J. (2018a). Processing of Xoconostle fruit (*Opuntia joconostle*) juice for improving its commercialization using membrane filtration. *J. Food Process Preserv.*, **42**, e13394.

21. Domingues, R. C. C., Ramos, A. A., Cardoso, V. L., and Reis, M. H. M. (2014). Microfiltration of passion fruit juice using hollow fibre membranes and evaluation of fouling mechanisms. *J. Food Eng.*, **121**, pp. 73–79.

22. Panigrahi, C., Karmakar, S., Mondal, M., Mishra, H. N., and De, S. (2018). Modeling of permeate flux decline and permeation of sucrose during microfiltration of sugarcane juice using a hollow-fiber membrane module. *Innov. Food Sci. Emerg. Technol.*, **49**, pp. 92–105.

23. Carrin, M. E., Ceci, L., and Lozano, J. E. (2000). Effect of pectinase immobilization during hollow fiber ultrafiltration of apple juice. *J. Food Process Eng.*, **23**, pp. 281–298.

24. Ghosh, P., Garg, S., Mohanty, I., Sahoo, D., and Pradhan, R. C. (2019). Comparison and storage study of ultra-filtered clarified jamun (*Syzygium cumini*) juice. *J. Food Sci. Technol.*, **56**, pp. 1877–1889.

25. Conidi, C., Tasselli, F., Cassano, A., and Drioli, E. (2017a). Quality of kiwifruit juice clarified by modified poly(ether ether ketone) hollow fiber membranes. *J. Membr. Sci. Res.*, **3**, pp. 313–319.

26. de Oliveira, R. C., Docê, R. C., and de Barros, S. T. D. (2012). Clarification of passion fruit juice by microfiltration: Analyses of operating parameters, study of membrane fouling and juice quality. *J. Food Eng.*, **111**, pp. 432–439.

27. de Barros, S. T. D., Andrade, C. M. G., Mendes, E. S., and Peres, L. (2003). Study of fouling mechanism in pineapple juice clarification by ultrafiltration. *J. Membr. Sci.*, **215**, pp. 213–224.

28. Laorko, A., Tongchitpakdee, S., and Youravong, W. (2013). Storage quality of pineapple juice non-thermally pasteurized and clarified by microfiltration. *J. Food Eng.*, **116**, pp. 554–561.

29. Cassano, A., Conidi, C., and Drioli, E. (2011). Clarification and concentration of pomegranate juice (*Punica granatum* L.) using membrane processes. *J. Food Eng.*, **107**, pp. 366–373.

30. Cassano, A., Conidi, C., and Ruby-Figueroa, R. (2014). Recovery of flavonoids from orange press liquor by an integrated membrane process. *Membranes*, **4**, pp. 509–524.

31. Conidi, C., Cassano, A., Caiazzo, F., and Drioli, E. (2017b). Separation and purification of phenolic compounds from pomegranate juice by ultrafiltration and nanofiltration membranes. *J. Food Eng.*, **195**, pp. 1–13.

32. Panigrahi, C., Mondal, M., Karmakar, S., Mishra, H. N., and De, S. (2020). Shelf life extension of sugarcane juice by cross flow hollow fibre ultrafiltration. *J. Food Eng.*, **274**, 109880.

33. Vladisavljevíc, G. T., Vukosavljevíc, P., and Veljovíc, M. S. (2013). Clarification of red raspberry juice using microfiltration with gas backwashing: A viable strategy to maximize permeate flux and minimize a loss of anthocyanins. *Food Bioprod. Process.*, **91**, pp. 473–480.

34. Salazar, F. N., De Bruijn, J. P. F., Seminario, L., Güell, C., and López, F. (2007). Improvement of wine cross-flow microfiltration by a new hybrid process. *J. Food Eng.*, **79**, pp. 1329–1336.

35. El Rayess, Y. E., Albasi, C., Bacchin, P., Taillandier, P., Raynal, J., Mietton-Peuchot, M., and Devatine, A. (2011). Cross-flow microfiltration applied to oenology: A review. *J. Membr. Sci.*, **382**, pp. 1–19.

36. Daufin, G., Escudier, J. E., Carrère, Bérot, S., Fillaudeau, L., and Decloux, M. (2001). Recent and emerging applications of membrane processes in the food and dairy industry. *Food Bioprod. Process.*, **79**, pp. 89–102.

37. Czekaj, P., López, F., and Güell, C. (2001). Membrane fouling by turbidity constituents of beer and wine: Characterization and prevention by means of infrasonic pulsing. *J. Food Eng.*, **49**, pp. 26–36.

38. Cassano, A., Mecchia, A., and Drioli, E. (2008). Analyses of hydrodynamic resistances and operating parameters in the ultrafiltration of grape must. *J. Food Eng.*, **89**, pp. 171–177.

39. Gan, Q., Howell, J. A., Field, R. W., England, R., Bird, M. R., O'Shaughnessy, C. L., and MeKechinie, M. T. (2001). Beer clarification by microfiltration: Product quality control and fractionation of particles and macromolecules. *J. Membr. Sci.*, **194**, pp. 185–196.

40. Atkinson, S. (2005). Beer clarification. *Membr. Technol.*, **7**, pp. 8–9.

41. Fillaudeau, L., Boissier, B., Moreau, A., Blanpain-Avet, P., Ermolaev, S., Jitariouk, N., and Gourdon, A. (2007). Investigation of rotating and vibrating filtration for clarification of rough beer. *J. Food Eng.*, **80**, pp. 206–217.

42. Noordman, T. R., Berghuis, O. A. E., Mol, M. N. M., Peet, C. J., Muller J. L. M., Broens, L., and Van Hoof, S. (1999). Membrane filtration for bright beer, an alternative to Kieselguhr filtration, *Proc. 27th EBC Congress* (Cannes, France), pp. 815–822.

43. Cimini, A. and Moresi, M. (2016a). Assessment of the optimal operating conditions for pale lager clarification using novel ceramic hollow-fiber membranes. *J. Food Eng.*, **173**, pp. 132–142.

44. Cimini, A. and Moresi, M. (2016b). Beer clarification by novel ceramic hollow-fiber membranes: Effect of pore size on product quality. *J. Food Sci.* **81**, pp. E2521–E2528.

45. Das, B., Sarkar, S., Sarkar, A., Bhattacharjee, S., and Bhattacharjee, C. (2016). Recovery of whey proteins and lactose from dairy waste: A step towards green waste management. *Process Saf. Environ. Protect.*, **101**, pp. 27–33.

46. Sofuwani, N. Z. A., Aslina, H. S., and Mazlina, S. M. K. (2016). Separation of lactose from raw goat's milk by cross-flow hollow fiber ultrafiltration membrane. *Int. Food Res. J.*, **23**, pp. S209–S219.

47. Novalin, S., Neuhaus, W., and Kulbe, K. D. (2005). A new innovative process to produce lactose-reduced skim milk. *J. Biotechnol.*, **119**, pp. 212–218.

48. Neuhaus, W., Novalin, S., Klimacek, M., Splechtna, B., Petzelbauer, I., Szivak, A., and Kulbe, K. D. (2006). Optimization of an innovative hollow-fiber process to produce lactose-reduced skim milk. *Appl. Biochem. Biotechnol.*, **134**, pp. 1–14.

49. Ye, H., Lei, J. J., Zhang, Y. Z., Li, H., Li, G. F., Cao, Z. P., and Xie, L. P. (2011). Investigation of microfiltration for pretreatment of whey concentration. *Desalin. Water Treat.*, **34**, 173–178.

50. Rektor, A. and Vatai, G. (2004). Membrane filtration of Mozzarella whey. *Desalination*, **162**, 279–286.

51. Ardisson-Korat, A. V. and Rizvi, S. S. H. (2004). Vatless manufacturing of low-moisture part-skim mozzarella cheese from highly concentrated skim milk microfiltration retentates. *J. Dairy Sci.*, **87**, pp. 3601–3613.

52. Marcelo, P. A. and Rizvi, S. S. H. (2008). Physicochemical properties of liquid virgin whey protein isolate. *Int. Dairy J.*, **18**, pp. 236–246.

53. Jaffrin, M. Y. (2008). Dynamic shear-enhanced membrane filtration: A review of rotating disks, rotating membranes and vibrating systems. *J. Membr. Sci.*, **324**, pp. 7–25.

54. Chai, M., Ye, Y., and Chen, V. (2017). Separation and concentration of milk proteins with a submerged membrane vibrational system. *J. Membr. Sci.* **524**, pp. 305–314.

55. Shahraki, M. H., Maskooki, A., and Faezian, A. (2017). Hollow fibers filtration and cleaning processes under ultrasound and gas bubbling combination. *J. Food Process Eng.*, **40**, e12325.

56. Schopf, R., Schork, N., Amling, E., Nirschl, H., Guthausen, G., and Kulozik, U. (2020). Structural characterisation of deposit layer during milk protein microfiltration by means of in-situ MRI and compositional analysis. *Membranes*, **10**, 59.

57. Beier, S. P., Guerra, M., Garde, A., and Jonsson, G. (2006). Dynamic microfiltration with a vibrating hollow fiber module: filtration of yeast suspensions. *J. Membr. Sci.*, **281**, pp. 281–287.

58. Girotto, F., Alibardi, L., and Cossu, R. (2015). Food waste generation and industrial uses: A review. *Waste Manage.*, **45**, pp. 32–41.

59. Lin, C. S. K., Koutinas, A. A., Stamatelatou, K., Mubofu, E. B., Matharu, A. S., Kopsahelis, N., Pfaltzgraff, L. A., Clark, J. H., Papanikolaou, S., Kwan, T. H., and Luque, R. (2014). Current and future trends in food waste valorization for the production of chemicals, materials and fuels: A global perspective. *Biofuels Bioprod. Biorefin.*, **8**, pp. 686–715.

60. Schieber, A., Stintzing, F. C., and Carle, R. (2001). By-products of plant food processing as a source of functional compounds-recent developments. *Trends Food Sci. Technol.*, **12**, pp. 401–413.

61. Castro-Muñoz, R., Conidi, C., and Cassano, A. (2019). Membrane-based technologies for meeting the recovery of biologically active compounds from foods and their by-products. *Crit. Rev. Food Sci. Nutr.*, **59**, pp. 2927–2948.

62. Castro-Muñoz, R., Barragán-Huerta, B. E., Fíla, V., Denis, P. C., and Ruby-Figueroa, R. (2018b). Current role of membrane technology: From the treatment of agro-industrial by-products up to the valorization of valuable compounds. *Waste Biomass Valorization*, **9**, pp. 513–529.

63. Dermeche, S., Nadour, M., Larroche, C., Moulti-Mati, F., and Michaud, P. (2013). Olive mill wastes: Biochemical characterizations and valorization strategies. *Process Biochem.*, **48**, pp. 1532–1552.

64. Russo, C. (2007). A new membrane process for the selective fractionation and total recovery of polyphenols, water and organic substances from vegetation waters (VW). *J. Membr. Sci.*, **288**, pp. 239–246.

65. Tundis, R., Conidi, C., Loizzo, M. R., Sicari, V., and Cassano, A. (2020). Olive mill wastewater polyphenol-enriched fractions by integrated membrane process: A promising source of antioxidant, hypolipidemic and hypoglycaemic compounds. *Antioxidants*, **9**, 602.

66. Cassano, A., Conidi, C., Giorno, L., and Drioli, E. (2013a). Fractionation of olive mill wastewaters by membrane separation techniques. *J. Hazard. Mater.*, **248–249**, pp. 185–193.

67. Gebreyohannes, A. Y., Mazzei, R., Curcio, E., Poerio, T., Drioli, E., and Giorno, L. (2013). Study on the in situ enzymatic self-cleansing of microfiltration membrane for valorization of olive mill wastewater. *Ind. Eng. Chem. Res.*, **52**, pp. 10396–10405.

68. Han, Y., Giorno, L., and Gugliuzza, A. (2017). Photoactive gel for assisted cleaning during olive mill wastewater membrane microfiltration. *Membranes*, **7**, 66.

69. Giacobbo, A., Do Prado, J. M., Meneguzzi, A., Bernardes, A. M., and De Pinho, M. N. (2015). Microfiltration for the recovery of polyphenols from winery effluents. *Sep. Purif. Technol.*, **143**, pp. 12–18.

70. Giacobbo, A., Meneguzzi, A., Bernardes, A. M., and De Pinho, M. N. (2017). Pressure-driven membrane processes for the recovery of antioxidant compounds from winery effluents. *J. Clean Prod.*, **155**, pp. 172–178.

71. Cassano, A., Bentivenga, A., Conidi, C., Galiano, F., Saoncella, O., and Figoli, A. (2019). Membrane-based clarification and fractionation of red wine lees aqueous extracts. *Polymers*, **11**, 1089.

72. Simone, S., Conidi, C., Ursino, C., Cassano, A., and Figoli, A. (2016). Clarification of orange press liquors by PVDF hollow fiber membranes. *Membranes*, **6**, 9.

73. Di Donna, L., Iacopetta, D., Cappello, A. R., Gallucci, G., Martello, E., Fiorillo, M., Dolce, E., and Sindona, G. (2014). Hypocholesterolaemic activity of 3-hydroxy-3-methyl-glutaryl flavanones enriched fraction from bergamot fruit (*Citrus bergamia*): "In vivo" studies. *J. Funct. Food*, **7**, 558–568.

74. Conidi, C., Cassano, A., and Drioli, E. (2011). A membrane based study for the recovery of polyphenols from bergamot juice. *J. Membr. Sci.*, **375**, pp. 182–190.

75. Cassano, A., Conidi, C., and Drioli, E. (2013b). A membrane-based process for the valorization of the bergamot juice. *Sep. Sci. Technol.*, **48**, pp. 537–546.

76. Wang, L., Wei, J., Zhao, K., and Wu, B. (2015). Preparation and characterization of high-hydrophilic polyhydroxy functional PP hollow fiber membrane. *Mater. Lett.*, **159**, pp. 189–192.

77. Li, H. B., Zeng, X. H., Shi, W. Y., Zhang, H. X., Huang, S. F., Zhou, R., and Qin, X. H. (2020). Recovery and purification of potato proteins from potato starch wastewater by hollow fiber separation membrane integrated process. *Innov. Food Sci. Emerg. Technol.*, **63**, 102380.

Chapter 7

Hollow Fibers for Reverse Osmosis and Nanofiltration

Jincai Cai[a] and Tai-Shung Chung[b,c]
[a]*School of Life Sciences & Chemical Technology, Ngee Ann Polytechnic, 535 Clementi Road, Singapore 599489, Singapore*
[b]*Department of Chemical and Biomolecular Engineering, National University of Singapore, 4 Engineering Drive 4, Singapore 117585, Singapore*
[c]*Graduate Institute of Applied Science and Technology, National Taiwan University of Science and Technology, Taipei, Taiwan*
chencts@mail.ntust.edu.tw

This chapter summarizes the advancement of hollow fiber membranes for reverse osmosis (RO) and nanofiltration (NF) technologies. The separation mechanisms of RO and NF are briefly discussed. The transition of commercial RO hollow fiber products and the challenges of further developing RO hollow fiber membranes are reviewed. The research and development activities around NF are presented in various aspects.

Hollow Fibers and Nanofibers in Membrane Science: Preparation, Characterization, and Applications
Edited by Alberto Figoli, Mir Saeed Seyed Dorraji, and Francesco Galiano

ISBN 978-981-4968-03-4 (Hardcover), 978-1-003-25663-2 (eBook)
www.jennystanford.com

7.1 Reverse Osmosis and Hollow Fiber Membranes

7.1.1 Mechanism of Reverse Osmosis

The origin of Reverse Osmosis (RO) is based on osmosis, which is a naturally occurring phenomenon in biological systems. Osmosis refers to the targeted diffusion of water through a semipermeable membrane, and water molecules spontaneously move from a solution of low solute concentration (low osmotic pressure π_1) into a solution of high solute concentration (high osmotic pressure π_2) without external forces exerted. Figure 7.1a shows how osmosis works [1, 2]. Semipermeable membranes have special properties. They are permeable to water molecules but impermeable to larger molecules such as salt particles. Water will continuously migrate toward the solution of a high solute concentration until the solute concentration is equalized on both sides of the semipermeable membrane (Fig. 7.1b). The difference in height between the two solutions corresponds to the osmotic pressure of the solution that is at equilibrium ($\Delta h = \Delta\pi = \pi_2 - \pi_1$) [3]. Osmosis can be slowed, stopped, or even reversed if sufficient pressure is applied to the membrane from the high solute concentration side. If the pressure applied on the more concentrated solution is greater than the osmotic pressure gradient ($\Delta h = P_2 - P_1 > \Delta\pi = \pi_2 - \pi_1$), the naturally occurring osmotic flow is reversed (Fig. 7.1c). In this manner, water is pushed through the membrane into the less concentrated solution, while dissolved solids are retained by the membrane. Hence, water on one side of the membrane is purified or demineralized, and the dissolved solutes on the other side of the membrane are concentrated or dewatered.

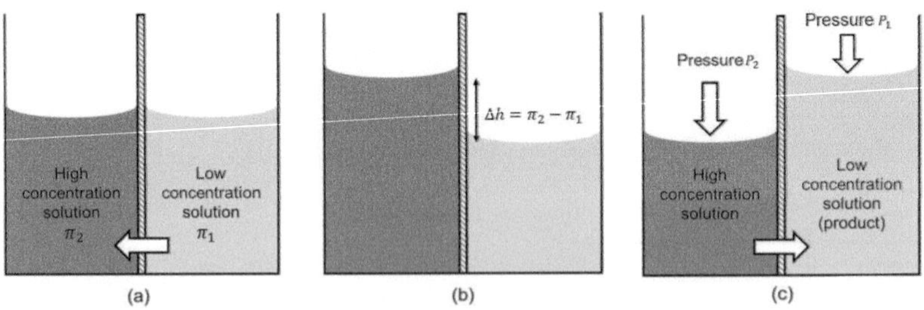

Figure 7.1 Schematic of (a) osmosis, (b) osmotic equilibrium, and (c) RO. Reprinted from Ref. [1] with permission of Elsevier.

The RO technique has developed greatly in recent decades and has progressed from an emerging technology to become a consolidated, efficient, and competitive demineralization process for the separation of dissolved solids from liquid solvents. It is being applied in water treatment, wastewater treatment, food and beverage processing, organic mixture separation, and many other applications [2]. Though many liquids may be applied, the primary application of RO is water-based separations. Hence, the following discussion on hollow fiber RO membranes is focused on water purification. Figure 7.2 shows the diagram of the simplest RO system. The RO system works by using a high-pressure pump to pressurize the feed solution of high solute concentration and force the water across the semipermeable RO membrane, leaving a majority of the dissolved solutes (e.g., salts) behind in the reject stream. The permeate is almost pure water and is collected as the product. Due to the added resistance of the membrane, the applied pressure required to achieve RO is normally significantly higher than the osmotic pressure gradient across the RO membrane. In actual industrial processes, the feed always contains a lot of contaminants such as suspended solids, organic materials, biological species, dissolved solids, etc. To avoid severe RO membrane fouling, the feed goes through intensive pretreatment operations such as coagulation–flocculation, media filtration, dissolved air flotation, disinfection, scale inhibition, microfiltration (MF), ultrafiltration (UF), nanofiltration (NF), etc. before it is introduced to the RO system [4, 5].

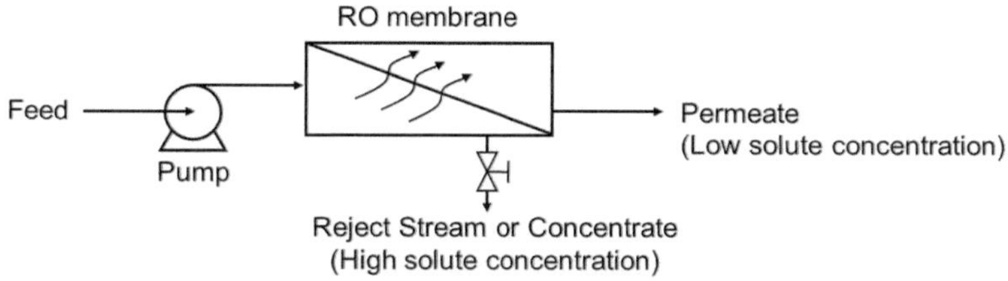

Figure 7.2 Diagram of an RO system.

RO can remove greater than 99% of the dissolved salts (ions), particles, colloids, organics, bacteria, and pyrogens from the feed. The excellent separation characteristics of an RO system result from the different transport rates of water, solvent, and other dissolved solutes through the semipermeable RO membrane. Many mechanisms have been proposed to describe the transport of mass through RO membranes; however, two schools of thoughts predominate [3, 6]. The first school of thought assumes that the separation layer of the RO membrane is non-porous and defect free. It believes that the molecules of interest dissolve in the separation layer of the RO membrane and then diffuse through the membrane. This is the solution-diffusion model, which is considered the leading theory on RO membrane transport. Dissolved salts are rejected because they cannot permeate through the dense layer. Rejection to organics is similar to that of salts, but the passage of organics of low molecular weight through the RO membrane may also occur. The second school of thought assumes that the separation layer of the RO membrane contains a multitude of micropores. Due to the physicochemical interactions between the solution and the membrane, water passes through the pores while salts are rejected. Whether organics can pass through the pores or are rejected depends on the size, by molecular sieving or screening principle.

The performance of an RO system is mainly described by water (or solvent) recovery, rejection of the dissolved salts and other contaminants, water (or solvent) flux, and antifouling properties. An ideal RO membrane should offer high water (or solvent) flux, high rejection as well as excellent mechanical strength and durability. These parameters depend largely on the properties of the membrane material, particularly, the chemical nature of the membrane polymer and the structure of the membrane.

7.1.2 Early Development of Hollow Fiber RO Membranes

Membranes for commercial RO applications are normally composite materials having a very thin and dense separation layer anchored on thick, strong, and porous support. Cellulose acetate (CA) is the first high-performance and commercially viable RO membrane material discovered [6, 7]. Figure 7.3 shows the chemical structure of CA. Early CA RO membranes were developed by casting CA solution onto fabric backing material. The main advantage of CA is its low price, less prone to fouling, and relatively good chlorine resistance. A typical CA RO membrane exhibits a water flux of 37.5 L/m^2h (LMH) at 425 psi pressure and an average NaCl rejection of 97.5% from a

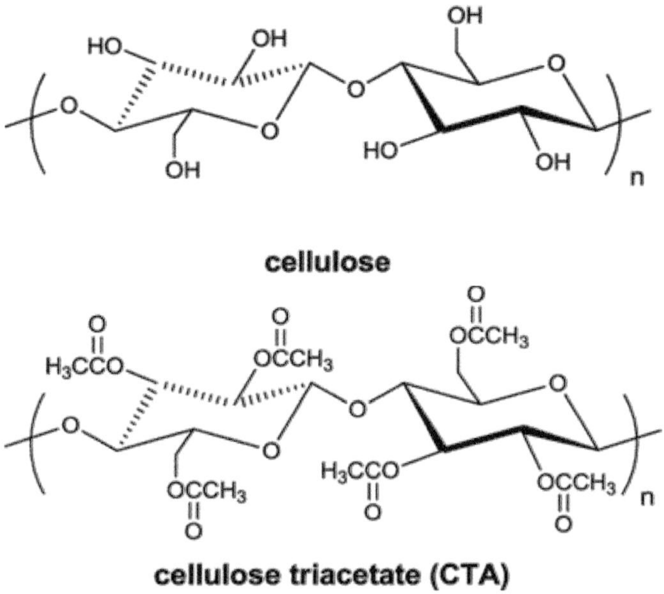

Figure 7.3 Chemical structure of CA and CTA. Reprinted from Ref. [7] with permission of Elsevier.

2000 mg/L NaCl solution [8]. The inherent weakness of CA material is its vulnerability to microorganisms, in addition to hydrolysis over time and sensitivity to operating temperature. A more successful RO membrane in commercial applications is a thin-film composite (TFC) polyamide (PA) membrane fabricated via interfacial polymerization method (Fig. 7.4). A typical TFC PA membrane exhibits a water flux of 50 LMH at 800 psi pressure and an average NaCl rejection of 99.5% from a 35,000 mg/L NaCl solution [8]. Compared with CA-based RO membranes, PA-based TFC RO membranes are obviously more permeable to water and more impermeable to salt. Besides, the structure of the latter makes them more robust at elevated operation pressures. The main disadvantage of PA material is its low tolerance to oxidizing reagents, such as chlorine and hypochlorite, which is commonly used for disinfection in water.

Though commercially available, RO membranes are predominantly flat-sheet composite membranes. Membranes in hollow fiber form have gained their broad recognition of certain advantages for seawater desalination, such as self-supporting, very robust to allow for high-pressure operation and high recovery, and large membrane area. The self-supporting nature lowers the requirement on hardware for fabrication, i.e., spacer and supports for flat-sheet membranes. The excellent robustness of hollow fibers makes it possible to operate at elevated pressure, e.g., 1200 psi, which allows enhanced recoveries in seawater desalination. The high membrane area of hollow fiber modules allows the operation at relatively lower permeate flux than flat-sheet membranes in order to achieve the same production rate. This reduces the severity of concentration polarization and the propensity of scaling. Hollow fiber membranes for RO processes normally have an inner diameter of less than 100 microns and an outer diameter less than 200 microns. An element with such hollow fibers has about 100 times more active surface area than a conventional plate-and-frame or tubular unit of the same volume. The earliest hollow fiber RO membranes might be cellulose triacetate (CTA) membrane developed by Dow Chemical Company, poly(furfuryl alcohol) (PFA) membranes developed by North Star, CA membrane developed by Toyobo, and aromatic polyamide membranes developed by DuPont [10, 11]. The composite hollow fiber RO membrane developed at FRL used PFA as the selective layer and polysulfone (PSf) hollow fibers as the substrate [11]. These PFA/PSf hollow fibers showed >99% rejection of salts (e.g., NaCl, KCl, and Na_2SO_4) and up to 13.6 LMH flux at 800 psi pressure.

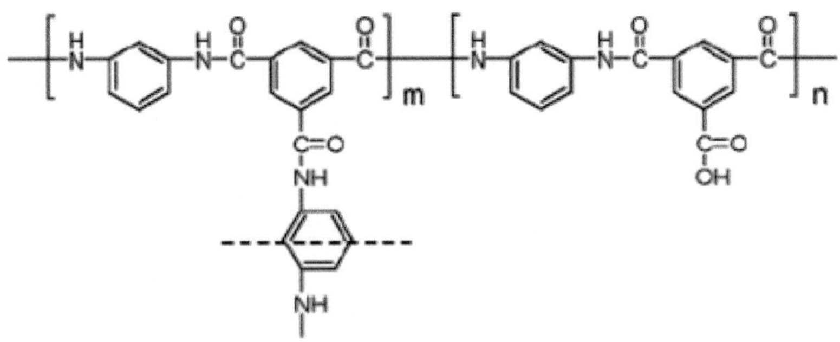

Figure 7.4 Chemical structure of PA. Reprinted from Ref. [9] with permission of the American Chemical Society.

The first non-cellulosic material for hollow fiber RO membrane is aromatic PA developed by Richter and Hoehn [12]. PA-based RO membranes were commercialized by DuPont under the trade name Permasep® hollow fiber (e.g., B-9 and B-10), for brackish water and seawater treatment. The B-10 hollow fibers are modified versions of B-9 fibers with thicker fiber walls and higher rejection with chemically treated fiber surface by tannic acid [13]. Abron and Osburn studied the performance of DuPont B-5 type hollow fiber RO membranes for the removal of dichlorodiphenyltrichloroethane (DDT) and Aldrin pesticides from aqueous solution [14]. The membranes were observed to only achieve 15% DDT and Aldrin recovery regardless of input concentrations. Sorption behavior disclosed that DDT was more strongly sorbed by the membrane and diffused at a slower rate than Aldrin. As organic materials, the transport of DDT and Aldrin was thought to be different from inorganic salts and it might be only pore flow, ordinary diffusion, or both. Figure 7.5 shows a hank of such hollow fibers, which have a very small diameter. Since then, DuPont continuously improved the design and performance of its hollow fiber RO elements and introduced new products such as B-10 TWIN® Permeator and Hollow fiber CartidgeTM to the market.

(a) (b)

Figure 7.5 A hank of DuPont aromatic PA hollow fiber RO membrane (a) and image and dimension of the fibers (b). Reprinted from Ref. [13] with permission of Elsevier.

Figure 7.6 shows the design of the Permasep® module. The fibers are operated at "outside-in" mode and the product passes into the fiber lumens, flowing down the lumens through the tube sheet and into the product header. It is believed that the "outside-in" flow mode is less demanding on the mechanical strength of the fibers than the "inside-out" mode. In addition, the pressure drop in the lumens is much lower when it carries very clean product water

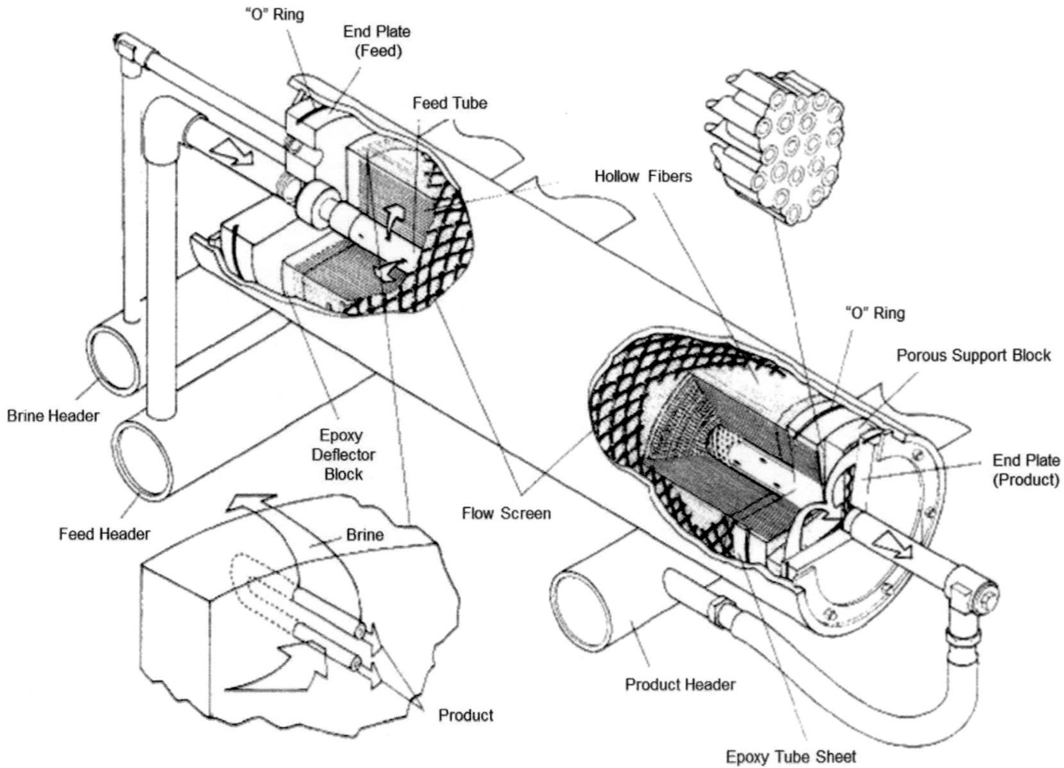

Figure 7.6 Schematic of Permasep® hollow fiber RO module. Reprinted from Ref. [13] with permission of Elsevier.

instead of the feed water. The typical performance parameters of the B-10 Permeator for seawater desalination were 30–50% recovery, energy consumption of 3.7–8.2 kWh/m^3, and permeate quality of <500 mg/L at 1000–1200 psi feed pressure [15]. There was experimental evidence showing that the B-10 hollow fibers were considerably compressed after desalination at high pressures [16]. Generally, these aromatic polyamide hollow fibers do not show more attractive transport properties than CA fibers though they also have a skinned asymmetric structure. However, the DuPont hollow fiber RO membranes based on linear aromatic PA were leading the RO market in the 1980s and early 1990s [17].

Al-Qahtany and Al-Bastaki continuously monitored the desalination performance of DuPont hollow fiber RO membrane for a period of 35,000 h [18]. Reduction in water permeation rate and increase in salt passage occurred due to the physical changes in the hollow fiber material resulting from continuous applied pressure, the influences from the chemicals, and regular chemical treatment. Butt et al. operated a hollow fiber and a spiral-wound element in parallel for brackish water desalination under identical scale control treatment and the same recovery [19]. After the 7500 h trial, the hollow fibers were more severely fouled. The scale deposits on the membrane surface were mainly contributed by organic matter and biomass, while calcium aluminosilicate clay was found to be the main source of inorganic scale. In another comparative study, seawater desalination performance of a twin hollow fine fiber (DuPont B-10) and a spiral-wound TFC membrane (Koch Fluid Systems®) element was monitored at totally identical operational aspects [20]. After 1 year of operation, a 0.4% salt rejection decrease was observed for the spiral-wound membrane, i.e., dropping from 99.8% to 99.4%. A more significant salt rejection loss of 3.4% occurred for the hollow fiber membrane, i.e., dropping from 99.8% to 96.4%. Slime formation was found in both elements, but it only apparently affected the performance of hollow fibers. This is the direct evidence that the spiral-wound RO module is superior to the twin hollow fiber module in resisting slime.

Aromatic PA hollow fibers were tested for the desalination of high-salinity water containing 20,000–53,000 mg/L total dissolved solid (TDS) [21]. At an operating pressure of 800 psi, decreasing recovery was observed at all the three operating temperatures, while the higher recovery was obtained at higher temperatures for each feed TDS. Interestingly, the recoveries at 24–26°C and 33–35°C were 34% and 42%, respectively, for a feed TDS of 35,000 mg/L and both are higher than 30%, which was reported by the manufacturer. This observation indicates the strong flexibility of the hollow fiber membrane to respond to a changing environment. The salt rejections observed for a wide range of feed TDS levels are all above 99%. The desalination of low-salinity water (TDS of 2600–3300) using PA hollow fibers was also studied [22]. The PA hollow fibers showed satisfactory performance even if subjected to low salinity at a low pressure of 100 psi. The membrane responding tendency to feed water TDS was found to be predictable.

Modeling work revealed the permeate flow profiles and permeate salinity at axial and radial directions in B-10 hollow fiber elements [23]. The permeate flow shows different trends along with the module, i.e., increasing at the inner radius but decreasing at the outer radius, due to the more significant influence of osmotic pressure at the outer radius where the salt concentration at the membrane surface is high (Fig. 7.7a). The salt concentration of the permeate is higher when the RO element is operated at low pressure. On increasing the operating pressure, the permeate salinity drops. Additionally, the salt concentration is lower in the inner radius, and it rises toward the outer radius (Fig. 7.7b).

Another breakthrough in the RO industry was the advancement of hollow fiber Cartridge™ at DuPont [24]. The design innovations of the new hollow fiber Cartridge™ include the improved permeate flow and salt rejection through "double-end" design, multiple assemblies, self-regulating feed bypass, and re-engineered bundle cross section. The new element design (Fig. 7.8) benefits the RO plant in terms of high recovery per cartridge, plant compactness, and long-term membrane performance (or lower cost). Field tests of the new hollow fiber Cartridge™ elements verified the improvements.

A major disadvantage of hollow fiber membranes is the propensity to foul and plug due to the small space between hollow fibers and the dead zone inside the module [25]. With more understanding of the constraints and shortcomings of hollow fiber membranes as well as the CA and CTA chemistry, Dow purchased FilmTec Corporation in 1985 and gained the technology to produce TFC PA membranes through interfacial polymerization. Hence, Dow started to produce

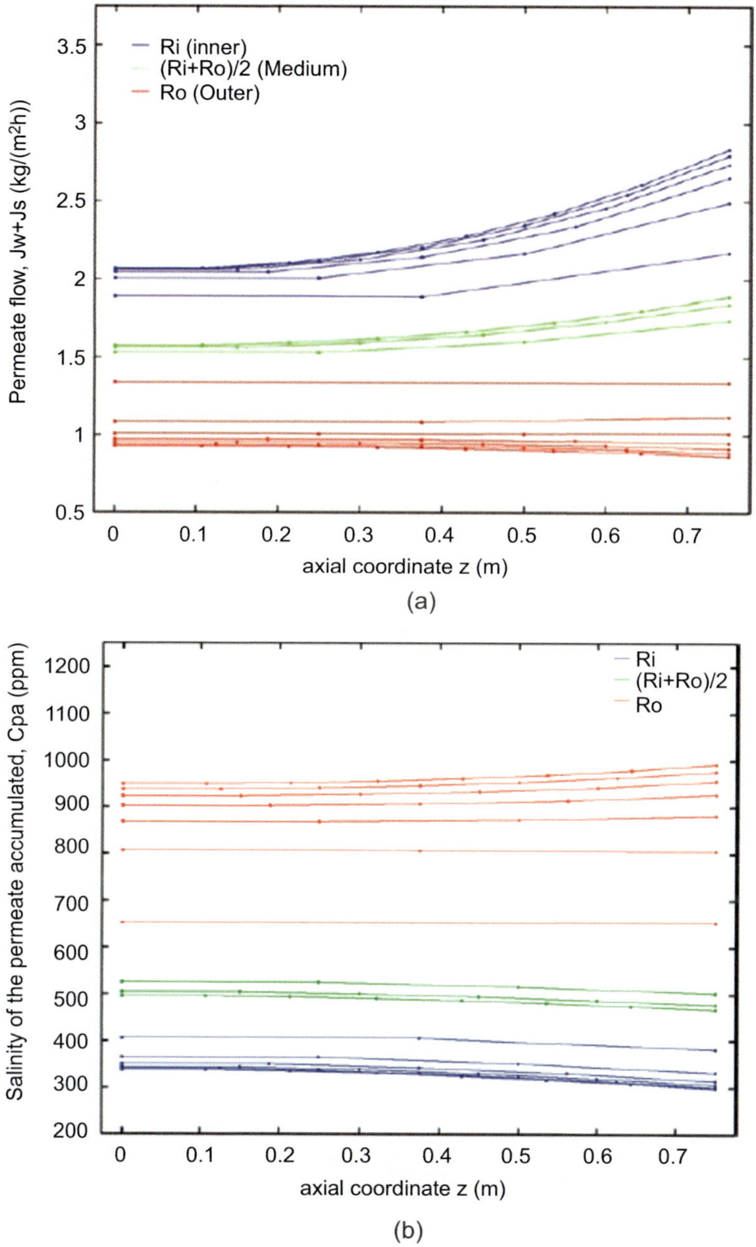

Figure 7.7 Profiles of the permeate flow (a) and permeate salinity for increasing mesh refinements. Reprinted from Ref. [23] with permission of Elsevier.

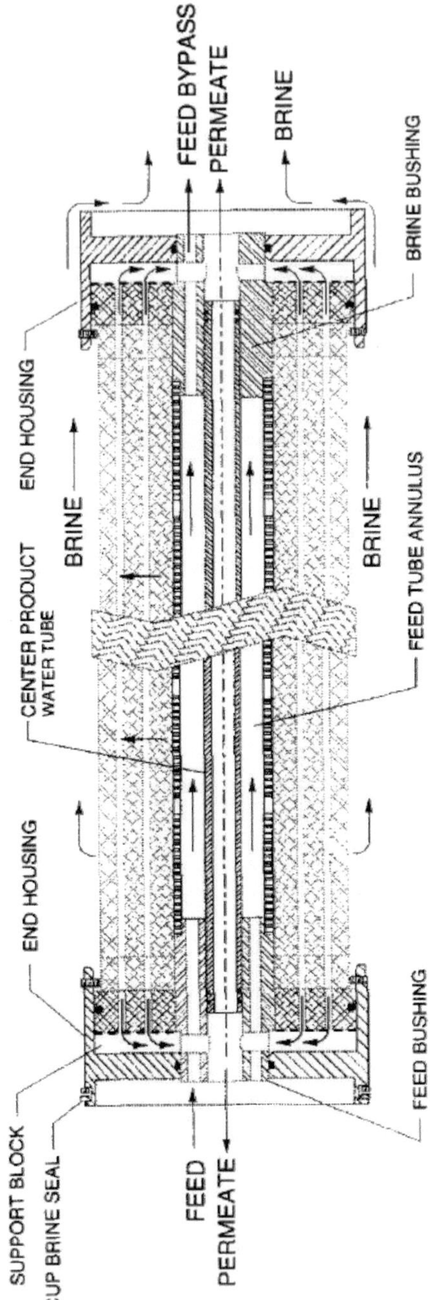

Figure 7.8 Schematic of the new hollow fiber Cartridge™. Reprinted from Ref. [24] with permission of Elsevier.

and provide spiral-wound RO elements and exited the hollow fiber market. With the advances of the TFC PA membrane and spiral-wound elements in the 1990s, the market share of hollow fiber RO elements shrank. Many more companies entered the RO membrane market and joined the competition with their spiral-wound elements. Many RO plants replaced the hollow fiber RO elements with spiral-wound elements. Conversion of hollow fiber to spiral-wound elements would help to reduce the operating costs of the plant, the membrane replacement costs, and the total water cost [26]. Nevertheless, there are still large old RO plants using hollow fiber elements, and Toyobo is the only remaining supplier of hollow fiber RO membranes.

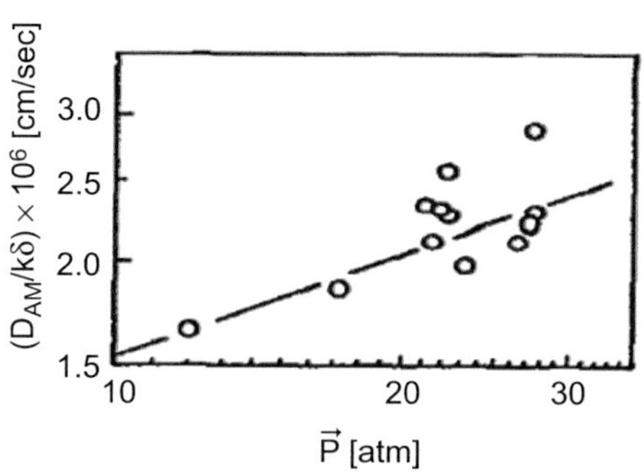

Figure 7.9 Effect of operating pressure on solute permeability constant ($D_{AM}/k\delta$). Reprinted from Ref. [28] with permission of Elsevier.

With PA- and cellulose-based hollow fiber RO membranes succeeding in commercial-scale desalination plants, research and development on other potential materials, fouling and cleaning, new applications, and mass transport phenomena continued. Chen and Petty studied the flow characteristics of RO hollow fibers and compared the coupled and uncoupled fluid models [27]. Based on the experiments on a large collection of hollow fiber membranes, they established the connection between observed hydrodynamic parameters and intrinsic membrane properties. They suggested the coupling between transport across the fiber wall and the flow through the fiber bore when describing the transport phenomenon for RO hollow fibers. Ohya et al. examined the pure water permeability (PWP) and reverse osmotic characteristics of B-9 hollow fiber modules using ultrapure water and 2775–3650 ppm NaCl solutions, respectively [28]. The PWP slightly dropped when increasing the pressure probably due to the mechanical compaction of the hollow fibers and the pressure drop inside the hollow fibers. It was about one-tenth of that obtained with CA membrane. The B-9 module showed increasing solute permeability constant at higher pressures, and the phenomenon was the reverse of that for the CA membrane (Fig. 7.9). The mean mass transfer coefficient was observed to vary with radial fluid velocity and its values were about one-half to one-tenth of the values obtained with a spiral-wound module. The deteriorated salt rejection might be caused by the imperfections on the B-9 hollow fiber surface.

Allegrezza et al. reported their work on the transfer of bench-scale flat composite membrane to a continuous process for composite hollow fiber RO membrane [11]. PSf was used to form the porous substrate, and furan resin was used to form the separation layer. The process included the spinning of PSf hollow fibers, non-solvent extraction of the nascent PSf fibers, coating with a furfuryl alcohol solution, and curing of the coated layer to complete polymerization (Fig. 7.10). The process produced dry hollow fiber membranes that could be directly subjected to the potting process without additional post-treatment. The transfer was very successful, and the composite hollow fiber RO membranes showed >99% rejection to NaCl and up to 13.6 LMH water flux at 800 psi pressure.

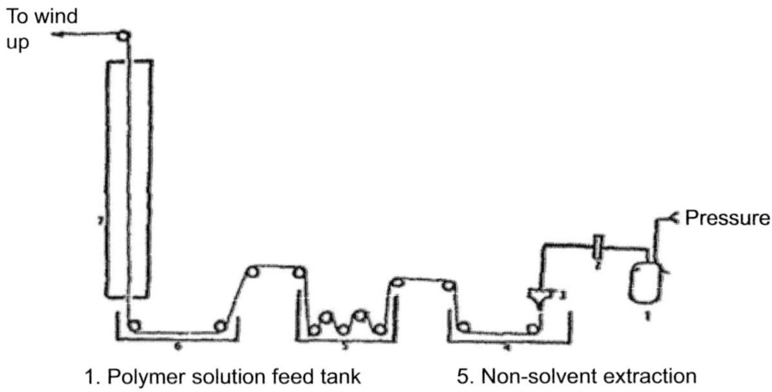

1. Polymer solution feed tank
2. Filter
3. Spinneret
4. Coagulation bath
5. Non-solvent extraction
6. Coationg formulation
7. Heated curing tower

Figure 7.10 Composite hollow fiber membrane manufacturing process. Reprinted from Ref. [11] with permission of Elsevier.

The success in the development of hollow fiber membranes for RO desalination also promoted studies in the direction of improved transport properties, better chemical resistance, higher mechanical strength, and more advanced spinning and fabrication methods [29]. Assisted by computer calculations, Churaev and Starov developed a theory to describe the separation by hollow fiber RO membranes [30]. The study disclosed the dependence of the concentration of the outflowing solution and outflowing rate on the fiber length, the fiber wall properties, and the pressure drop. The model established could be used to predicate the optimum conditions applicable to hollow fiber membranes of different parameters as well as configurations, i.e., dead-end or U-shape. In 1985, Lin reported the preparation of CTA hollow fiber RO membranes by gel-spinning process with sulfolane and polyethylene glycol (PEG) as plasticizers [31]. The change in the performance was thought to be greatly affected by the change in the molecular network structure in the hollow fiber skins. The change in the microstructure was formed mainly during the cold draw step. The melt-draw ratio would also influence the membrane performance, but its influence was not more significant than other parameters such as spinning temperature, gelation bath, environmental temperature, and humidity.

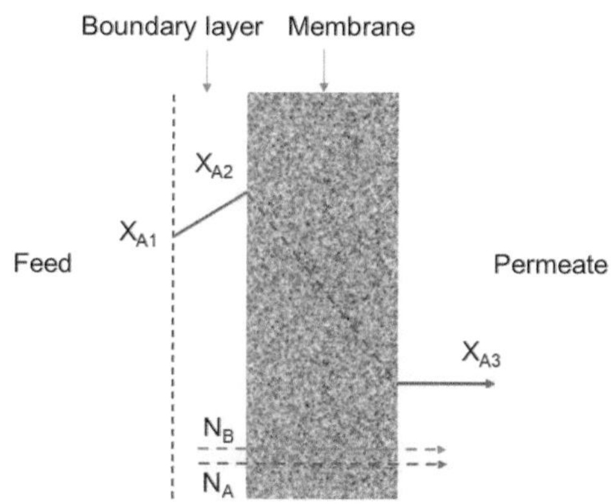

Figure 7.11 Concentration distribution in the boundary layer and the RO membrane.

One of the earliest works on RO transport, i.e., Kimura–Sorirajan analysis, was published in 1967 [32]. The hypothesis was that both the solute and water would permeate through the membrane, while a large part of the solute would diffuse back to the bulk feed through the boundary layer due to the low solute diffusivity. The following equations were developed to relate *PWP*, the permeate rate of water (N_B), the solute transport parameter ($\bar{D}_s / k\delta$), and the mass transfer coefficient (k) on the feed side of the membrane [32]:

$$A = \frac{PWP}{M_w \times S \times 3600 \times P} \tag{7.1}$$

$$N_B = A\left[P - \pi(X_{A2}) + \pi(X_{A3})\right] \tag{7.2}$$

$$N_B = \frac{\bar{D}_s}{k\delta}\left(\frac{1 - X_{A3}}{X_{A3}}\right)\left(c_2 X_{A2} - c_3 X_{A3}\right) \tag{7.3}$$

$$N_B = kc_1\left(1 - X_{A3}\right)\ln\left(\frac{X_{A2} - X_{A3}}{X_{A1} - X_{A3}}\right) \tag{7.4}$$

where A is the permeability constant of pure water, π is the osmotic pressure, S is the effective membrane surface area, P is the operating pressure, X_A is the mole fraction of solute, M_W is the molecular weight of water, δ is the membrane thickness, and c is the molar density of the solution. Abdel-Jawad and Darwish modified the Kimura–Sorirajan model for the prediction of the performance of RO hollow fibers [33]. The new models were developed based on the assumption of negligible concentration polarization due to the low permeation flux as well as the insignificance of the salt rejection and the pressure drop in the permeate side. The simplifications lowered the complexity of calculations without apparently affecting the accuracy.

Hollow fiber membranes were used to separate ternary NaCl/acetic acid aqueous solutions in an RO process [34]. The permeability of NaCl was found to be an order of magnitude lower than that of the acetic acid. However, the presence of the salt inhibited the transport of acetic acid, while the acetic acid tended to enhance the transport of NaCl. The experimental results on binary and ternary mixtures were found to support the complete mixing assumption in the shell side of the hollow fiber modules. Futselaar et al. calculated the product cost of transverse flow and longitudinal flow hollow fiber RO modules for desalination [35]. The permeate flux (J) was described using the following equation:

$$J = A\left\{\Delta P_{\text{mem}} - b^{*}\left[C_{\text{f}} - \frac{B \cdot C_{\text{f}}\exp(J/k)}{J + B \cdot \exp(J/k)}\right]\exp(J/k)\right\} \tag{7.5}$$

where ΔP_{mem} is the transmembrane pressure difference, b^* is the osmotic pressure coefficient, B is the solute permeability, and C_{f} is the feed concentration. The total product cost per unit volume permeate $(J \cdot S)$ was determined by

$$k_{\text{t}} = \frac{k_{\text{c}}}{J} + \frac{k_{\text{p}}}{J \cdot S}\left(\Delta P_{\text{pro}} Q_{\text{f}} + \Delta P_{\text{mod}} Q_{\text{c}}\right) \tag{7.6}$$

in which k_{t} is the total product cost, k_{p} is the cost of pumping energy delivered to the fluid, k_{c} is the cost of capital per unit operational time per unit membrane area, ΔP_{pro} is the process pressure, ΔP_{mod} is the pressure difference over the module, Q_{f} is the flow rate of the feed, and Q_{c} is the flow rate along the membrane surface. The RO permeate flow rate was more or less independent of the superficial velocity. Both modules showed maximal productivity at a superficial velocity with negligible circulation flow, indicating that the only way to increase productivity would be the use of more permeable membranes. The hollow fiber permeation properties and the pressurization cost of the feed dominated the product cost. The transverse flow module showed higher product cost than that of the longitudinal flow module, and the cost increase was more significant at higher velocities due to the strong pressure drop over the module. The study suggested that the transverse flow module would be only considered when the permeation resistance is determined by the fluid boundary layers instead of the membrane itself.

Sekino reported a precise analytical model for the description of mass transfer in Toyobo CTA Hollosep® hollow fiber RO modules [36]. The structure of Toyobo Hollosep® module shown in Fig. 7.13 is different from the DuPont Permasep® module (Fig. 7.6). In this module, about 1×10^5 to 1×10^6 pieces of CTA hollow fibers with an outer diameter of 160 μm are cross-wound around a porous central tube. The modules are also operated as "outside-in" mode with the permeate flowing through the bore side of the fibers. The author applied the friction-concentration-polarization (FCP) model to two different modules, i.e., one with high-flux fibers and another with high-rejection fibers. For either brackish water or seawater desalination, the test data agreed well with the calculated product rate and salt passage at different recoveries, feed concentrations, and operating pressures. Starov et al. developed more complicated models to predict the hollow fiber RO membrane performance, i.e., productivity and selectivity, for three cases, including "outside-in" with feed and permeate in concurrent flow, "outside-in" with feed and permeate in countercurrent flow, and "inside-out" mode [37, 38]. The influences of fluid pressure, fluid velocity, and fiber geometries on the performance were studied. One interesting finding was that the fiber length and diameter could be optimized to render the desired balance between productivity and selectivity. According to a more recent study, the concentration polarization on the hollow fiber surface could not be neglected [39]. For these CTA fibers, the near-surface dense layer with the smallest pore sizes considerably contributed to the water permeation property [40]. The CTA hollow fibers from Toyobo also displayed a good rejection to trace organic chemicals [41]. Hydrophobic interaction played a greater role than electrostatic interaction in the rejection of such chemicals. The separation performance of the CTA hollow fibers was comparable to PA RO membranes.

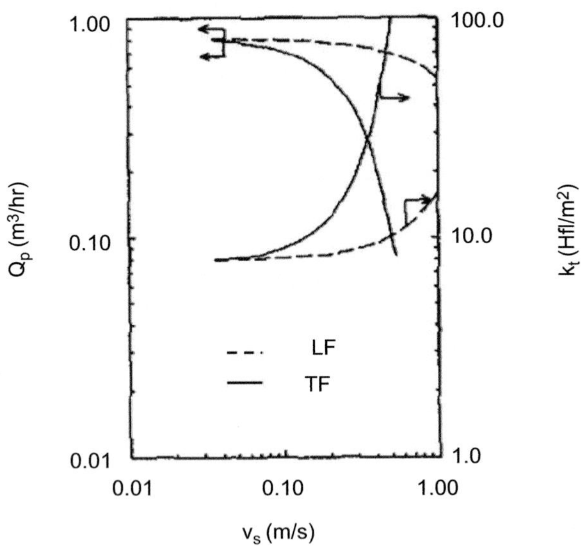

Figure 7.12 RO permeate flow rate and product cost as a function of the superficial velocity. Reprinted from Ref. [35] with permission of Elsevier.

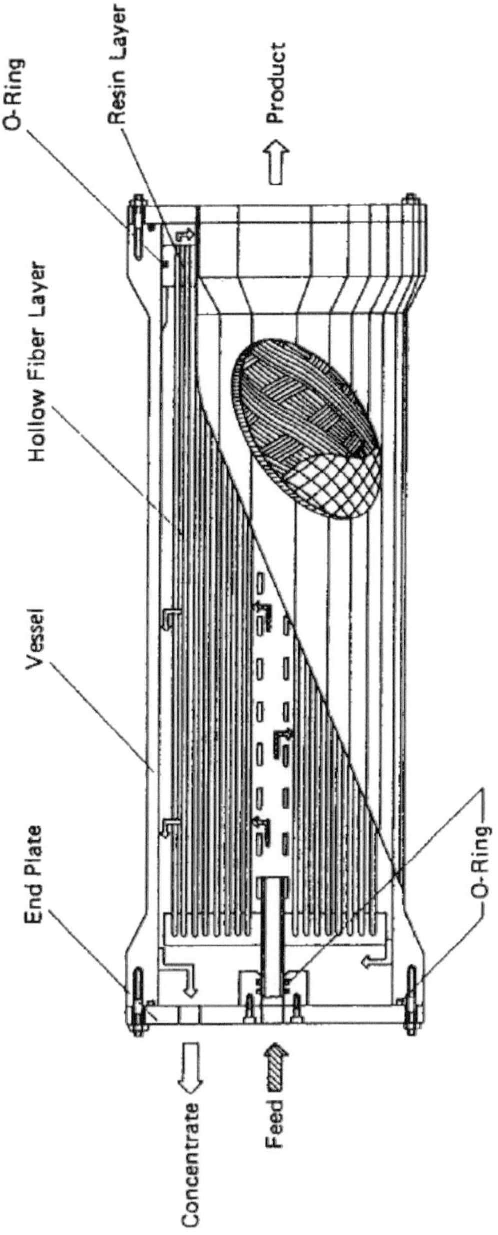

Figure 7.13 Structure of Toyobo Hollosep® hollow fiber RO membrane module. Reprinted from Ref. [36] with permission of Elsevier.

7.1.3 Recent Developments in Hollow Fiber RO Membranes

With the spiral-wound PA TFC membranes dominating the RO desalination market, the work on RO hollow fibers has slowed down in recent years. The research has focused on mass transfer, antifouling, and new applications with reduced effort on the development of better RO hollow fibers. Idris et al. fabricated CA hollow fibers in their lab and studied the effect of dope extrusion rates (DERs) on membrane performance in the RO process [42]. The CA hollow fibers have an outer diameter in the range of 500 and 598 μm, which is much bigger than commercial products. At the lowest DER of 2.5 mL/min, the resulted CA hollow fibers showed 0.77 L/m^2 per day water flux and 86.7% salt rejection. On increasing the DER to 4.0 mL/min, the salt rejection is enhanced and reached the maximum value of 94.4% with a water flux of 1.58 L/m^2 per day. Further increase in the DER led to a decrease in salt rejection. The increase in salt rejection indicates enhanced polymer molecule orientation due to a higher shear rate and stretching. In addition, the skin thickness is also reduced due to the coupling effect of shear deformation and phase inversion. However, a critical shear rate might exist, beyond which surface defects could form upon overstretching. A rheological study of CA solution confirms the influence of shear rate [43]. Analysis of the membrane pore structure through mathematic modeling reveals that low DER and thus low shear rate would create relatively larger pores and thick fiber skins [44]. The membrane pores tend to reduce, and the active skin layer becomes thinner until a critical level of shear is reached. Beyond the critical shear rate, the pore size and the active layer no longer reduce but remain in the same range.

A very interesting work on RO CTA hollow fiber membrane development was reported in 2019 [45]. In this process (Fig. 7.14), twisted poly(p-phenyl eneterephthalamide) (PPTA) fiber bundle was used as the supporting layer and the separation layer was formed by coating melt CTA on the outer surface of the PPTA bundle (Fig. 7.14). The CTA fibers with different thicknesses in the separation layer can be easily controlled by using spinnerets with different dimensions. Without the PPTA fiber bundle as the supporting layer, the CTA hollow fibers would collapse at operating pressure above

40 bar. The CTA hollow fibers with PPTA bundle support could still perform even at 60 bar, which is the maximum pressure applied in the performance tests. The incorporation of the twisted fiber bundle did not influence the separation performance at all the pressures tested.

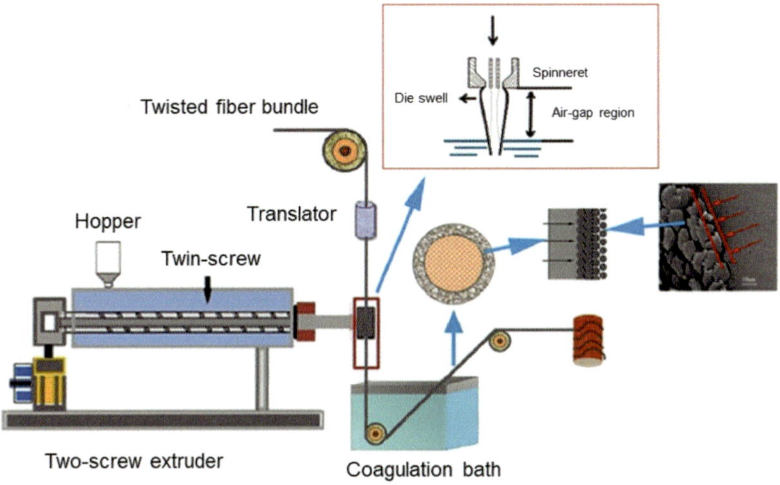

Figure 7.14 Process for the fabrication of twisted fiber bundle-reinforced CTA hollow fibers. Reprinted from Ref. [45] with permission of Elsevier.

A lot of effort has been put to the development of TFC PA hollow fibers since 2010 for nanofiltration, forward osmosis, and pressure retard osmosis processes. These fibers indeed have a separation layer almost as dense and selective as existing RO membranes. Important advancements have been made recently. Gai et al. developed thin-film nanocomposite (TFN) hollow fiber membranes comprising Na^+-functionalized carbon quantum dots (CQDs). Polyethersulfone (PES) hollow fiber was used as the substrate, and the PA separation layer was formed on the inner surface via amended interfacial polymerization, where CQDs or Na-CQDs synthesized in house were added to the m-Phenylenediamine (MPD) solution. With CQDs, the PA layer is well integrated with the PES substrate (Fig. 7.15). The acidic properties of CQDs result in thinner and smoother PA layer due to their influence on the interfacial polymerization by reacting with the amine groups of MPD as well as inhibiting the reaction between MPD and TMC. However, the addition of Na-CQDs into the

MPD solution makes the PA layer thicker and rougher. It is possible that these Na-CQDs induce additional interstitial space among the nascent PA network and promote the diffusion of MPD toward the TMC phase. Na-CDQs based TFN membranes show enhanced water permeability with acceptable salt rejection.

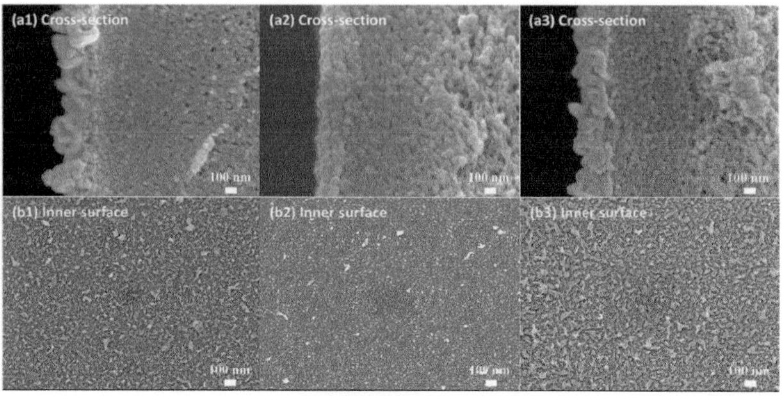

Figure 7.15 Morphology of the PA layers of (1) TFC membrane, (2) and (3) TFN membranes incorporated with CQDs and Na-CQDs, respectively. Reprinted from Ref. [46] with permission of Elsevier.

Zhang et al. developed fouling-resistant hollow fiber RO membranes fabricated through interfacial polymerization on the inner surface of a PES substrate [47]. For antifouling modifications, they synthesized copolymers and grafted the copolymers onto the PA layer (Fig. 7.16). The surface modification significantly enhanced the hydrophilicity of the PA layer, as seen from the water contact angle reducing from about 34° to below 20°. The modified hollow fibers showed excellent resistance to bovine serum albumin (BSA), $CaCl_2$, and sodium alginate foulants. In a 10-day test with real wastewater retentate, the modified hollow fibers showed constant water flux at 35 LMH without apparent fouling tendency.

By incorporating metal-organic frameworks (MOFs), a new class of TFN hollow fiber membrane was developed for low-pressure RO desalination [48]. The "outside-in" mHKUST-1@PA/PES hollow fibers were prepared through interfacial polymerization adding polydopamine-modified HKUST-1 MOF nanofillers into MPD solution. The water contact angle reduced from 51° for pristine PA to 34° for mHKUST-1@PA/PES, indicating enhanced hydrophilicity.

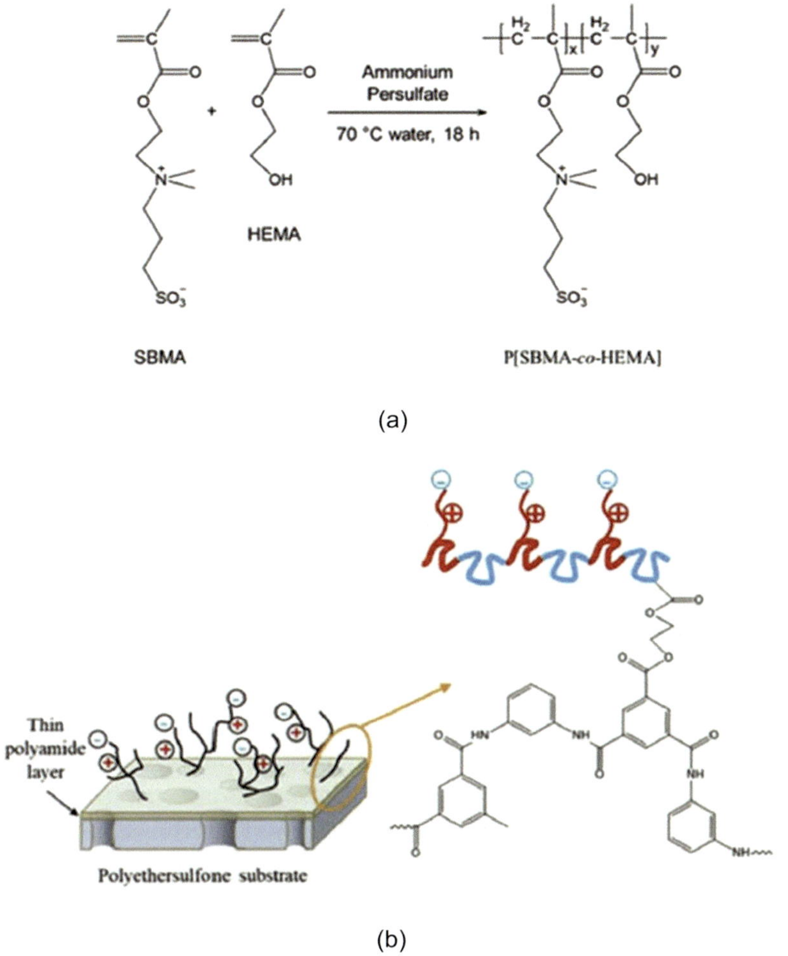

Figure 7.16 (a) The synthesis route of the copolymer p[SBMA-co-HEMA] and (b) surface grafting by p[SBMA-co-HEMA] onto the surface of polyamide layer. Reprinted from Ref. [47] with permission of Elsevier.

At the same time, the nanofillers made the fibers more negatively charged. An increase of 24% in PWP was observed for the mHKUST-1@PA/PES hollow fiber RO membrane with 97.4% salt rejection at 4 bar operating pressure. The increase in PWP was due to the low intrinsic resistance of nanoporous channels provided by the mHKUST-1 nanofillers as well as the increased hydrophilicity. The stronger negative charge should have contributed to the enhanced salt rejection. A 30-day desalination test showed that the newly developed TFN hollow fibers were very stable, exhibiting a steady performance of 6.94 LMH/bar PWP and 97.8% salt rejection.

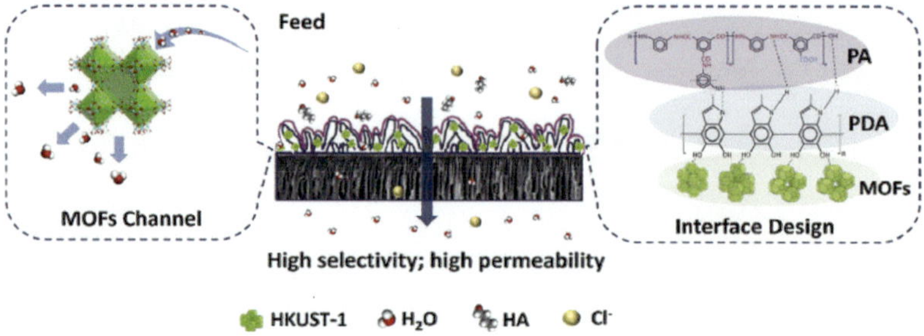

Figure 7.17 Schematic of the mHKUST-1@PA/PES membrane and salt removal in low-pressure RO application. Reprinted from Ref. [48] with permission of Elsevier.

Usta et al. proposed an interesting design of hollow fiber RO membrane modules based on their modeling work [49]. As shown in Fig. 7.19, the convoluted module geometry was obtained by twisting the straight module to achieve a 360° rotation. Based on their modeling, such a twisted module would induce swirling flow structures that increase the momentum mixing throughout the module. As a result, the concentration polarization was reduced by transporting regions of high salt concentration away from fiber surfaces. A performance increase of 70% could be achieved with the twisted module.

For seawater and brackish water desalination, RO is currently the most reliable and economic state-of-the-art technique. With the maturity of PA TFC spiral-wound RO elements, the research on hollow fibers has become less and less. On one hand, existing CTA hollow fibers do not perform better than PA TFC flat-sheet membranes. On the other hand, the technology for PA TFC hollow fiber RO membranes is not ready for commercial-scale production. Since the last decade, the interest of transferring the interfacial polymerization from flat-sheet membrane to hollow fiber configuration has been growing. However, success is mostly in a small scale and good reproducibility is not easy to attain. Moreover, the trials are mainly on the fiber inner surface and the TFC hollow fibers can only operate at "inside-out" mode, which is not desirable for desalination. It is currently well

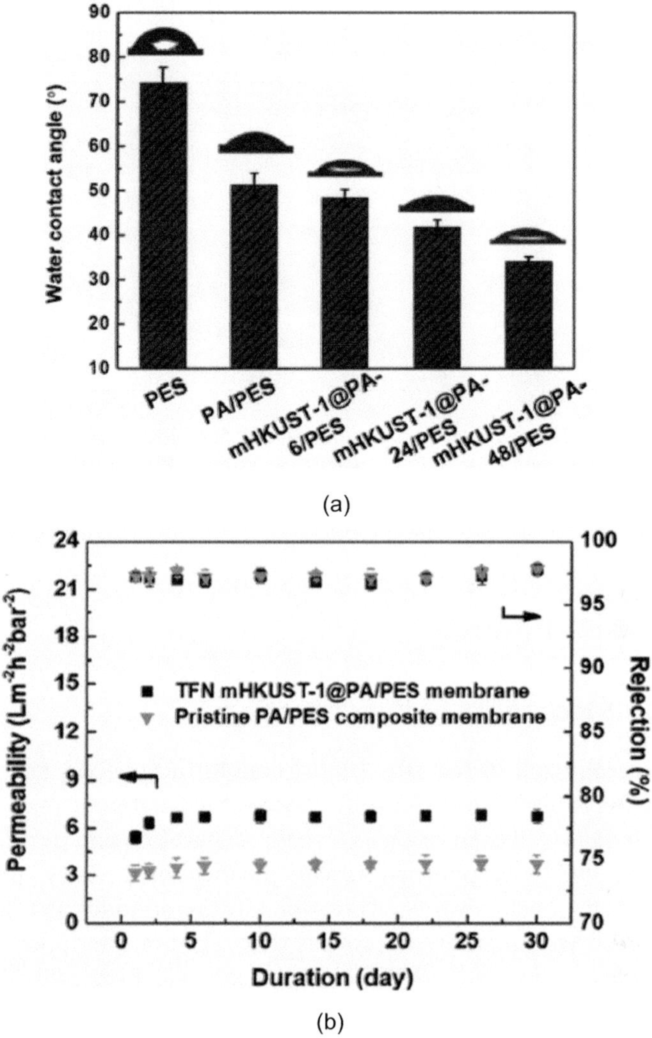

Figure 7.18 Comparison of (a) surface hydrophilicity and (b) long-term performance of water permeability and NaCl rejection. Reprinted from Ref. [48] with permission of Elsevier.

recognized that interfacial polymerization on curving hollow fiber surface is much more challenging than a flat surface. To make defect-free "outside-in" hollow fibers for the RO process, it may need a completely different technique in order to form the thin, dense, and defect-free separation layer on the fiber surface.

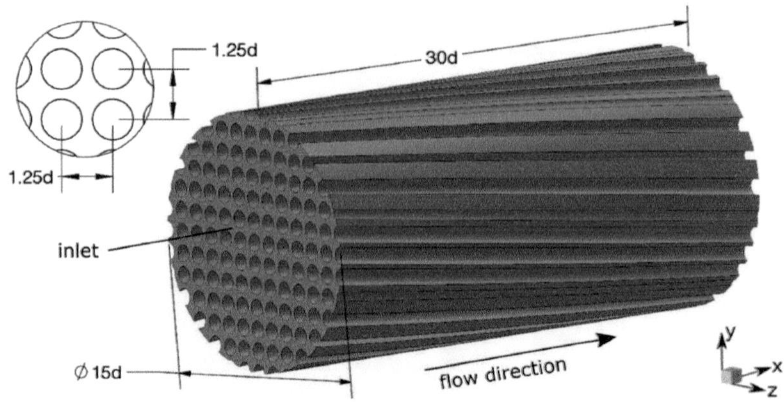

Figure 7.19 Schematic of the computational domain for proposed twisted geometry. Reprinted from Ref. [49] with permission of Elsevier.

7.2 Nanofiltration and Hollow Fiber Membranes

7.2.1 Mechanism of Nanofiltration

Since recognized in the late 1980s, nanofiltration (NF) has found a growing interest in water and wastewater treatment (e.g., water softening) as well as in aqueous/organic separation and purification applications (e.g., pharmaceutical purifications, dairy product purifications, and solvent purifications in the oil industry). NF is a pressure-driven membrane process that takes in the upper end of RO and the lower end of UF, i.e., molecular weight cutoff (MWCO) of 200–1000 Da and pore sizes of 0.5–2.0 nm. The transport of the solute and solvent through an NF membrane is induced by the chemical potential gradient between the two streams partitioned by the membrane. The separation property of NF membranes relies on both the size and the charge of the solutes. Therefore, the transport and separation through NF membranes may be explained by a combination of at least three mechanisms, i.e., solution-diffusion, molecular sieving, and electrostatic interactions between the electrically charged ions and the charged membrane surface [50, 51].

The transport of uncharged (or neutral) solutes is via the steric mechanism, i.e., size exclusion, so the retention of such solutes depends on the small pore size of the NF membrane. The rejection of charged solutes depends on the charge of the membrane originated from the dissociation of ionizable groups at the membrane surface and from the membrane pore structure. The ions, upon dissociation of the charged solutes, are repelled and excluded by the fixed ions of the same charge on the membrane surface [52]. This effect is well known as Donnan exclusion. On increasing the concentration of the charged solutes, the retention ability of the membranes would drop because of the cation's shield effect on the membrane's negatively charged groups. For a contacting solution containing monovalent and polyvalent anions, the NF membrane would preferentially reject the polyvalent anions.

7.2.2 Development of Hollow Fiber NF Membranes

In terms of NF membranes, the dominant areas of research have been manufacturing and modification to enhance water flux, improve separation efficiency, reduce cost, increase fouling resistance, and increase lifetime. In general, NF membranes are made in three geometries, i.e., flat sheet, hollow fiber, and tubular, and are packed into modules of different configurations. In the NF membrane market, flat-sheet TFC membranes and corresponding spiral-wound modules are dominant. With the broadening of NF applications, spiral-wound modules are severely suffering from fouling caused by suspended solids and they do not permit back-flushing. Tubular membranes are less susceptible but have a low packing density. Consequently, the development of high-performance hollow fiber NF membranes has attracted a lot of attention.

Methods for the fabrication of hollow fiber NF membranes fall into one of six categories, as shown in Fig. 7.20 [53]. Among them, phase inversion and interfacial polymerization are used most often. Phase inversion is normally used to fabricate the integrally skinned asymmetric NF hollow fibers using the same or different polymers. The phase-inversion technique is easy to apply, and the production cost is low, but the resulted hollow fibers are generally not competitive in terms of permeate flux and separation efficiency. Interfacial polymerization is inherited from RO TFC membrane fabrication processes and is applied on porous hollow fibers fabricated via the phase-inversion method. The process parameters of interfacial polymerization can be adjusted to obtain NF membranes with attractive flux and satisfactory rejection. However, conducting interfacial polymerization on curving hollow fiber surface is very challenging let alone continuous fabrication.

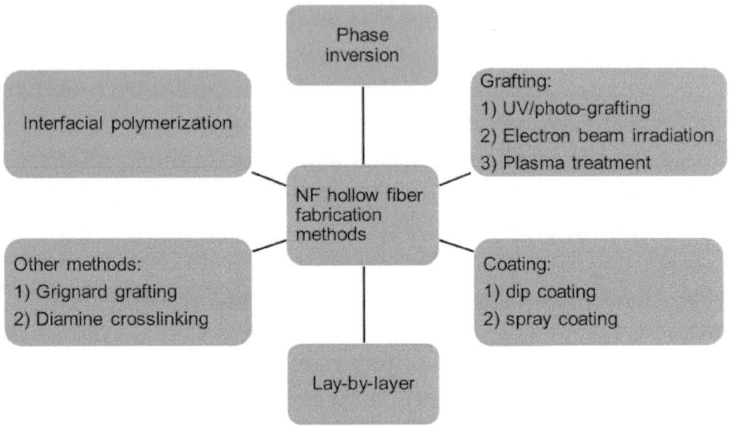

Figure 7.20 Methods of fabrication of hollow fiber nanofiltration membranes.

7.2.2.1 Phase-inversion NF hollow fibers

The formation of asymmetric hollow fibers consists of seven steps: (1) dope preparation, (2) degas, (3) metering, (4) spinning, (5) evaporation (in the air gap region), (6) coagulation, and (7) solvent exchange (Fig. 7.21). Spinning is a non-isothermal process, and there are at least three forces (stresses) applied upon the as-spun polymeric spinning solution, i.e., shear and elongation stresses within the spinneret, gravity induced by the fiber's own weight, and stresses induced by the take-up unit and coagulation bath.

Two coagulations take place at the internal and external surfaces during the hollow fiber spinning. There is simultaneous solvent outflow to coagulant and coagulant inflow, and the ratio of the two flows is defined as *k* in Fig. 7.22. The difference in solubility parameter between the spinning solution and internal or external coagulant affects the coagulation rate. The phase diagram can be taken as the qualitative understanding of phase inversion. The *k* value determines the macroscopic membrane porosity. Depending on the dope composition and the *k* values, the precipitation path may occur via nuclei growth or spinodal decomposition. One can also control the fiber outer skin morphology by changing the external coagulant chemistry or adjusting the coagulation conditions. It is commonly accepted that coagulation with solvent/water mixture tends to induce delayed demixing and form more porous fiber skin.

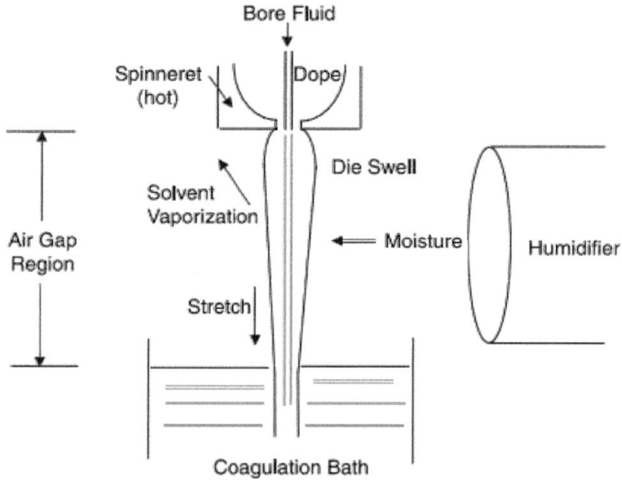

Figure 7.21 Schematic of dry-jet wet-spinning of hollow fibers. Reprinted from Ref. [54] with permission of John Wiley and Sons.

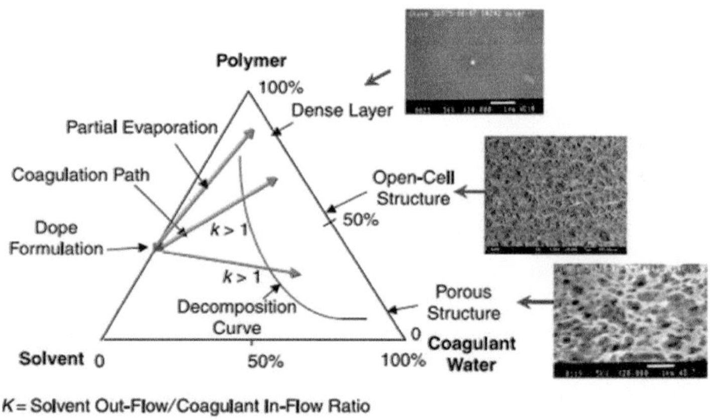

Figure 7.22 A qualitative description of phase separation kinetics and membrane morphology. Reprinted from Ref. [54] with permission of John Wiley and Sons.

Through the direct phase-inversion method, single- and dual-layer asymmetric NF hollow fibers have been developed with many different materials. In 2006, polybenzimidazole (PBI)-based single-layer NF hollow fiber was reported [55]. The PBI hollow fibers have an obviously asymmetric structure with the outer skin denser than the inner (Fig. 7.23). Close to the outer layer, there are finger-like macrovoids probably formed by water-induced intrusion during phase inversion. Delayed phase separation induced by the solvent-enriched internal coagulant leads to the sponge-like structure near the inner skin. The imidazole group within PBI molecules makes the hollow fibers have an isoelectric point near pH 7 and show different rejection performance to phosphate, arsenate, arsenite, and borate ions depending on the pH levels [56]. Yang et al. used poly(phthalazinone ether sulfone ketone) (PPESK) to fabricate inner-selective NF hollow fibers [57]. Non-solvent additives were found to affect the structure of the resulted fibers. With an MWCO of 600 Da, the resulted PPESK hollow fibers show a water flux of 211 LMH at 3 bar pressure.

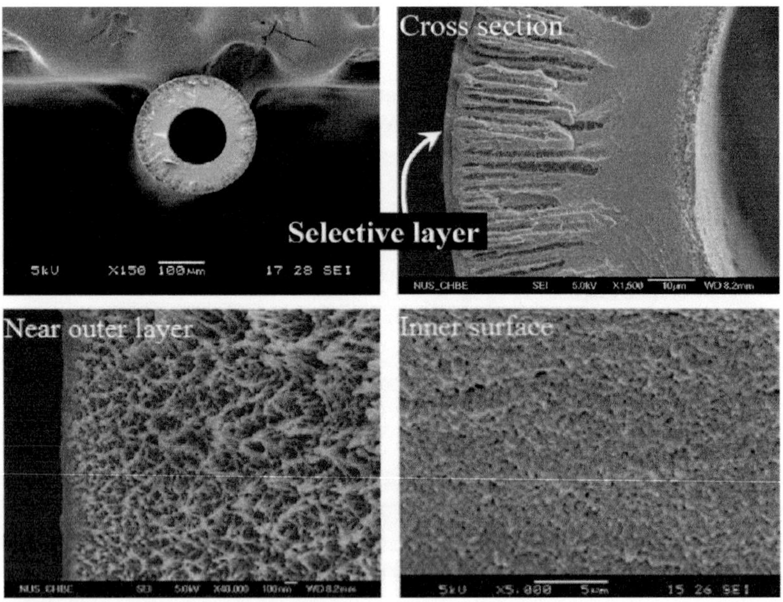

Figure 7.23 Morphology of PBI single layer NF hollow fiber. Reprinted from Ref. [55] with permission of Elsevier.

With the addition of charged surface modifying macromolecules (cSMM), PES NF hollow fibers were developed [58]. The PES hollow fibers had typical outer and inner diameters of 575 and 318.2 μm, respectively. The mean pore size and PWP of the PES hollow fibers were 1.22 nm and 11.08 LMH/bar. The PES hollow fibers with cSMM were associated with sulfonic groups and were thus negatively charged. The negative charge enabled the fibers to show twice better removal of bisphenol A (BPA), which was also negatively charged. Loose inner-selective PES hollow fiber NF membranes were fabricated through a one-step spinning process by Gao et al. [59]. The addition of sulfonated polysulfone (SPSf) to the dope solution significantly increased the dope viscosity due to hydrogen bonding between the sulfonic acid groups of SPSf and the solvent and the additives, resulting in retarded water intrusion into the nascent fiber during the spinning and suppression of macrovoid formation. Furthermore, the hydrophilic nature of SPSf favors delayed demixing, which may also hinder the formation of macrovoids. The adverse impact of including SPSf was that the mechanical strength was reduced. With the optimal quantity of about 2.5 wt.% SPSf addition to the dope, the SPSf/PES NF hollow fibers showed a 70% rejection to PEG1000 and a PWP of 12.8 LMH/bar. Loose outer-selective PES NF hollow fibers were also developed in the same research group [60]. The dope solution was also based on PES and SPSf, but LiBr was used as an additive. The increase in the percentage of SPSf in the dope solution tended to widen the fiber surface pores and enhanced the negative surface charge. Lower SPSf content favored the formation of fiber outer skin. Though having a MWCO of 1500 Da, the outer-selective SPSf/PES NF membrane achieved 94.9% rejection of Indigo carmine dye whose molecular weight is 466 g/mol.

Loose hollow fiber NF membranes were also prepared by incorporating graphene oxide (GO) into the PSf dope solution and using air as bore fluid for the spinning [61]. GO is very hydrophilic, and it induces faster water intrusion and formation of finger-like macrovoids. The GO/PSf hollow fibers presented dense outer skin with pores of 2–3 nm and a loose inner surface. With GO, these loose NF hollow fibers showed a negative charge in a broad pH range of 3 to 10. With relatively loose selective layer and negative charge, the best performing GO/PSf hollow fibers showed very attractive fluxes of up to 95.4, 98.7, and 101 LMH for Congo Red, Direct Yellow 24, and Acid Orange 10 solutions with dye rejections of 99.9%, 99.8%, and 78.5%, respectively, at 2 bar operating pressure. All the hollow fibers showed low rejection of NaCl (<5%) and moderate rejection of Na_2SO_4 (<40%). Wang et al. fabricated poly(m-phenylene isophthalamide) (PMIA) NF hollow fibers for water recovery from perfluorooctane sulfonate (PFOS)-containing wastewater [62]. The natively charged PMIA hollow fibers were outer-selective, with a mean effective pore diameter of 0.404 nm and an MWCO of 904 Da. High rejection of PFOS (91.17–97.49%) was obtained owing to a combined steric hindrance and Donnan exclusion effect. Zhao et al. incorporated carboxylic multi-walled carbon nanotubes (C-MWCNT) into PMIA dope for the spinning of hollow fiber NF membranes [63]. The addition of 0.1 wt.% C-MWCNT resulted in an almost 100% increase in the water flux. In addition, the mechanical strength and elongation were enhanced by 28.9% and 29.5%, respectively. The rejection was adversely affected if more C-MWCNT was added. Ibrahim et al. blended tannic acid-functionalized halloysite nanotubes (THNTs) with PSf dope and fabricated loose hollow fiber NF membranes [64]. With 2 wt.% THNTs, the resulted hollow fibers achieved rejections of >99% to reactive black 5 and >97.5% to reactive orange 16 with permeation fluxes of 43.5 LMH and 44 LMH, respectively.

Chu et al. developed reinforced PES loose NF hollow fibers [64]. Instead of using an internal coagulant, the authors guided the twisted polyester (PET) fiber bundle through the core of the spinneret and coated PES onto the outer surface of the PET bundle. Representative PET-reinforced PES hollow fibers (PST-1) had a mean effective pore size of 1.702 nm and an MWCO of 9353 Da. These reinforced PES fibers showed very high fractionation efficiency for dye/salt mixture solution, i.e., 99% rejection to Congo Red dye and <7% rejection to NaCl with a high permeate flux of 52.3 LMH at 6 bar operating pressure over 100 h. The most remarkable property of these hollow fibers was the mechanical strength, i.e., tensile strength up to 185.7 MPa. With the robustness and fractionation for dye and salt, the PET-reinforced PES hollow fiber NF membranes are potentially applicable for textile wastewater treatment.

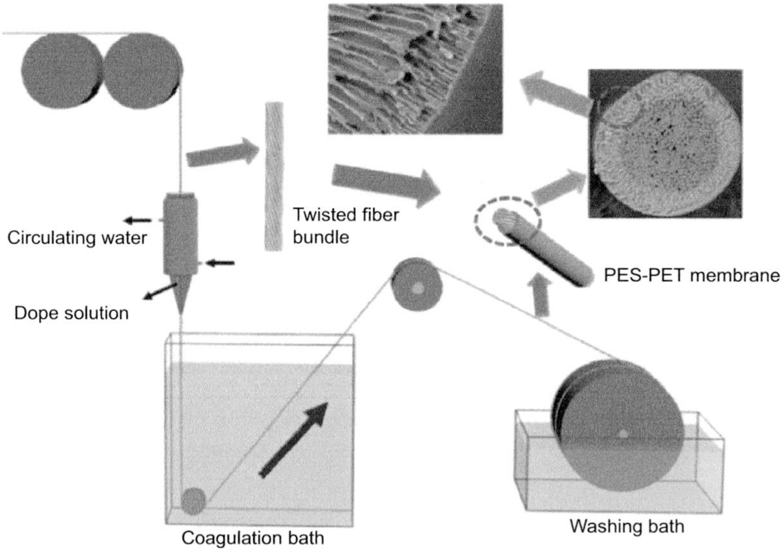

Figure 7.24 Fabrication process of the reinforced PES NF hollow fibers. Reprinted from Ref. [65] with permission of Elsevier.

Co-extrusion of two dope solutions is a technical breakthrough in spinning that makes it possible to prepare skinned composite hollow fiber membranes in one step. Figure 7.25 shows the typical design of the dual-layer spinneret [66]. For the dual-layer hollow fibers, the inner layer is usually porous and serves as a mechanical support for the selective outer layer, while the outer layer made from a high-performance polymer contains a selective dense skin. By way of co-extrusion, one can make good NF hollow fibers with eliminating the complex post-treatment process, maximizing the performance by only engineering the structure and chemistry of the selective layer, and drastically reducing the cost on the functional materials of high performance as the thin selective layer. He et al. reported their work on dual-layer hollow fibers by co-extruding sulfonated polyethersulfone (SPES) and PSf [67]. As SPES is hydrophilic and PSf is hydrophobic, the delamination problem of the two layers can occur but can be avoided at high SPES concentrations (e.g. >25 wt%). The SPES/PSf hollow fibers showed low retention for inorganic salts and high retention for charged organic dye molecules. Therefore, these hollow fibers were categorized as loose NF or tight UF hollow fibers.

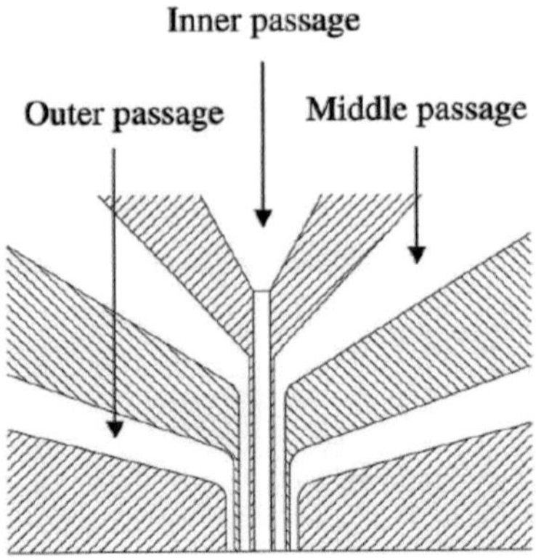

Figure 7.25 Diagram of dual-layer spinneret. Reprinted from Ref. [66] with permission of Elsevier.

Sun et al. reported dual-layer NF hollow fibers by simultaneous co-extrusion of Torlon® polyamide-imide (PAI) and CA in a phase-inversion process. To improve the integration of the two layers, pre-mixing of the outer dope (PAI) and the inner dope (CA) before exiting the spinneret was obtained by using a specially designed spinneret with an indented pin. As seen from Fig. 7.26, the CA support layer is highly porous while the selective PAI layer is thin and dense. The authors studied the influence of alcohol additives (methanol, ethanol, and isopropanol) on the PAI layer. Relatively stronger non-solvent (e.g., methanol) would facilitate the mass exchange with the external coagulant and produce a thinner but more closely packed nodule-like selective layer. Weak non-solvent favors delayed demixing, resulting in thick selective layer and macrovoid-free and delamination-free structure. PAI was also used to form the selective layer of PAI/PES dual-layer NF hollow fibers [69]. PAI/PES UF hollow fibers were first fabricated with a PWP of 170 LMH/bar and an MWCO of 21,000 Da. The PAI-PES UF hollow fibers were then immersed into polyallylamine (PAAm) solutions at 70°C for crosslinking. The concentration and molecular weight of PAAm and reaction time significantly influenced the selective layer and separation performance. With an MWCO of <6000 Da and positive charge, the developed NF hollow fibers were able to achieve a salt water permeability of 15.8 LMH/bar, and rejections of 94.2% and 92.3% to Mg^{2+} and Ca^{2+}, respectively, at 2 bar operating pressure using 3000 ppm TDS solution. PBI was used to form the selective layer of PBI/PES dual-layer hollow fiber NF membranes [70]. The PBI/PES hollow fibers presented a very narrow pore size distribution with a mean effective pore radius of 0.32 nm and an MWCO of 249 Da. The amphoteric charge property of PBI enabled an enhanced Donnan exclusion effect, which contributed to the high rejections of 98% and 93% to $Cr_2O_7^{2-}$ and Pb^{2+}, respectively.

Figure 7.26 Morphology of dual-layer PAI/CA dual-layer hollow fiber. Reprinted from Ref. [68] with permission of Elsevier.

Another method to form dual-layer NF hollow fibers is to coat a functional polymer onto the UF hollow fiber substrate. Sulfonated polymers, due to their hydrophilicity and charge, are popular coating materials in many studies. He et al. coated SPEEK on PES hollow fibers to form composite membranes [71]. Low SPEEK concentration (e.g., 0.5 wt.%) led to the penetration of the polymer solution into the support layer. The coating thickness was in the range of 0.19–1.5 mm with 0.5–5 wt.% SPEEK coating solutions. The surface pores were between 1.7 and 4.0 nm. The PWP values were in the range of 0.8–30 LMH/bar. The SPEEK/PES fibers showed 90% rejection of sulfate anions and 10% rejection of calcium cations [72]. When applied for the filtration of glyphosate at 5 bar operating pressure and pH 11, more than 90% glyphosate rejection and less than 20% NaCl rejection were obtained. Unfortunately, SPEEK/PES NF hollow fibers suffered for reduced rejection after extended contact with water, probably due to the loss of SPEEK from the top skin. To make a more stable selective layer, Song et al. coated a mixture of SPEEK and polyvinyl alcohol (PVA) onto the inner surface of PES UF hollow fibers and crosslinked PVA with glutaraldehyde (GA) [73]. The resulted SPEEK/PVA network significantly reduced the SPEEK chain reptation, and thus improved the stability of the selective layer and the NF performance. In the study of Han et al., sulfonated poly(phthalazinone ether sulfone) (SPPES) polymer was dip-coated on poly(phthalazinone ether sulfone) (PPES) hollow fibers to form composite NF membranes [74]. The SPPES/PPES NF hollow fibers were outer-selective and could be backwashed after fouling, showing good temperature resistance. About 4–6% of SPPES in the coating solution was the optimum, resulting in fibers with high Na_2SO_4 rejection and acceptable flux.

7.2.2.2 TFC and TFN NF hollow fibers

Interfacial polymerization on phase-inversion hollow fibers has been widely used to fabricate high-performance TFC NF membranes. Similar to the formation of PA selective layer of RO membranes, two highly reactive monomers, i.e., a diamine (in an aqueous solution) and acyl chloride (in an organic solvent), are brought into contact at the substrate surface for polymerization. Verissimo et al. used N,N'-diaminopiperazine (DAP) and TMC as monomers for the interfacial polymerization on the lumen side of PEI hollow fibers [75]. This is one of the earliest TFC hollow fiber NF membranes. The best membranes showed Na_2SO_4 rejection of 87% and a PWP of 6 LMH/bar. Using piperazine (PIP) and TMC monomers, Wei et al. fabricated PA thin film on the inner surface of PSf/PES UF hollow fibers [76]. The resulted PA/PSf/PES NF hollow fibers had a relatively hydrophilic surface, an MWCO of 520 Da, a rejection of 96.2% to Na_2SO_4, and a flux of 47.5 LMH at 4 bar operating pressure. Wei et al. also conducted interfacial polymerization on the inner surface of PES hollow fibers [77]. They introduced a PEI solution to the fibers' inner surface and then brought in TMC/n-Hexane solution after purging the lumen side with pure nitrogen gas. The PA layer formed on the fibers' inner surface after heat treatment in an oven. The optimized NF hollow fibers presented an MWCO of 500 Da, a water flux of 35.3 LMH, and

$MgCl_2$ rejection of 97.3% at 4 bar operating pressure. Using PEI and TMC monomers and PES substrate, Fang et al. prepared TFC hollow fibers with much higher water flux (17 LMH/bar) and an MWCO of 500 Da [78]. By using PEI/PIP mixture for the aqueous phase in interfacial polymerization, the water permeability was further enhanced to 18.2 LMH/bar with an even lower MWCO of 380 Da [79]. Mixed diamines of PIP and 2,2'-bis(1-hydroxyl-1-trifluoromethyl-2,2,2-triflutoethly)-4,4'-methylene-dianniline (BHTTM) were also used for the interfacial polymerization with TMC to prepare outer-selective TFC hollow fibers with enhanced chlorine tolerance [80].

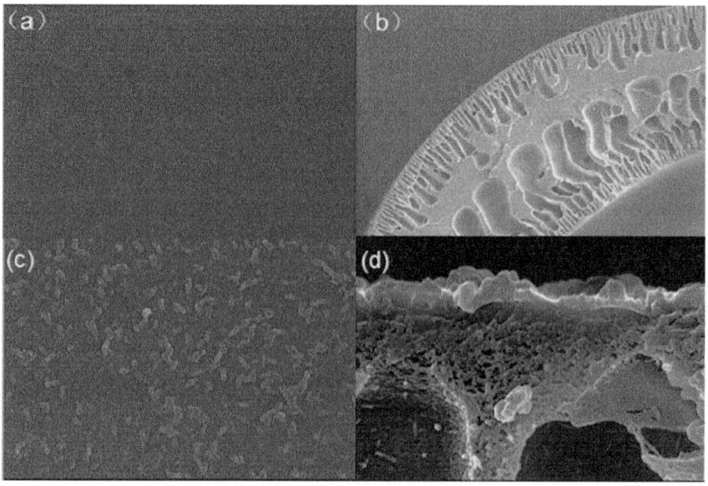

Figure 7.27 Morphology of PMIA substrate and PA/PMIA NF TFC hollow fibers. (a) Outer surface of PMIA substrate, (b) cross section of PMIA substrate, (c) top surface of PA/PMIA membrane, (d) cross section of PA/PMIA fiber. Reprinted from Ref. [82] with permission of Elsevier.

TFC hollow fiber NF membranes were also prepared by polymerizing PIP and TMC on PSf and PMIA substrates [81, 82]. Figure 7.27 shows the morphology of the PMIA hollow fiber substrate and PA/PMIA composite membrane. A typical ridge-and-valley structure of PA thin film can be clearly seen on the membrane surface. At an operating pressure of 6 bar, the PA/PMIA NF hollow fibers showed Na_2SO_4 rejection of 98.45% and water flux of 104.13 LMH. Positively charged PA/PAN TFC hollow fibers were fabricated by the interfacial polymerization of 1,4-Bis(3-aminopropyl) piperazine (DAPP) and TMC [83]. Commercial PAN hollow fibers were hydrolyzed first in 1.0 M NaOH solution to generate carboxyl groups before interfacial polymerization on the fibers' outer surface. Due to low reaction activity, the DAPP content had a great impact on the performance of the PA/PAN fibers. The rejection difference between $MgCl_2$ and LiCl was stable at 47.5%.

Rajaeian et al. developed PA TFN NF hollow fibers based on ceramic substrates [84]. Before interfacial polymerization, the home-made Al_2O_3 hollow fibers were cleaned in an ultrasonic water bath, then treated in boiling H_2O_2 to introduce OH groups on the surface, soaked with 1-methyl-2-pyrrolidone (NMP) solvent, and coated with PES. The PA thin film was formed via interfacial polymerization of MPD and TMC, while functionalized TiO_2 nanoparticles were dispersed in MPD and TMC solutions. Figure 7.28 shows the schematic of the TiO_2-incorporated TFC membrane. By incorporating functionalized TiO_2 into the PA layer, the water flux was increased up to twofold compared with the pure PA membrane with negligible rejection loss. A similar PA/TiO_2 TFN membrane was also developed through interfacial polymerization of PEI and TMC on ceramic hollow fibers [85]. These TFN hollow fibers offer a PWP of 26.4 LMH/bar and $MgCl_2$ rejection of 95.5%. As well, the membrane showed excellent antifouling properties and stability.

Based on PES/PVDF dual-layer hollow fibers, SAPO-34-incorporated TFN NF hollow fibers were fabricated via interfacial polymerization of PIP and TMC monomers [86]. The SAPO-34 nanoparticles were brought into the PA layer through the TMC solution and dioxane was used as a co-solvent additive for TMC. With pores of 0.4 nm, the SAPO-34 nanoparticles on the membrane surface provided additional water channels. The hydroxyl groups of the nanoparticles made the membrane surface more hydrophilic, further enhancing the permeation properties. A high PWP of 20.1 LMH/bar was observed for the membranes with high separation capability for polarized micropollutants. Li et al. reported TFN hollow fibers with reduced GO-NH_2 [87]. The GO-NH_2 powders were brought into the PA layer by the PIP solution. The reduced GO-NH_2 could polymerize with TMC and create a network structure within the PA layer. In addition, the hydrophilicity and electrostatic repulsion were enhanced. The best performing TFN hollow fibers showed a water flux of 38.57 LHM and a rejection of 98.21% to Na_2SO_4 at 2 bar operating pressure.

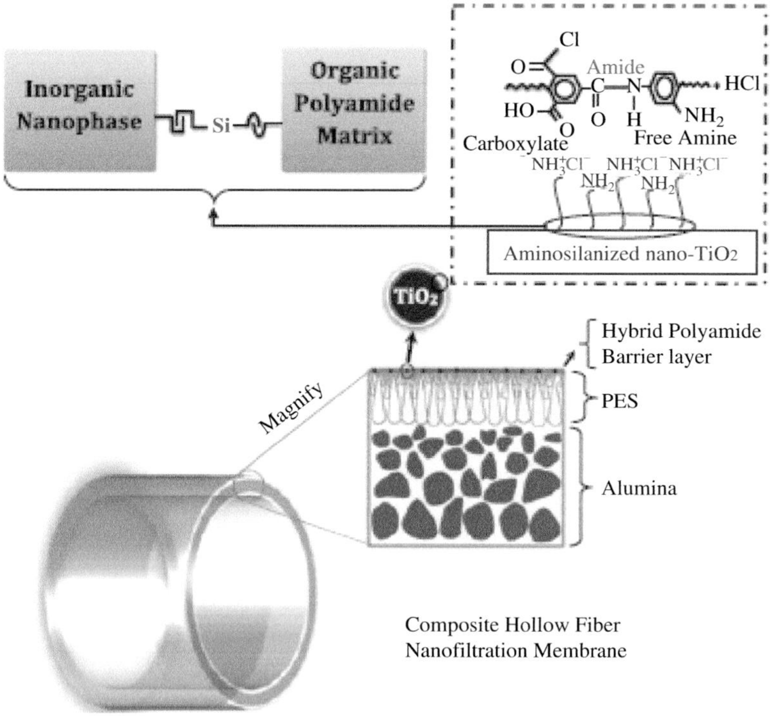

Figure 7.28 Schematic of TiO_2-incorporated PA TFC NF hollow fiber. Reprinted from Ref. [84] with permission of Elsevier.

With the same monomers of PIP and TMC, Kong et al. formed PA thin film on PVC hollow fibers [88]. Interestingly, the PVC hollow fiber substrate with medium pore size and porosity resulted in a selective PA layer with good permeability, superior rejection to $MgSO_4$, and good long-term stabilities. In another work, hydroxyl-terminated hyperbranched polyester (HPE) was used as a macromonomer and mixed with PIP for the interfacial polymerization with TMC [89]. The presence of HPE reduced the PIP/acyl chloride ratio at the interface and decreased the degree of network crosslinking of the PA skin layer, leading to a loose PA skin layer with smaller nodules on the membrane surface. When the HPE content was higher than 60%, the PA layer was too loose and the rejection of $MgSO_4$ significantly drops. The optimum HPE/PIP ratio was 3:2, corresponding to a permeate flux of 45.6 LMH and $MgSO_4$ rejection of 98.1% at 4 bar operating pressure.

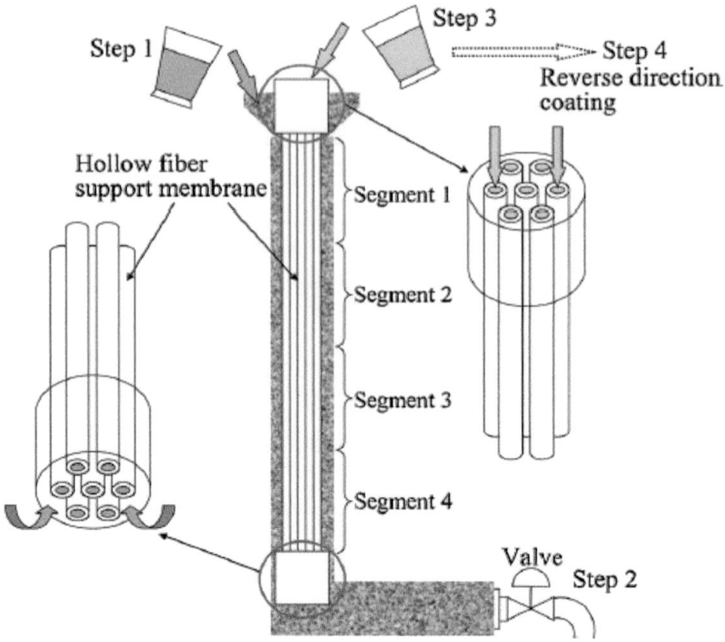

Figure 7.29 Diagram of dip-coating process for the preparation of TFC NF hollow fibers. Reprinted from Ref. [91] with permission of Elsevier.

Echaide-Gorriz et al. reported a bi-layered polyamide-metal organic framework (PA-MOF) TFC membranes on P84 hollow fibers [90]. The MOF layer was first formed on the fiber surface and its thickness was in the range of 0.83–1.33 μm depending on the synthesis time. Subsequently, the interfacial polymerization of MPD and TMC was conducted on the MOF layer. The PA-MOF TFC hollow fibers were less permeable compared with other TFC membranes.

Li et al. reported a dip-coating and reverse direction coating method for the interfacial polymerization of PIP and TMC on PSf hollow fiber inner surface [91]. The procedure is presented in Fig. 7.29. PSf hollow fibers were immersed in the PIP solution to allow PIP wet inner and outer surfaces. After the TMC solution was introduced to the lumen side, the fibers were immediately reversed upside down and the TMC solution was introduced for the second time. The reverse direction TMC coating made it possible to form a uniform PA layer. An optimum TMC coating time, i.e., 20 min dip-coating and reverse direction coating of 60 seconds, led to TFC hollow fiber membrane with 99.5% $MgSO_4$ rejection and a PWP of 6.2 LMH/bar. Chong and Wang conducted interfacial polymerization of PEI and PIP on the lumen side of microfiltration ceramic hollow fibers [92]. The excess PEI solution was removed by purging cyclohexane before pumping the TMC solution to the lumen side. Good integration was obtained between the PA layer and the ceramic substrate. With the porous ceramic substrate and loose PEI-based PA network, the resulted membranes showed excellent PWP of 16–18 LMH/bar and an MWCO of 250 Da. Even at an elevated temperature of 80°C, the PA/ceramic membranes showed high rejections to $MgCl_2$ and sucrose.

Through the polymerization of MPD and TMC on the inner surface of PES hollow fibers, Zhu et al. formed the PA layer first and then modified the PA layer by grafting polyamidoamine (PAMAM) dendrimer on its surface [93]. The PAMAM grafting formed an additional layer on the PA surface, not only shrinking the surface pores, but also making the membrane positively charged by bringing in functional groups such as tertiary amino groups (RH_3H^+ and R_3HN^+). With the PAMAM grafting, the PWP of the membrane reached 3.6 LMH/bar with >99% rejections to the heavy metal ions such as Pb^{2+}, Cu^{2+}, Ni^{2+}, Cd^{2+}, Zn^{2+}, and As^{5+}.

7.2.2.3 Organic solvent NF hollow fibers

Organic solvent nanofiltration (OSN) is a very hot topic in the area of NF research. Different from conventional NF membranes, membranes for OSN applications are expected to have high stability in organic solvents in addition to flux and separation efficiency. In many studies, newly developed functional materials are usually incorporated into the membrane matrix or selective layer through methods beyond those discussed earlier. Thus, hollow fiber membranes in the category of OSN application are separately discussed here. Darvishmanesh et al. prepared OSN hollow fibers using polyphenylsulfone (PPSU) [94]. After immersion in n-hexane for 10 days, the elastic modulus of the PPSU hollow fibers remained approximately the same and the elongation at break was increased. Toluene tended to plasticize PPSU, leading to a significant drop in the elastic modulus and the overall strength of the fibers. The PPSU OSN hollow fibers showed good stability in the filtration of mild organic solvents such as alcohols and alkanes. Xu et al. proposed a synergetic covalent-crosslinking theory and used PEI and 1,6-hexanediamine (HAD) to crosslink polyimide (PI) hollow fibers [95]. The method efficiently enhanced the affinity between the PI substrate and PEI coating layer. Furthermore, they used GA to enhance the solvent resistance and improve dye rejection. Having a mean pore diameter of 0.69 nm, the OSN hollow fibers showed an ethanol permeability of 0.447 LMH/bar and a rejection of 96.5% to Rhodamine B. These hollow fibers showed stable OSN performance after being immersed in normal polar solvents for 7 days.

Goh et al. fabricated TFC OSN hollow fibers by conducting the interfacial polymerization of PEI and TMC on the HAD-crosslinked PI substrate [96]. Upon crosslinking, the inner and outer diameters of PI hollow fibers decreased, and the fiber wall became thinner. Besides that, the microstructure near the fiber surface was denser and the membrane became slightly more brittle. The crosslinked PI hollow fibers showed enhanced stability in IPA and DMF and negligible weight loss after soaking with the solvents (Table 7.1). The TFC hollow fibers achieved very stable and attractive acetone and IPA permeability of 11.6 and 4.5 LMH/bar with more than 90% rejection of acid fuchsin.

Table 7.1 Swelling factor and weight loss for the crosslinked membrane

	Solvent	Before immersion	After immersion	Difference %
Mass (mg)	DMF	74.4 ± 2.0	74.2 ± 3.6	0.3
	Acetone	74.1 ± 3.1	74.2 ± 2.5	0
	IPA	74.3 ± 2.6	74.3 ± 2.9	0
Length (cm)	DMF	10.0 ± 0.2	10.6 ± 0.1	6
	Acetone	10.0 ± 0.1	10.1 ± 0.2	1
	IPA	10.0 ± 0.1	10.0 ± 0.2	0

Note: Reprinted from Ref. [96] with permission of Elsevier.

After crosslinking with (3-aminopropyl)trimethoxysilane, PAI hollow fibers became stable in aggressive solvents like DMF due to the formation of aminetethered inorganic networks [97]. The best membrane achieved permeability of 6.4 LMH/bar in IPA and 0.9 LMH/bar in DMF with rejections over 97% and 98%, respectively. Farahani and Chung incorporated amine-functionalized carbon nanotubes (NH_2-MWCNTs) into P84 polyimide and obtained OSN hollow fibers [98]. The addition of NH_2-MWCNTs into the P84 matrix allowed the amidation reaction between the amine groups of NH_2-MWCNTs and the imide groups of P84, resulting in narrower pore structure and enhancing mechanical and thermal stabilities. The NH_2-MWCNTs/P84 hollow fiber membranes exhibited great permeabilities of 4.31, 2.26, 1.45, and 1.17 LMH/bar for acetone, methanol, ethyl acetate, and ethanol, respectively. These OSN hollow fibers also showed excellent rejections of 99.8% to methylene blue.

Tashvigh and Chung reported OSN hollow fibers based on H_2SO_4-doped PBI [99]. The acid doping not only reduced the MWCO of the PBI fibers from 2000 Da to 500 Da but also significantly enhanced the membrane's chemical stability and flexibility. The PBI-H_2SO_4 hollow fibers could effectively separate both tetracycline/methanol and L-α-lecithin/hexane mixtures with rejections higher than 98%. Using polyamic acid as the precursor, poly(4,4′-oxydiphenylene pyromellitimide) (PMDA-ODA) polyimide OSN hollow fibers were developed [100]. Polyamic acid hollow fibers were first fabricated and subjected to a series of treatments, including solvent exchange, chemical imidization in acetic anhydride and trimethylamine mixed solvent, and heat treatment at 100°C, and washing with IPA before the OSN hollow fibers were obtained. The obtained PMDS-ODS polyimide hollow fibers exhibited stable OSN performance, such as DMF permeability of 2.5 LMH/bar and 96.7% of RB rejection. Incorporating TiO_2 nanoparticles to the membrane could accelerate the phase inversion in the spinning process, increase the membrane hydrophilicity, widen the mean pore sizes, and result in sharper pore size distribution.

Aba et al. prepared a GO dispersion with GO flakes of 5–10 μm and coated GO onto ceramic hollow fiber substrates (Al_2O_3 and YSZ) [101]. The GO/Al_2O_3 hollow fiber membrane was unstable in air due to the shrinkage-induced stress in the GO membrane. The GO hollow fibers were stable when stored in water, having an MWCO <300 Da and attractive organic solvent fluxes. Wang et al. dip-coated aluminum on γ-Al_2O_3/α-Al_2O_3 UF hollow fibers to form OSN composite membranes [102]. The aluminum layer and the membrane performance were dependent on the coating conditions, especially the soaking time. By only increasing the soaking time from 5 s to 30 s, the MWCO was reduced from 14,000 Da to 1400 Da.

Multiple coating reduced the mean pore size, but it also significantly increased the thickness of the aluminum layer. The γ-Al_2O_3/α-Al_2O_3 OSN hollow fibers with a mean effective pore size of 1.61 nm showed a surprisingly high permeability of 17.4 LMH/bar and high retention to multivalent ions. Mahalingam et al. also developed GO OSN hollow fiber membranes [103]. They fabricated PEI hollow fiber substrate first and used hexamethylene diamine (HMD) to crosslink the imide groups of PEI. Subsequently, a layer of GO was formed by spray-coating GO dispersion on crosslinked PEI fiber surface. The crosslinking made the hollow fibers stable in organic solvents, while the GO coating could reject over 90% of Rose Bengal (RB) in acetone.

7.2.2.4 NF hollow fibers prepared from other methods

Crosslinking, photografting, plasma treatment, etc. are also commonly used in the development of NF hollow fiber membranes. Zheng et al. fabricated positively charged composite membrane by coating and crosslinking PVA on to commercial polypropylene (PP) MF hollow fibers [104]. The PP hollow fibers had very large surface pores while no pores were visible with crosslinked PVA (Fig. 7.30). The PVA skin layer presented a thickness of <400 nm, and the membrane surface was hydrophilic. The composite hollow fiber membrane exhibited a PWP of 8.6 LMH/bar and an MWCO of 650 Da. In the submerged filtration tests, rejections of 92.8% to $CaCl_2$ and 35% to NaCl were obtained. Removals to cationic dyes Brilliant Green, Victoria Blue B, and Crystal Violet were 99.8%, 99.8%, and 99.2%, respectively. Through a similar procedure, this research group also tried to coat and crosslink sodium carboxymethyl cellulose (CMCNa) on PP hollow fibers [105]. The resulted composite hollow fibers were more permeable (PWP = 10.8 LMH/bar) with an MWCO of 760 Da.

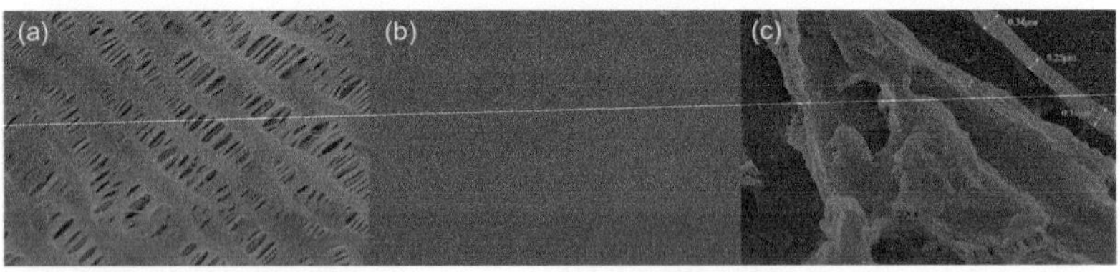

Figure 7.30 Images of (a) PP hollow fiber, (b) PVA-PP NF composite membrane, and (c) cross section of the composite membrane. Reprinted from Ref. [104] with permission of Elsevier.

Tham et al. prepared PAN UF hollow fibers and crosslinked the fibers at 70°C using hydrazine to obtain OSN membranes [106]. The cross linked PAN fibers showed an impressive rejection of >99.9% to Remazol Brilliant Blue R and ethanol permeance of 2.32 LMH/bar. Goh et al. reported NF hollow fibers with GO as the selective barrier [107]. The PAI hollow fibers were crosslinked with PEI and were positively charged. Then, negatively charged GO nanosheets were electrostatically deposited onto the outer surface of PAI/PEI hollow fibers. As seen from Fig. 7.31, the deposited GO nanosheets were able to reduce the surface pore diameter and narrow down the pore size distribution. With the PAI substrate, NF hollow fibers with boron-bridged GO as the selective layer were also developed [108]. The fabrication process included four steps: (1) fabrication of PAI hollow fiber substrate, (2) crosslinking of PAI with hyperbranched polyethylenimide (HPEI) at 70°C, (3) coating with a solution containing GO/boron compound, and (4) thermal annealing at 50–80°C. Introducing structurally rigid borate orthoester bonds with the GO laminates generated a much stable selective layer with stable water flux and rejection. Thermal annealing tended to improve the density of the GO laminate structure. The best performing fibers exhibited a PWP of 11.7 LMH/bar and high retention rates of >97% toward the dyes with molecular weights of 300–1000 Da. These authors also tried layer-by-layer (LBL) method to form the GO framework layer on HPEI-crosslinked PAI hollow fiber substrate [109]. The resulted GO/PAI-PEI composite hollow fiber membranes showed >95% rejections to Pb^{2+}, Ni^{2+}, and Zn^{2+} with relatively lower PWP of 4.7 LMH/bar.

Through LBL coating, Rajabzadeh et al. developed positively charged NF hollow fibers by alternating poly(sodium 4-styrene sulfonate) (PSS) and a poly-cation, i.e., PEI, poly (allylamine hydrochloride) (PAH) or poly(diallyldimethylammonium chloride) (PDADMAC) on UF PES hollow fibers [110]. The resulted membrane with two bilayers of PSS/PAH exhibited a water flux of 12 LMH/bar and 94% Mg^{2+} rejection. The same research group reported NF hollow fibers from inner surface LBL deposition [111]. The inner-selective LBL NF hollow fibers were superior to outer-selective alternatives as they could withstand high operation pressure and offer higher hardness rejection. Crosslinking with GA resulted in densification of the polyelectrolyte bilayers, thus enhancing the rejection to hardness [112].

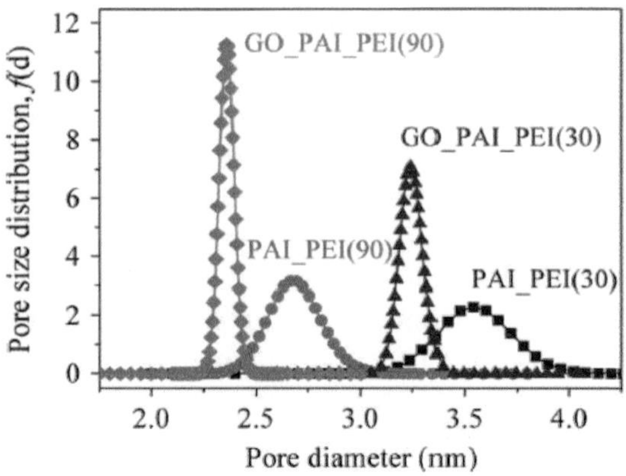

Figure 7.31 Pore size distribution of (■) PAI_PEI(30); (▲) GO_PAI_PEI(30); (●) PAI_PEI(90); and (♦) GO_PAI_PEI(90) hollow fiber membranes. Reprinted from Ref. [107] with permission of Elsevier.

After crosslinking with PEI, P84 polyimide hollow fibers were further modified by chelating polymers containing negatively charged functional groups such as poly(acrylic acid-co-maleic acid) (PAM), poly(acrylic acid) (PAA), and poly(dimethylamine-co-epichlorohydrin-co-ethylenediamine) (PDMED) [113]. The chelating polymers were not only able to change the membrane surface charge and pore size, but also provide an extra mean to remove heavy metal ions through adsorption in addition to traditional steric effect and Donnan exclusion. With the same polyelectrolytes of PSS and PDADMAC, Menne et al. developed backwashable LBL NF hollow fibers through static and dynamic depositions [114]. The retention layer of bilayers built at dynamic conditions was denser and smoother than the bilayers assembled under static conditions. The thickness of the bilayers and the retention could be precisely tuned (Fig. 7.32). The resulted LBL NF membrane was flux- and retention competitive to existing commercial membranes and it was stable toward backwashing.

Using the LBL method, Ilyas et al. formed PAH and PAA multilayer on the inner surface of commercial UF hollow fibers and obtained NF membranes [115]. The hydrophilicity of the resulted membranes could be controlled with PAA or PAH terminal layer. The flux and retention of charged micropollutants could be adjusted by tuning the coating conditions. In a most recent study, poly(2-methacryloxyeth yltrimethylammonium bromide) (PMOTAB)/PSS or PDADMAC/PSS bilayers were deposited on the lumen side of SPES hollow fibers with dynamic LBL method [116]. Filtration tests showed that the resulted hollow fibers were in loose NF or tight UF range. However, PMOTAB/PSS multilayer was very sensitive to the oxidative cleaning reagent sodium hypochlorite, and the performance deteriorated very fast due to the ester or methacryloxy groups of PMOTAB. PDADMAC/PSS membrane shows much better stability without a significant drop in the performance until approximately 50,000 ppm·h hypochlorite treatment.

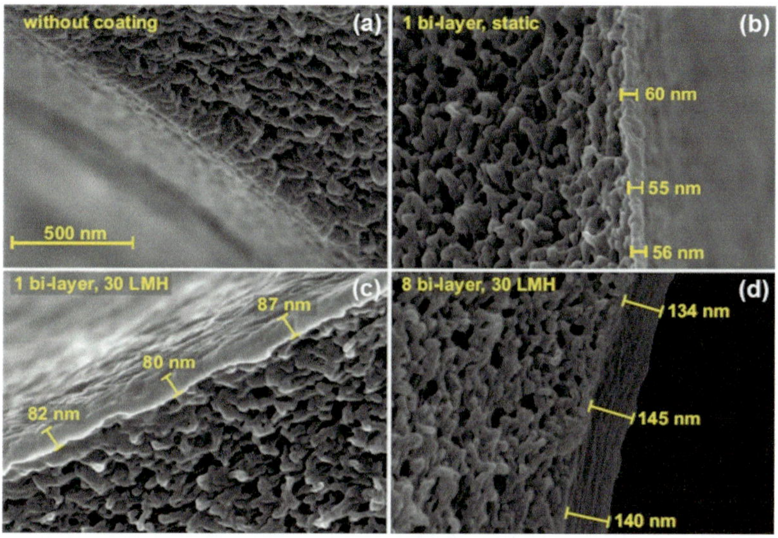

Figure 7.32 Morphology of LBL NF membranes: (a) without coating, (b) one bi-layer coated in static mode, (c) one bi-layer coated in dynamic mode, and (d) eight bi-layers coated in dynamic mode. Reprinted from Ref. [114] with permission of Elsevier.

Akbari et al. reported NF hollow fibers by polymerizing vinyl monomer on UF hollow fibers under photografting, as shown in Fig. 7.33 [117]. Under UV illumination, chemical bonds in the PSf fibers were cleaved to create radical sites. In the presence of vinyl monomer, free radical polymerization occurred at the radical sites and thus polymer chains were grafted to the fiber surface by a covalent bond. For photografting purposes, the UF hollow fibers were fabricated in another process and needed to be collected or transferred to the bobbin.

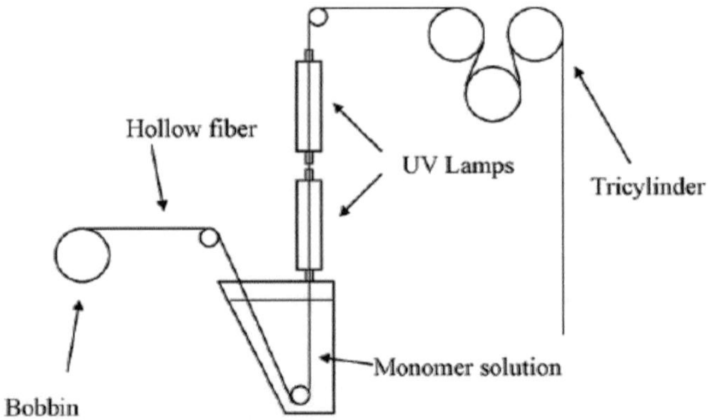

Figure 7.33 Schematic of a photoreactor setup. Reprinted from Ref. [117] with permission of Elsevier.

Wang et al. reported negatively charged NF hollow fibers by grafting and polymerizing 2-acrylamido-2-methylpropanesulfonic acid (AMPS) on PSf substrate [118]. Low-temperature plasma reactor was used in the study, and the grafting time varied from 0 to 110 s. Suitable plasma pretreatment made PSf hollow fibers more hydrophilic, and this step had a great impact on the rejection performance of the final NF membranes. At a pretreatment time of 30 s, AMPS monomer concentration of 5%, and a grafting reaction time of 80 s, the NF hollow fibers demonstrated 80.8% Na_2SO_4 rejection with a permeate flux of 6.4 LMH at 2 bar operating pressure. The same research group also applied simultaneous electron beam radiation for the grafting of AMPS [119]. Though not very tight in pore structure, the resulted membranes showed good rejection of Cr(VI) (95.1%) due to the high density of negative charges.

7.2.3 Challenges in NF Hollow Fiber Membranes Fabrication

Significant efforts have been put on the development of NF hollow fiber membranes, and delightful progress has been made. Among the membranes developed, TFC hollow fibers fabricated via interfacial polymerization are obviously superior to those obtained from the traditional phase-inversion method in terms of permeate flux and selectivity. The newly developed methods, such as UV or photografting, plasma grafting, LBL deposition, and electron beam irradiation, are able to produce membranes with attractive performance, but these methods have the same limitation of implementation on commercial scale. Incorporating new functional nanomaterials has been proved an effective way to tune the membrane structure, morphology, and effect on solute transport. However, since these nanomaterials could incur severe safety and environmental issues, in the case they are leached out from the membrane matrix, further investigations are needed [52].

Despite success in lab-scale fabrication of TFC NF hollow fibers, there are many challenges to scale up the technique. Applying interfacial polymerization on the surface of a bundle of fibers is extremely challenging due to the difficulty in handling solution flow. There is no way to completely remove the excess liquid carrying the amine-type monomer; hence, it is almost impossible to form a defect-free PA film on the hollow fiber substrate with existing methods. Furthermore, the formed PA film would not adhere properly to the hollow fiber substrate. Producing outer-selective TFC hollow fibers is even more difficult than producing inner-selective hollow fibers because the fibers usually stick close to each other. By and large, TFC hollow fiber membranes are still in the early stage and need more in-depth research and development, especially to find the feasible and scalable methods to do interfacial polymerization on fiber bundles, improve the integration between the PA film and the substrate, and enhance the chlorine tolerance [53].

Acknowledgments

Prof. Chung likes to thank the Yushan Scholar Program supported by the Ministry of Education, Taiwan.

References

1. Qaslm M., Badrelzaman M., Darwish N. N., Darwish N. A., and Hilal N. (2019). Reverse osmosis desalination: A state-of-the-art review, *Desalination*, **459**, pp. 59–104.

2. Wenten I. G. and Khoiruddin. (2016). Reverse osmosis applications: Prospect and Challenges, *Desalination*, **391**, pp. 112–125.

3. Eisenberg T. N. and Middlebrooks E. J. (1986) *Reverse Osmosis Treatment of Drinking Water* (Butterworth Publishers).
4. Jamaly S., Darwish N. N., Ahmed I., and Hasan S. W. (2014). A short review on reverse osmosis pretreatment technologies, *Desalination*, **354**, pp. 30–38.
5. Anis S. F., Hashaikeh R., and Hilal N. (2019). Reverse osmosis pretreatment technologies and future trends: A comprehensive review, *Desalination*, **452**, pp. 159–195.
6. Kucera J. (2015) *Reverse Osmosis-Industrial Processes and Applications*, 2nd Edition (Scrivener Publishing LLC).
7. Sanders D. F., Smith Z. P., Guo R., Robeson L. M., McGrath J. E., Paul D. R., and Freeman B. D. (2013). Energy-efficient polymeric gas separation membranes for a sustainable future: A review, *Polymer*, **54**, pp. 4729–4761.
8. Bao C. (2010) Design of advanced reverse osmosis and nanofiltration membranes for water purification, PhD Thesis.
9. Kim S. H., Kwak S-Y., and Suzuki T. (2005). Positron annihilation spectroscopic evidence to demonstrate the flux-enhancement mechanism in morpholoty-controlled thin-film-composite (TFC) membrane, *Environ. Sci. Technol.*, 39, pp. 1764–1770.
10. Johnson J. and Busch M. (2010). Engineering aspects of reverse osmosis module design, *Desalin. Water Treat.*, **15**, pp. 236–248.
11. Allegrezza A. E., Burchesky R. D., Gotz G., Davis R. B., and Coplan M. J. (1977). Hollow fiber composite reverse osmosis membrane, *Desalination*, **20**, pp. 87–94.
12. Hoehn H. H. and Richter J. W. (1980). Aromatic polyimide, polyester and polyamide separation membranes, US Patent, USRE30351E.
13. Longsdale H. K. (1982). The growth of membrane technology, *J. Membr. Sci.*, **10**, pp. 81–181.
14. Abron L. A. and Osburn J. O. (1973). A transport mechanism in hollow fiber reverse osmosis membranes for the removal of DDT and Aldrin from water, *Water Res.*, **7**, pp. 461–477.
15. Jason J. and Busch M. (2012). Engineering aspects of reverse osmosis module design, *Desalination Water Treat.*, **15**, pp. 236–248.
16. Hanbury W. T. and Yuceer A. (1985). Constriction of B10 hollow fiber bore diameters under operating conditions, *Desalination*, **54**, pp. 27–44.
17. Hailemariam R. H., Woo Y. C., Damtie M. M., Kim B. C., Park K.-D., and Choi J.-S. (2020). Reverse osmosis membrane fabrication and

modification technologies and future trends: A review, *Adv. Colloid Interface Sci.*, 276, pp. 1002100.1-22.

18. Al-Qahtany H. I. and Al-Bastaki N. M. S. (1995). Effect of aging on the performance of RO hollow fiber membranes in a section of an RO plant, *Desalination*, **101**, pp. 177–183.
19. Butt F. H., Rahman F., and Baduruthamal U. (1997). Hollow fine fibers vs. spiral-wound reverse osmosis desalination membranes, Part 2: Membrane autopsy, *Desalination*, **109**, pp. 83–94.
20. Bou-Hamad S., Abdel-Jawad M., Al-Tabtabaei M., and Al-Shammari S. (1998). Comparative performance analysis of two seawater reverse osmosis plants: Twin hollow fine fiber and spiral wound membranes, *Desalination*, **120**, pp. 95–106.
21. Khan W. Z. and Hamad E. Z. (1996). Performance of an aromatic polyamide hollow fiber membrane subjected to high salinity water, *Desalination*, **105**, pp. 191–197.
22. Khan W. Z. (2009). Desalination of raw water using a polyamide hollow fiber membrane, *Desalination*, **244**, pp. 59–65.
23. Marcovecchio M. G., Scenna N. J., and Aguirre P. A. (2010). Improvements of a hollow fiber reverse osmosis desalination model: analysis of numerical results, *Chem. Eng Res. Des.*, **88**, pp. 789–802.
24. Reibersdorfer G., Shields C. P., and Stranz J. W. (1998). The hollow fiber Cartridge^{TM}: Operational experience with an improved reverse osmosis device/technology, *Desalination*, **117**, pp. 311–321.
25. Gorenflo A., Redondo J. A., and Reverberi F. (2005). Basic options and two case studies for retrofitting hollow fiber elements by spiral-wound RO technology, *Desalination*, **178**, pp. 247–260.
26. Polasek V., Talo S., and Sharif T. (2003). Conversion of hollow fiber to spiral technology in large seawater RO systems-process design and economics, *Desalination*, **156**, pp. 239–247.
27. Chen C. and Petty C. A. (1973). Flow characteristics of semipermeable hollow fiber undergoing reverse osmosis, *Desalination*, **12**, pp. 281–293.
28. Ohya H., Nakajima H., Takagi K., Kagawa S., and Negishi Y. (1977). An analysis of reverse osmotic characteristics of B-9 hollow fiber module, *Desalination*, **21**, pp. 257–274.
29. Soltanieh M. and Gill W. (1984). An experimental study of the complete-mixing model for radial flow hollow fiber reverse osmosis systems, *Desalination*, **49**, pp. 57–88.

30. Churaev N. V. and Starov V. M. (1982). The theory of reverse osmosis separation of solutions on hollow fiber membranes, *J. Colloid Interface Sci.*, **89**, pp. 77–85.
31. Lin S. Q. (1985). The structure of reverse osmosis hollow fiber membrane of cellulose triacetate (CTA), *Desalination*, **56**, pp. 395–403.
32. Kimura S. and Sorirajan S. (1967). Analysis of data in reverse osmosis with porous cellulose acetate membranes used, *AIChE J.*, 13, pp. 497–503.
33. Adbel-Jawad M. and Darwish M. A. (1989). Simplified analysis of transport in reverse osmosis (RO) hollow fibers (HF) membranes, *Desalination*, **75**, pp. 97–116.
34. Sapienza F. J., Gill W. N., and Soltanieh M. (1990). Separation of ternary salt/acid aqueous solutions using hollow fiber reverse osmosis, *J. Membr. Sci.*, **54**, pp. 175–189.
35. Futselarr H., Zoontjes R. J. C., Reight T., and Racz I. G. (1993). Economic comparison of transverse and longitudinal flow hollow fiber membrane modules for reverse osmosis and ultrafiltration, *Desalination*, **90**, pp. 345–361.
36. Sekino M. (1993). Precise analytical model of hollow fiber reverse osmosis modules, *J. Membr. Sci.*, **85**, pp. 241–252.
37. Starov V. M., Smart J., and Lloyd D. R. (1995). Performance optimization of hollow fiber reverse osmosis membranes, Part I. Development of theory, *J. Membr. Sci.*, **103**, pp. 257–270.
38. Smart J., Starov V. M., and Lloyd D. R. (1996). Performance optimization of hollow fiber reverse osmosis membranes, Part II. Comparative study of flow configurations, *J. Membr. Sci.*, **119**, pp. 117–128.
39. Kumano A., Sekino M., Matsui Y., Fujiwara N., and Matsuyama H. (2008). Study of mass transfer characteristics for a hollow fiber reverse osmosis module, *J. Membr. Sci.*, **324**, pp. 136–141.
40. Hagihara H., Ito Kenji, Oshima N., Yabuuchi A., Suda H., and Yanagishita H. (2014). Depth profiling of the free-volume holes in cellulose triacetate hollow-fiber membranes for reverse osmosis by means of variable-energy positron annihilation lifetime spectroscopy, *Desalination*, **344**, pp. 86–89.
41. Fujioka T., Khan S. J., McDonald J. A., and Nghiem L. D. (2015). Rejection of trace organic chemicals by a hollow fiber cellulose triacetate reverse osmosis membrane, *Desalination*, **368**, pp. 69–75.
42. Idris A., Noordin M. Y., Ismail A. F., and Shilton S. J. (2002). Study of shear rate influence on the performance of cellulose acetate reverse osmosis hollow fiber membranes, *J. Membr. Sci.*, **202**, pp. 205–215.

43. Idris A., Ismail A. F., Gordeyev S. A., and Shilton S. J. (2002). Rheology assessment of cellulose acetate spinning solution and its influence on reverse osmosis hollow fiber membrane performance, *Polym Testing*, **22**, pp. 319–325.

44. Idris A., Ismail A. F., Shilton S. J., Roslina R., and Musa M. (2002). The deduction of fine structural studies of reverse osmosis hollow fiber membranes using surface force-pore flow model, *Sep. Purif. Technol.*, **29**, pp. 217–227.

45. Chen K., Xiao C., Liu H., Ling H., Chu Z., and Hu Z. (2019). Design of robust twisted fiber bundle-reinforced cellulose triacetate hollow fiber reverse osmosis membrane with thin separation layer for seawater desalination, *J. Membr. Sci.*, **578**, pp. 1–9.

46. Gai W., Zhao D. L., and Chung T.-S. (2019). Thin film nanocomposite hollow fiber membranes comprising Na+-functionalized carbon quantum dots for brackish water desalination, *Water Res.*, **154**, pp. 54–61.

47. Zhang Y., Yang L., Promoda K. P., Gai W., and Zhang S. (2019). Highly permeable and fouling-resistant hollow fiber membranes for reverse osmosis, *Chem. Eng. Sci.*, **207**, pp. 903–910.

48. Lin Y., Chen Y., and Wang R. (2019). Thin film nanocomposite hollow fiber membranes incorporated with surface functionalized HKUST-1 for highly efficient reverse osmosis desalination process, *J. Membr. Sci.*, **589**, pp. 1–13.

49. Usta M., Morabito M., Anqi A., Alrehili M., Hakim A., and Oztekin A. (2018). Twisted hollow fiber membrane modules for reverse osmosis-driven desalination, *Desalination*, **441**, pp. 21–34.

50. Wang K. Y. (2005) Fabrication and characterization of ultrafiltration and nanofiltration membranes, PhD Thesis.

51. Sun S. (2011). Fabrication of nanofiltration hollow fiber membranes for sustainable pharmaceutical manufacture, PhD Thesis.

52. Mohammad A. W., Teow Y. H., Ang W. L., Chung Y. T., Otaley-Radcliffe D. L., and Hilal N. (2015). Nanofiltration membranes review: recent advances and future prospects, *Desalination*, **356**, pp. 226–254.

53. Urper G. M., Sengur-Tasdemir R., Turken T., Genceli E. A., Tarabara V. V., and Koyuncu I. (2017). Hollow fiber nanofiltration membranes: A comparative review of interfacial polymerization and phase inversion fabrication methods, *Sep. Sci. Technol.*, **52**, pp. 2120–2136.

54. Chung T.-S. (2008), *Advanced Membrane Technology and Applications*, eds, Norman N. Li, Fane A. G., Ho W. S. W., and Matsuura T., Chapter

1 "Fabrication of hollow-fiber membranes by phase inversion," (John Wiley & Sons, Inc) pp. 821–835.

55. Wang K. Y. and Chung T.-S. (2006). Fabrication of polybenzimidazole (PBI) nanofiltration hollow fiber membranes for removal of chromate, *J. Membr. Sci.*, **281**, pp. 307–315.
56. Lv J., Wang K. Y., and Chung T.-S. (2008). Investigation of amphoteric polybenzimidazole (PBI) nanofiltration hollow fiber membrane for both cation and anions removal, *J. Membr. Sc.*, **310**, pp. 557–566.
57. Yang Y., Jian X., Yang D., Zhang S., and Zou L. (2008). Poly(phthalazinone ether sulfone ketone) (PPESK) hollow fiber asymmetric nanofiltration membranes: Preparation, morphologies and properties, *J. Membr. Sci.*, **270**, pp. 1–12.
58. Bolong N., Ismail A. F., Salim M. R., Rana D., Matsuura T., and Tabe-Mohammadi A. (2010). Negatively charged polyethersulfone hollow fiber nanofiltration membrane for the removal of bisphenol A from wastewater, *Sep. Purif. Technol.*, **73**, pp. 92–99.
59. Gao J., Thong Z., Wang K. Y., and Chung T.-S. (2017). Fabrication of loose inner-selective polyethersulfone (PES) hollow fibers by one-step spinning process for nanofiltration (NF) of textile dyes, *J. Membr. Sci.*, **541**, pp. 413–424.
60. Thong Z., Gao J., Lim J. X. Z., Wang K. Y., and Chung T.-S. (2018). Fabrication of loose outer-selective nanofiltration (NF) polyethersulfone (PES) hollow fibers via single-step spinning process for dye removal, *Sep. Purif. Technol.*, **192**, pp. 483–490.
61. Ji D., Xiao C., An S., Zhao J., Hao J., and Chen K. (2019). Preparation of high-flux PSf/GO loose nanofiltration hollow fiber membranes with dense-loose structure for treating textile wastewater, *Chem. Eng. J.*, **363**, pp. 33–42.
62. Wang T., Zhao C., Li P., Li Y., and Wang J. (2015). Fabrication of novel poly(m-phenylene isophthalamide) hollow fiber nanofiltration membrane for effective removal of trace amount perfluorooctane sulfonate from water, *J. Membr. Sci.*, **477**, pp. 74–85.
63. Zhao C., Yang B., Han J., Meng Y., Yu L., Hou D., Wang J., Zhao Y., Zhai Y., Wang S., and Sun X. (2018). Preparation of carboxylic multiwalled-carbon-nanotube-modified poly(m-phenylene isophthalamide) hollow fiber nanofiltration membranes with improved performance and application for dye removal, *Appl. Surf. Sci.*, **453**, pp. 502–512.
64. Ibrahim G. P. S., Isloor A. M., Moslehyani A., and Ismail A. F. (2017). Bio-inspired, fouling resistant, tannic acid functionalized halloysite nanotube reinforced polysulfone loose nanofiltration hollow fiber

membranes for efficient dye and salt separation, *J. Water Process Eng.*, **20**, pp. 138–148.

65. Chu Z., Chen K., Xiao C., Ji D., Ling H., Li M., and Liu H. (2020). Improving pressure durability and fractionation property via reinforced PES loose nanofiltration hollow fiber membranes for textile wastewater treatment, *J. Taiwan Inst. Chem. Eng.*, **108**, pp. 71–81.
66. Li D. F., Chung T.-S., Wang R., and Liu Y. (2002). Fabrication of fluoropolyimide/polyethersulfone (PES) dual-layer asymmetric hollow fiber membranes for gas separation, *J. Membr. Sci.*, **198**, pp. 211–223.
67. He T., Mulder M. H. V., Strathmann H., and Wessling. (2002). Preparation of composite hollow fiber membranes: co-extrusion of hydrophilic coatings onto porous hydrophobic support structures, *J. Membr. Sci.*, **207**, pp. 143–156.
68. Sun S. P., Wang K. Y., Peng N., Hatton T. A., and Chung T.-S. (2010). Novel polyamide-imide/cellulose acetate dual-layer hollow fiber membranes for nanofiltration, *J. Membr. Sci.*, **363**, pp. 232–242.
69. Setiawan L., Shi L., and Wang R. (2014). Dual layer composite nanofiltration hollow fiber membranes for low-pressure water softening, *Polymer*, **55**, pp. 1367–1374.
70. Zhu W. P., Sun S. P., Gao J., Fu F.-J., and Chung T.-S. (2014). Dual-layer polybenzimidazole/polyethersulfone (PBI/PES) nanofiltration (NF) hollow fiber membranes for heavy metals removal from wastewater, *J. Membr. Sci.*, **456**, pp. 117–127.
71. He T., Frank M., Mulder M. H. V., and Wessling M. (2008). Preparation and characterization of nanofiltration membranes by coating polyethersulfone hollow fibers with sulfonated poly(ether ether ketone) (SPEEK), *J. Membr. Sci.*, **307**, pp. 62–72.
72. Song J., Li X.-M., Figoli A., Huang H., Pan C., He T., and Jiang B. (2013). Composite hollow fiber nanofiltration membranes for recovery of glyphosate from saline wastewater, *Water Res.*, **47**, pp. 2065–2074.
73. Song J., Li X.-M., Li Z., Zhang M., Yin Y., Zhao B., Kong D., Chen G., and He T. (2015). Stabilization of composite hollow fiber nanofiltration membranes with a sulfonated poly(ether ether ketone) coating, *Desalination*, **355**, pp. 83–90.
74. Han J., Yang D., Zhang S., Wang J., and Jian X. (2014). Preparation and performance of SPPES/PPES hollow fiber composite nanofiltration membrane with high temperature resistance, *Desalination*, **350**, pp. 95–101.

75. Verissimo S., Peinemann K.-V., and Bordado J. (2005). New composite hollow fiber membrane for nanofiltration, *Desalination*, **184**, pp. 1–11.

76. Wei X., Kong X., Sun C., and Chen J. (2013). Characterization and application of a thin-film composite nanofiltration hollow fiber membrane for dye desalination and concentration, *Chem. Eng. J.*, **223**, pp. 171–182.

77. Wei X., Wang S., Shi Y., Xiang H., Chen J., and Zhu B. (2014). Characterization of a positively charged composite nanofiltration hollow fiber membrane prepared by a simplified process, *Desalination*, **350**, pp. 44–52.

78. Fang W., Shi L., and Wang R. (2013). Interfacially polymerized composite nanofiltration hollow fiber membranes for low-pressure water softening, *J. Membr. Sci.*, **430**, pp. 129–139.

79. Fang W., Shi L., and Wang R. (2014). Mixed polyamide-based composite nanofiltration hollow fiber membranes with improved low-pressure water softening capability, *J. Membr. Sci.*, **468**, pp. 52–61.

80. Zhou B.-W., Zhang H.-Z., Xu Z.-L., and Tang Y.-J. (2016). Interfacial polymerization on PES hollow fibers using mixed diamines for nanofiltration removal of salts containing oxyanions and ferric ions, *Desalination*, **394**, pp. 176–184.

81. Li H., Shi W., Wang W., and Zhu H. (2015). The extraction of sericin protein from silk reeling wastewater by hollow fiber nanofiltration membrane integrated process, *Sep. Purif. Technol.*, **146**, pp. 342–350.

82. Wang T., He X., Li Y., and Li J. (2018). Novel poly(piperazine-amide) (PA) nanofiltration membrane based poly(m-phenylene isophthalamide) (PMIA) hollow fiber substrate for treatment of dye solutions, *Chem. Eng. J.*, **351**, pp. 1013–1026.

83. Li X., Zhang C., Zhang S., Li J., He B., and Cui Z. (2015). Preparation and characterization of positively charged polyamide composite nanofiltration hollow fiber membrane for lithium and magnesium separation, *Desalination*, **369**, pp. 26–36.

84. Rajaeian B., Rahimpour A., Tade M. O., and Liu S. (2013). Fabrication and characterization of polyamide thin film composite (TFN) nanofiltration membrane impregnated with TiO_2 nanoparticles, *Desalination*, **313**, pp. 176–188.

85. Li Y.-X., Cao Y., Wang M., Xu Z.-L., Zhang H.-Z., Liu X.-W., and Li Z. (2018). Novel high-flux polyamide/TiO_2 composite nanofiltration membranes on ceramic hollow fiber membranes, *J. Membr. Sci.*, **565**, pp. 322–330.

86. Liu T.-Y., Liu Z.-H., Zhang R.-X., Wang Y., der Bruggen B. V., and Wang X.-L. (2015). Fabrication of a thin film composite hollow fiber nanofiltration membrane for wastewater treatment, *J. Membr. Sci.*, **488**, pp. 92–102.

87. Li X., Zhao C., Yang M., Yang B., Hou D., and Wang T. (2017). Reduced graphene oxide-NH_2 modified low pressure nanofiltration composite hollow fiber membranes with improved water flux and antifouling capabilities, *Appl. Surf. Sci.*, **419**, pp. 418–428.

88. Kong X., Zhou M.-Y., Lin C.-E., Wang J., Zhao B., Wei X.-Z., and Zhu B.-K. (2016). Polyamide/PVC based composite hollow fiber nanofiltration membranes: Effect of substrate on properties and performance, *J. Membr. Sci.*, **505**, pp. 231–240.

89. Kong X., Zhang Y., Zeng S.-Y., Zhu B.-K., Zhu L.-P., Fang L.-F., and Matsuyama H. (2016). Incorporating hyperbranched polyester into cross-linked polyamide layer to enhance both permeability and selectivity of nanofiltration membrane, *J. Membr. Sci.*, **518**, pp. 141–149.

90. Echaide-Gorriz C., Zapata J. A., Etxeberria-Benavides M., Tellez C., and Coronas J. (2020). Polyamide/MOF bilayered thin film composite hollow fiber membranes with tuned MOF thickness for water nanofiltration, *Sep. Purif. Technol.*, **236**, pp. 116265:1-8.

91. Li H., Shi W., Zhang H., Zhou R., and Qin X. (2020). Preparation of internally pressurized polyamide thin-film composite hollow fiber nanofiltration membrane with high ions selectivity by a facile coating method, *Prog Org. Coat.*, **139**, pp. 105456:1-13

92. Chong J. Y. and Wang R. (2019). From micro to nano: Polyamide thin film on microfiltration ceramic tubular membranes for nanofiltration, *J. Membr. Sci.*, **587**, pp. 117161:1-12.

93. Zhu. W.-P., Gao J., Sun S.-P., Zhang S., and Chung T.-S. (2015). Poly(amidoamine) dendrimer (PAMAM) grafted on thin film composite (TFC) nanofiltration (NF) hollow fiber membranes for heavy metal removal, *J. Membr. Sci.*, **487**, pp. 117–126.

94. Darvishmanesh S., Tesselli F., Jansen J. C., Tocci E., Bazzarelli F., Bernardo P., Luis P., Degreve J., Drioli E., and der Bruggen B. V., Preparation of solvent stable polyphenylsulfone hollow fiber nanofiltration membranes, *J. Membr. Sci.*, **384**, pp. 89–96.

95. Xu S., Wang Z. Y., Li S., Tian L., and Su B. (2020). Fabrication of polyimide-based hollow fiber membrane by synergetic covalent-crosslinking strategy for organic solvent nanofiltration (OSN) application, *Sep. Purif. Technol.*, **241**, 116751:1-10.

96. Goh K. S., Chong J. Y., Chen Y., Fang W., Bae T.-H., and Wang R. (2020). Thin-film composite hollow fiber membrane for low pressure organic solvent nanofiltration, *J. Membr. Sci.*, **597**, pp. 117760:1-10.

97. Lim S. K., Setiawan L., Bae T.-H., and Wang R. (2016). Polyamide-imide hollow fiber membranes crosslinked with amine-appended inorganic networks for application in solvent-resistant nanofiltration under low operating pressure, *J. Membr. Sci.*, **501**, pp. 152–160.

98. Farahani M. H. D. A. and Chung T.-S. (2018). Solvent resistant hollow fiber membranes comprising P84 polyimide and amine-functionalized carbon nanotubes with potential applications in pharmaceutical, food, and petrochemical industries, *Chem. Eng. J.*, **345**, pp. 174–185.

99. Tashvigh A. A. and Chung T.-S. (2019). Robust polybenzimidazole (PBI) hollow fiber membranes for organic solvent nanofiltration, *J. Membr. Sci.*, **572**, pp. 580–587.

100. Li Y., Cao B., and Li P. (2019). Effects of dope compositions on morphologies and separation performances of PMDA-ODA polyimide hollow fiber membranes in aqueous and organic solvent systems, *Appl. Surf. Sci.*, **473**, pp. 1038–1048.

101. Aba N. F. D., Chong J. Y., Wang B., Mattevi C., and Li K. (2015). Graphene oxide membranes on ceramic hollow fibers: Microstructural stability and nanofiltration performance, *J. Membr. Sci.*, **484**, pp. 87–94.

102. Wang Z., Wei Y.-M., Xu Z.-L., Cao Y., Dong Z.-Q., and Shi X.-L. (2016). Preparation, characterization and solvent resistance of γ-Al_2O_3/α-Al_2O_3 inorganic hollow fiber nanofiltration membrane, *J. Membr. Sci.*, **503**, pp. 69–80.

103. Mahalingam D. K., Falca G., Upadhya L., Abou-Hamad E., Batra N., Wang S., Musteata V., da Costa P. M., and Nunes S. P. (2020). Spray-coated graphene oxide hollow fibers for nanofiltration, *J. Membr. Sci.*, **606**, pp. 118006:1-9.

104. Zheng Y., Yao G., Cheng Q., Yu S., Liu M., and Gao C. (2016). Positively charged thin-film composite hollow fiber nanofiltration membrane for the removal of cationic dyes through submerged filtration, *Desalination*, **328**, pp. 42–50.

105. Yu S., Zheng Y., Zhou Q., Shuai S., Lv Z., and Gao C. (2012). Facile modification of polypropylene hollow fiber microfiltration membranes for nanofiltration, *Desalination*, **298**, pp. 49–58.

106. Tham H. M., Wang K. Y., Hua D., Japip S., and Chung T.-S. (2017). From ultrafiltration to nanofiltration: Hydrazine cross-linked

polyacrylonitrile hollow fiber membranes for organic solvent nanofiltration, *J. Membr. Sci.*, **542**, pp. 289–299.

107. Goh K., Setiawan L., Wei L., Si R., Fane A. G., Wang R., and Chen Y. (2015). Graphene oxide as effective selective barriers on a hollow fiber membrane for water treatment process, *J. Membr. Sci.*, **474**, pp. 244–253.

108. Zhang Y., Japip S., and Chung T.-S. (2017). Thermally evolved and boron bridged graphene oxide (GO) frameworks constructed on microporous hollow fiber substrates for water and organic matters separation, *Carbon*, **123**, pp. 193–204.

109. Zhang Y., Zhang S., Gao J., and Chung T.-S. (2016). Layer-by-layer construction of graphene oxide (GO) framework composite membranes for highly efficient heavy metal removal, *J. Membr. Sci.*, **515**, pp. 230–237.

110. Rajabzadeh S., Liu C., Shi L., and Wang R. (2014). Preparation of low-pressure water softening hollow fiber membranes by polyelectrolyte deposition with two bilayers, *Desalination*, **344**, pp. 64–70.

111. Liu C., Shi L., and Wang R. (2015). Enhanced hollow fiber membrane performance via semi-dynamic layer-by-layer polyelectrolyte inner surface deposition for nanofiltration and forward osmosis applications, *React. Funct. Polym.*, **86**, pp. 154–160.

112. Liu C., Shi L., and Wang R. (2015). Crosslinked layer-by-layer polyelectrolyte nanofiltration hollow fiber membrane for low-pressure water softening with the presence of SO42- in feed water, *J. Membr. Sci.*, **486**, pp. 169–176.

113. Gao J., Sun S.-P., Zhu W.-P., and Chung T.-S. (2014). Chelating polymer modified P84 nanofiltration (NF) hollow fiber membranes for high efficiency heavy metal removal, *Water Res.*, **63**. pp. 252–261.

114. Menne D., Kamp J., Wong J. E., and Wessling M. (2016). Precise tuning of salt retention of backwashable polyelectrolyte multilayer hollow fiber nanofiltration membranes, *J. Membr. Sci.*, **499**, pp. 396–405.

115. Ilyas S., Abtanhi S. M., Akkilic N., Roesink H. D. W., and de Vos W. M. (2017). Weak polyelectrolyte multilayers as tunable separation layers for micro-pollutant removal by hollow fiber nanofiltration membranes, *J. Membr. Sci.*, **537**, pp. 220–228.

116. Laakso T., Pihlajamaki A., and Manttari M. (2018). Effect of polycation structure on the fabrication of polyelectrolyte multilayer hollow fiber membrane for loose nanofiltration applications, *Sep. Purif. Technol.*, **194**, pp. 141–148.

117. Akbari A., Desclaux S., Rouch J. C., and Remigy J. C. (2007). Application of nanofiltration hollow fiber membranes, developed by photografting, to treatment of anionic dye solutions, *J. Membr. Sci.*, **297**, pp. 243–252.

118. Wang X.-L., Wei J.-F., Dai Z., Zhao K.-Y., and Zhang H. (2012). Preparation and characterization of negatively charged hollow fiber nanofiltration membrane by plasma-induced graft polymerization, *Desalination*, **286**, pp. 138–144.

119. Xu H.-M., Wei J.-F., and Wang X.-L. (2014) Nanofiltration hollow fiber membranes with high charge density prepared by simultaneous electron beam radiation-induced graft polymerization for removal of Cr(VI), *Desalination*, **346**, pp. 122–131.

Chapter 8

Fiber-Based Membrane Distillation for Desalination and Wastewater Treatment

F. Macedonio and A. Criscuoli
Institute on Membrane Technology, National Research Council of Italy (ITM-CNR), via P. Bucci, Cubo 17/C Rende (CS), Italy
f.macedonio@itm.cnr.it; a.criscuoli@itm.cnr.it

Water scarcity is becoming one of the biggest issues for humanity, as the countries under water stress are continuously increasing. Therefore, there is an urgent need for a wise management of available water sources, as well as of implementing a recycling strategy of wastewater. Often, wastewater contains useful compounds that can be concentrated/recovered during the treatment, with further benefits from an economic point of view. Another source of fresh water can be obtained from the sea, by applying desalination technologies. The aim of this chapter is to report about the efficiency of membrane distillation for the treatment of both wastewater and seawater. In particular, results obtained with polymeric fibers, both commercial and lab made, are presented and discussed.

Hollow Fibers and Nanofibers in Membrane Science: Preparation, Characterization, and Applications
Edited by Alberto Figoli, Mir Saeed Seyed Dorraji, and Francesco Galiano

ISBN 978-981-4968-03-4 (Hardcover), 978-1-003-25663-2 (eBook)
www.jennystanford.com

8.1 Introduction

Membrane distillation (MD) is an expanding membrane technology that has aroused great interest among membranologists around the world. In fact, both theoretical and experimental research on MD has significantly intensified in recent decades. Figure 8.1 shows the number of articles published in the peer-reviewed journals in the reverse osmosis (RO) and MD sectors from 1960 to today. Thus, we can observe a significant increase in articles related to MD in recent years, despite the number of articles published each year on MD being much lower than those relating to RO. This testifies the strong interest worldwide in seeking alternative and more advantageous membrane processes compared to the conventional RO process, especially for desalination.

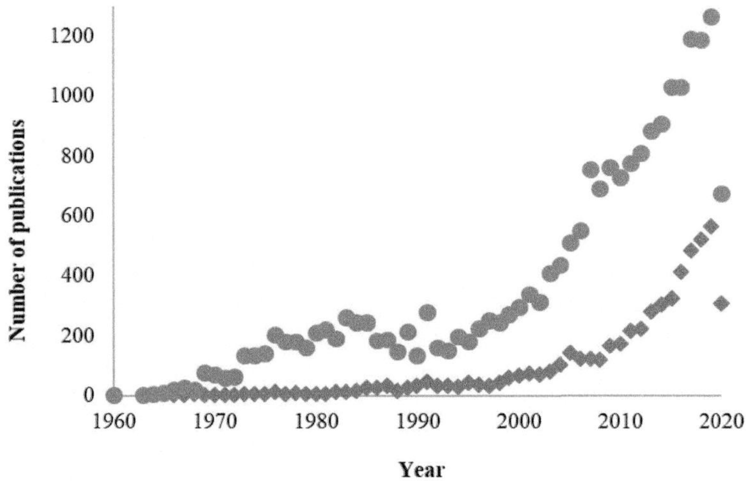

Figure 8.1 Number of research publications related to RO (circles) and MD (rhombus) from 1960 to 2020 (till June).

Most of the articles published until now on membrane distillation focused on the use of flat-sheet commercial membranes made of polytetrafluorethylene (PTFE) (e.g., Gore) and polypropylene (PP) (e.g., Celgard, Accurel) [1]. However, hollow fibers are receiving attention and are increasingly being explored. PTFE and PP cannot be used for the manufacture of hollow fibers through the phase-inversion technique due to their low solubility at room temperature, at least in most common solvents. For this reason, poly(vinylidene fluoride) (PVDF) was the preferred polymer for hollow fibers to be used in MD, as it is soluble in polar solvents and has a higher hydrophobicity than other polymers frequently used for membranes. In particular, efforts focused on (1) optimizing fiber morphology by integrating additives, changing solvent and polymer concentration, (2) manufacturing dual-layer fibers, and (3) developing innovative fiber designs. Except for mixing of polymeric or oligomeric additives that modify the surface, no further efforts were devoted to the development of new polymers tailored to the requirements of MD. This chapter illustrates the application of both commercial and lab-made polymeric fibers for the treatment of wastewater and for desalination by different configurations of membrane distillation.

8.2 Membrane Distillation Configurations

The typical four-membrane distillation configurations were investigated for carrying out tests on membranes with tubular

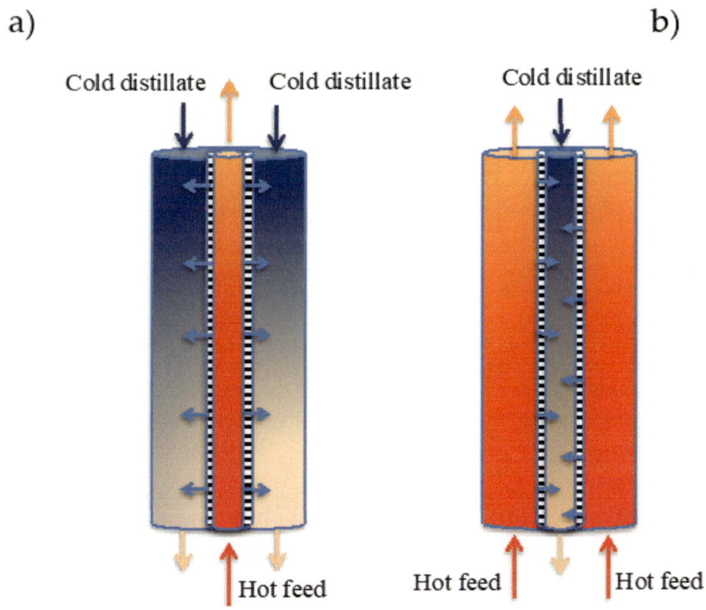

Figure 8.2 DCMD: (a) feed in the lumen side and (b) feed in the shell side.

geometry: direct contact membrane distillation (DCMD), air gap membrane distillation (AGMD), vacuum membrane distillation (VMD), and sweep gas membrane distillation (SGMD). In DCMD, the hot feed to be treated can be sent in both lumen and shell side, while the cold distillate flows at the shell and lumen side, respectively (Fig. 8.2).

In the AGMD configuration, the hot liquid is usually sent inside the lumen of a microporous membrane, while a cold stream is sent in the lumen of a dense tube, on whose external surface the vapor condensation occurs, as shown in Fig. 8.3.

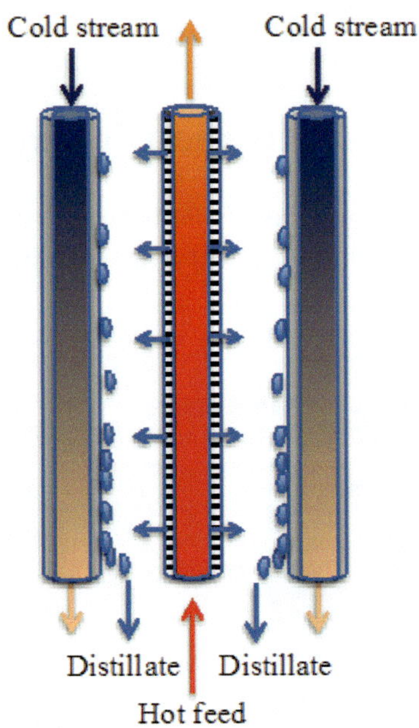

Figure 8.3 AGMD. Feed in the lumen side of the membrane and condensation of the water vapor on the external surfaces of dense tubes surrounding the membrane.

In VMD and SGMD, like in DCMD, the hot feed can be sent either in the lumen and in the shell (see Fig. 8.4 and Fig. 8.5, respectively).

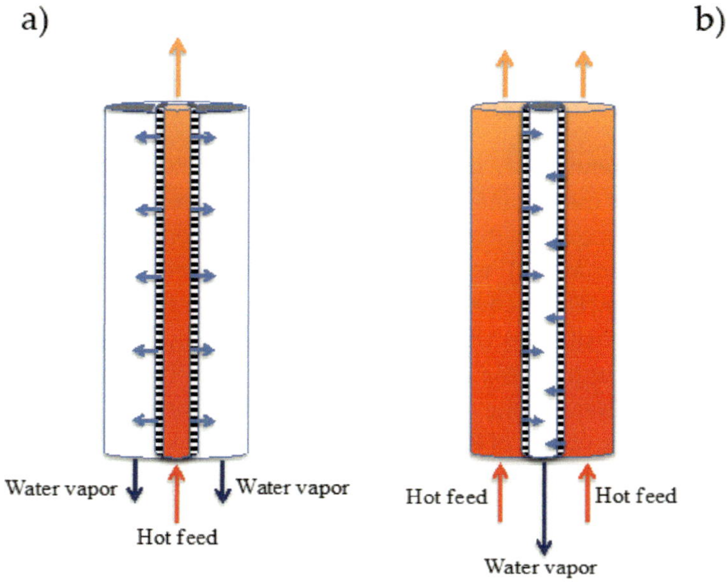

Figure 8.4 VMD: (a) feed in the lumen side with the shell under vacuum and (b) feed in the shell side with the lumen under vacuum.

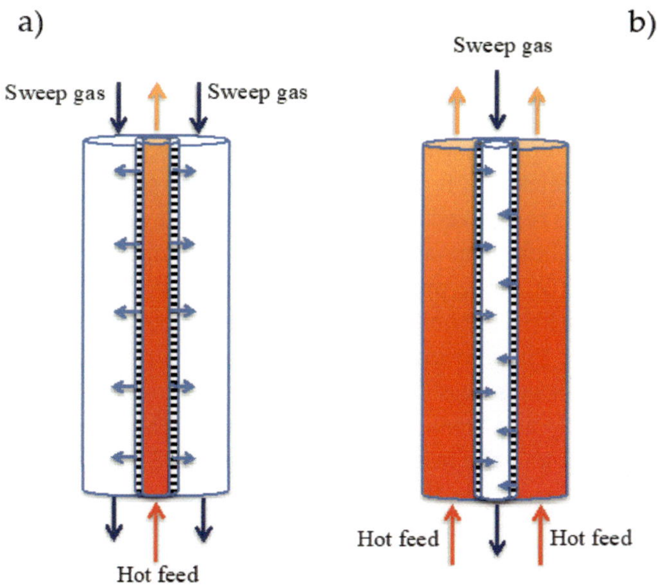

Figure 8.5 SGMD: (a) feed in the lumen side and (b) feed in the shell side.

8.3 Wastewater Treatment

8.3.1 Commercial

Vacuum membrane distillation based on tubular/capillary/hollow fiber membranes was used to concentrate various types of wastewaters, coming from different industrial productions.

Zhang et al. [2] tested VMD to concentrate lignocellulosic hydrolyzates. The membranes employed were in PVDF with a pore size and porosity of 0.18 μm and 85%, respectively, and a solar heat exchanger supplied the needed heat to the feed. The feed was sent inside the fibers, while vacuum was applied at the shell side of the module. Flux values of 8.46 L/m^2h and glucose rejections of 99.5% were obtained re-circulating the feed (glucose, 5 g/L) at 1 m/s and 65°C. As the glucose content increased, the flux value reduced to 5.55 L/m^2h and 3.30 L/m^2h for a glucose concentration of 35.96 g/L (glucose amount after enzymatic hydrolysis) and of 136.07 g/L (final value obtained after 270 min of VMD test), respectively. Nevertheless, the glucose rejection was always higher than 98%.

The same equipment was used by Chen et al. [3] to remove two by-products—furfural and acetic acid—from diluted lignocellulosic hydrolyzates. Both transmembrane flux and removals increased with temperature. However, higher removals were obtained for the furfural: at 70°C, the flux was of 6.42 L/m^2h and furfural removal was 96.25% versus the 24.79% measured for the acetic acid. Nevertheless, an increase of 17.8% in the ethanol production efficiency was registered after the selective removal step.

Synthetic heavy-metal wastewaters, containing Zn^{2+}, Ni^{2+}, Cu^{2+}, Ca^{2+}, and EDTA, were prepared and treated by VMD in a module equipped with PTFE hollow fibers provided by Zhejiang Dongda Environment Engineering Co., China [4]. The nominal pore size of the membranes was 0.243 μm, their thickness 440 μm, their porosity 40%, and their contact angle 101°. The membrane area was 180 cm^2, and the tests were carried out at 57°C. The permeate flux did not depend on the feed pH nor the EDTA content, whereas decreased at higher values of calcium content. On the contrary, the permeate conductivity was not affected by the calcium concentration of the feed, but significantly increased at pH lower than 2 and in the presence of EDTA.

A liquid solution containing methylene blue dye (18.5 ppm) was concentrated by VMD using a commercial module (Enka-Microdyn, North Carolina, USA) with an effective membrane area of 360 cm^2 [5]. The module was equipped with three tubular membranes made in polypropylene having a nominal pore size and thickness of 0.2 μm and 1.5 mm, respectively. The feed was sent in the lumen of the membranes. The produced distillate did not contain dyes, but a continuous decrease in flux with time (at 50°C, from 30 kg/m^2h to about 1.6 kg/m^2h) was observed due to fouling.

The potential of VMD for treating dye solutions was also studied by Criscuoli et al. [6]. A lab-made module was equipped with three commercial capillary polypropylene membranes having a pore size of 0.2 μm and a porosity of 75% (Membrana, Germany). The effective membrane area was of 28 cm^2, and tests were carried out by sending the dye solutions in the membrane lumen. Five dyes were considered (initial concentration of 50 ppm, temperature of 50°C), and in all cases pure water was collected as distillate. During the first 30 min of tests, the permeate fluxes decreased, due to membrane fouling, then, asymptotic values (always higher than 20 kg/m^2h) were reached. Some swelling was also observed, due to the interaction between dyes and the membrane material. However, original fluxes were restored after cleaning and prolonged VMD tests with distilled water.

A similar type of membrane was used by Carnevale et al. [7] to concentrate the olive mill wastewater. Both direct contact (DCMD) and vacuum membrane configurations were applied by sending the feed stream inside the capillaries at 70 L/h. At 50°C and at a feed concentration of 2500 ppm, rejection values of 99.9% and 99.6% were obtained with DCMD and VMD, respectively, while the corresponding transmembrane fluxes were around 6.5 kg/m^2h and 19 kg/m^2h. Moreover, the original water flux was recovered after cleaning and membrane distillation tests with deionized water. In prolonged DCMD tests, it was possible to reduce the feed volume by 99%, with a rejection of 99.6%. In this case, a water flux reduction of 8% was registered after the cleaning step.

Tomaszewska et al. investigated the HCl recovery from industrial effluent by DCMD [8], using a capillary module with an effective membrane area of 120 cm^2. The membranes were in polypropylene

(Accurel, Membrana, Germany) and had a thickness of 0.4 mm. The real pickling solution was fed at 70°C inside the capillaries, and the cold distillate (20°C) was sent in countercurrent at the shell side. It was found that the HCl flux increased (up to 350 mol/m^2 per day) with the salt concentration of the feed, due to the desalting effect. The HCl concentration in the distillate was in the range of 100–150 g/dm^3, whereas the salt was completely retained in the feed.

A membrane module equipped with capillary membranes in polypropylene (Accurel PP S6/2, Membrana, Germany) was used by Tomaszewska and Mientka [9] to separate HCl from HCl-H_2SO_4 solutions by DCMD. The capillary membranes had a thickness of 0.4 mm and a pore size of 0.2 μm, and the effective membrane area was 112 cm^2. For HCl content in the feed of 51.72 g/dm^3 and an H_2SO_4 concentration increasing from 50 to 150 g/dm^3, the HCl flux moved from 2.27 to 20.12 mol/m^2d and from 31.22 to 80.12 mol/m^2d at 60°C and 70°C, respectively.

The concentration of diluted hydrochloric acid solutions was studied by Liu et al. [10] who employed hollow fibers to fabricate an AGMD module with internal heat recovery. In the unit, the stream to be treated is first sent inside dense fibers, where it receives the heat of vapor condensation and is pre-heated; then it is heated up to the desired temperature (90°C) in an external heat exchanger and, finally, it is fed inside porous fibers, where it evaporates. The produced vapor migrates inside the air gap between porous and dense fibers, and it is condensed on the outer surface of the dense walls. Polypropylene dense and porous fibers (Chembrane Research and Engineering, Inc., Tianjin, China) were used. In particular, two porous fibers with a different mean pore size (PP48: 0.1 μm and PP60: 0.2 μm), porosity (PP48: 30% and PP60: 68%), and thickness (PP48: 0.05 mm and PP60: 0.125 mm) were tested. The two modules differed in terms of length, number of porous fibers, and ratio of the number of porous to dense fibers, but had similar membrane areas (PP48: 0.67 m^2 and PP60: 0.63 m^2) and packing densities (PP48: 35.7 % and PP60: 32.7 %). PP60 fibers resulted to be more suitable, leading to higher fluxes (the maximum value was 4.3 kg/m^2h for a 2 wt% HCl feed) and to a distillate with an HCl content of 8.7 wt% when the feed was concentrated up to 18 wt%. Moreover, PP60 fibers were stable in long-term tests (up to 30 days).

The same setup was used by Li et al. [11] to concentrate dilute sulfuric acid 2 wt% up to 40 wt% at 90°C. The corresponding transmembrane flux ranged from 5.3 and 1.9 L/m^{2}h. The module was stable for >30 days, leading to a distillate conductivity always lower than 150 μS/cm.

He et al. [12] also employed a similar AGMD setup, but equipped with polypropylene hollow fiber microporous membranes produced by Membrana (Germany) (mean pore size: 0.2 μm; porosity: 65%; thickness: 150 μm), for the treatment of the wastewater coming from an urea synthesis plant. The membrane area was of 0.708 m^2, and tests were carried out on solutions containing 1–10wt% of urea. The highest water flux was of 6.8 L/m^{2}h at 95°C with urea rejections of 99.99%. Scale-up tests were also carried out on real urea streams containing more than 200 mg/L of ammonia, by using a 3.7 m^2 module for 60 days. The average registered transmembrane flux at 85°C was around 4.7 L/m^{2}h, and the urea rejection was always above 99.99%, while the final ammonia content in the retentate was about 97.9 mg/L.

The removal of chloroform by VMD from a dilute aqueous stream was studied by Urtiaga et al. [13]. The membrane module used (membrane area of 389 cm^2), which was produced by Enka-Microdyn (MD 020 TO 2N), contained three capillaries in polypropylene, with a nominal pore size of 0.2 μm, a thickness of 1.55 mm, and a porosity of 75%. The liquid stream was sent in the fiber lumen. It was found that higher feed flow rates and operating temperatures improved chloroform removal. For instance, at 0.50 L/min and at a chloroform content of 500 mg/L, by increasing the feed temperature from 5 to 44°C, the time required to remove 50% of the chloroform reduced from 4500 s to 2100 s, respectively. At 5 L/min and 43°C, almost all chloroform was removed in 3500 s.

Liqui-Cel SuperPhobic hollow fibers (Alting, Hoerdt, France) were used to concentrate (from 30% to 50%) aqueous solutions of the dipolar aprotic solvent 1,3-dimethyl-2-imidazolidinone (DMI), by SGMD [14]. The polypropylene membranes were characterized by a 0.04 μm pore size, 40% porosity, a 40 μm thickness, and the module membrane area was 1.2 m^2. During experiments, air was sent at 21–26°C while the feed temperature was varied between 20°C and 67°C. The maximum vapor flux was around 0.12 L/m^{2}h and 99.2% of DMI was recovered.

Table 8.1 Commercial membranes utilized in MD for wastewater treatment

	Liquid Feed Composition	MD Configuration	Membrane Material	Membrane Geometry	Main Membrane Properties	Refs.
Concentration of lignocellulosic hydrolyzates	Glucose, 5–35.96 g/L	VMD	PVDF	Hollow fiber	Pore size: 0.18 μm Porosity: 85%	Zhang et al. [2]
Removal of inhibitors from lignocellulosic hydrolyzates	Glucose, 35.96 g/L; xylose, 5.96 g/L; acetic acid, 2.62 g/L; furfural, 0.72 g/L	VMD	PVDF	Hollow fiber	Pore size: 0.18 μm Porosity: 85%	Chen et al. [3]
Treatment of heavy-metal wastewaters	Zn^{2+}, 600–3000 ppm; Ni^{2+}, 200–1000 ppm; Cu^{2+}, 400–2000 ppm; Ca^{2+}, 0–500 ppm; EDTA, 0–300 ppm	VMD	PTFE	Capillary	Pore size: 0.243 μm Porosity: 40% Thickness: 0.440 mm	Ji et al. [4]
Treatment of dye solutions	Methylene blue, 18.5 ppm	VMD	PP	Capillary	Pore size: 0.2 μm Thickness: 1.5 mm	Banat et al. [5]
Treatment of dye solutions	Remazol brilliant blue R, 25–500 ppm; Reactive black 5, 50 ppm; Indigo vat blue, 50 ppm; Acid red 4, 50 ppm; Methylene blue, 50 ppm	VMD	PP	Capillary	Pore size: 0.2 μm Porosity: 75%	Criscuoli et al. [6]

	Liquid Feed Composition	MD Configuration	Membrane Material	Membrane Geometry	Main Membrane Properties	Refs.
Treatment of olive mill wastewaters	Polyphenols, 150–2500 ppm	DCMD and VMD	PP	Capillary	Pore size: 0.2 μm Porosity: 70%	Carnevale et al. [7]
HCl recovery from metal pickling solutions	Real solution: HCl, 85.5–126.8 g/dm^3	DCMD	PP	Capillary	Thickness: 0.4 mm	Tomaszewska et al. [8]
HCl separation from HCl-H_2SO_4 solutions	HCl, 51.72–111 g/m^3; H_2SO_4, 50–183 g/m^3	DCMD	PP	Capillary	Pore size: 0.2 μm Thickness: 0.4 mm	Tomaszewska and Mientka [9]
Concentration of diluted HCl solutions	HCl, 1–2 wt%	AGMD	PP	Hollow fiber	PP48 Pore size: 0.1 μm; Porosity: 30%; Thickness: 0.05 mm PP60 Pore size: 0.2 μm; Porosity: 68%; Thickness: 0.125 mm	Liu et al. [10]
Concentration of H_2SO_4 solutions	H_2SO_4, 2–20 wt%	AGMD	PP	Hollow fiber	PP48 Pore size: 0.1 μm; Porosity: 30%; Thickness: 0.05 mm PP60 Pore size: 0.2 μm; Porosity: 68%; Thickness: 0.125 mm	Li et al. [11]

(Continued)

Table 8.1 *(Continued)*

	Liquid Feed Composition	MD Configuration	Membrane Material	Membrane Geometry	Main Membrane Properties	Refs.
Concentration of urea solutions	Urea, 1–10 wt%	AGMD	PP	Hollow fiber	Pore size: 0.2 μm Porosity: 65% Thickness: 150 μm	He et al. [12]
Removal of chloroform from aqueous solutions	Chloroform, 212–2012 ppm	VMD	PP	Capillary	Pore size: 0.2 μm Porosity: 75% Thickness: 1.55 mm	Urtiaga et al. [13]
Concentration of 1,3-dimethyl-2-imidazolidinone (DMI) in aqueous solutions	DMI, 0–30 wt%	SGMD	PP	Hollow fiber	Pore size: 0.04 μm; Porosity: 40%; Thickness: 40 μm	Abejon et al. [14]
Treatment of organic wastewater	Glucose, 3.8 g/L; peptone, 1.8 g/L; $NaHCO_3$, 0.067 g/L; $MgSO_4$, 0.047 g/L; KH_2PO_4, 0.24 g/L; K_2HPO_4, 0.36 g/L; $CaCl_2$, 0.047 g/L; TOC, 2603 ppm	DCMD	PVDF	Capillary	Pore size: 0.22 μm; Thickness: 0.65 mm	Phattaranawik et al. [15]
Removal of strontium from simulated radioactive wastewater	Sr^{2+}, 10 ppm	VMD	PP	Capillary	Pore size: 0.18 μm Porosity: 60%	Jia et al. [16]

Membranes in PVDF (Pall Corporation and Asahi, Japan) were used by Phattaranawik et al. [15, 16] to develop a new concept of membrane bioreactor: the membrane distillation bioreactor (MDBR), as a potential alternative to the more classical integration of membrane bioreactor and reverse osmosis. The membranes' pore size, thickness, and length were 0.22 μm, 0.65 mm, and 40 cm, respectively, and the effective membrane area was 0.62 m^2. The membrane module was submerged into the stream to purify that was heated at the desired temperature of 58.5°C, while the cold permeate was recirculated inside the lumen fibers at 3 L/min and at an average temperature of 45°C. Ceramic air diffusers located at the bottom of the reactor ensured a fine air bubbling of the organic wastewater stream. Starting from a synthetic solution with a TOC of 2603 ppm, a permeate with very low TOC (1–1.7 ppm) was produced. Furthermore, a quite constant average transmembrane flux of 1.84 kg/m^2h was registered over 1-month operation.

VMD was applied by Jia et al. [16] to treat simulated radioactive wastewater containing strontium (Sr^{2+}, 10 ppm). Polypropylene hollow fiber membranes produced by Wochi, WHPP96-21, China (pore size, 0.18 μm; porosity, 60%) were assembled into a tubular module (140 fibers; effective area, 620 cm^2), and tests were carried out sending the feed inside the lumen fiber, while vacuum was applied at the shell side. At 70°C and 41.8 L/h, the Sr^{2+} rejection was >99.6% and a stable flux of 6.71 L/m^2h was registered.

8.3.2 Research

In addition to the hydrophobic microporous commercial fiber membranes, studies were conducted on the preparation of "specifically designed" membranes to be used in membrane distillation. This is because the existing commercial membranes were mainly developed for microfiltration purposes and do not present features that are typically needed for membrane distillation applications, like high hydrophobicity, high porosity, high liquid entry pressure, etc. Coking wastewaters were concentrated in an air-blowing vacuum membrane distillation (AVMD) by Zhao and Sun [17]. In their study, lab-made PVDF fibers with a pore size of 0.160 μm and a porosity of 85% were assembled in a module leading to an effective membrane area of 250 cm^2. The wastewater circulated in the fibers' lumen and vacuum was applied at the shell side. The air blowing resulted to be effective in improving the flux. Since the organic matter content in the feed was high, a pretreatment was requested to reduce the decline in flux, and at 70°C, the flux was kept at 18.6 kg/m^2h when the feed was concentrated by five times. The produced permeate had a conductivity of 10 μS/cm and a COD content ranging from 2 to 4 mg/L. The ammonia removal from water was investigated in a modified direct contact membrane distillation (MDCMD) unit, where 0.01 mol/L sulfuric acid solution was used in the permeate side to accelerate the removal [18]. The module was equipped with 50 capillary PVDF membranes produced by the authors (thickness, 0.1 mm; porosity, 80%; average pore radius, 0.22 μm; contact angle, 87°) and presented an effective membrane area of 141.3 cm^2. The same module was used to carry out the classical DCMD, sending only distilled water at the permeate side, and ammonia extraction by membrane contactor, sending at ambient temperature a 0.01 mol/L sulfuric acid receiving solution. In all cases, the feed was sent in the lumen fibers. When compared with the classical DCMD and the membrane contactor, the MDCMD showed the highest efficiency, leading to an ammonia removal of 99.5% within 105 min at 50°C, 0.5 m/s feed velocity, and 12.2 feed pH. At the same operating conditions (with the exception of the feed temperature for the membrane contactor, which was the room one), removals of 88% and 52% were achieved with the membrane contactor and the DCMD, respectively.

Shao et al. [19] applied three types of PP hollow fibers prepared by themselves to concentrate a solution of NMP by vacuum membrane distillation. Membranes were different in thickness (PP-1: 0.050 mm; PP-2: 0.040 mm, PP-3: 0.049 mm), porosity (PP-1: 62%; PP-2: 53%, PP-3: 48%) and the realized modules had membrane areas of 1.543×10^{-2} m^2, 0.870×10^{-2} m^2 and 0.718×10^{-2} m^2, respectively. During tests, the feed was sent at the shell side and the vapor was removed from the lumen side. The highest transmembrane fluxes of water vapor and NMP were achieved by using PP-1 membranes, due to their higher porosity. At 80°C, for a 10 wt% NMP feed, the water vapor flux was 9.5 kg/m^2h, while the NMP flux and rejection were about 0.020 kg/m^2h and 98%, respectively. An increase in the

NMP concentration in the feed led to a reduction in its rejection, moving from 97.5% to 97% for a 10 wt% and 40 wt%, respectively. The removal of water from aqueous lithium bromide solutions by VMD was investigated by Wang et al. [20]. Experiments were carried out on a capillary membrane module realized at the Key Laboratory of Hollow Fiber Membrane Materials and Membrane Process of Ministry of Education, Tianjin Polytechnic University, China. The module contained 300 PVDF membranes (mean pore size: 0.16 μm; porosity: 85%; thickness: 0.15 mm) and the total area was 0.3 m^2. The feed was recirculated inside the hollow fibers at a concentration of 50 wt%. In all tests, a high-purity distillate was obtained with transmembrane fluxes ranging from 0.7 to 1.9 kg/m^2h when the feed temperature increased from 65°C to 88°C, respectively.

Asymmetric PVDF hollow fibers were prepared by phase inversion, using dimethylacetamide (DMAc) as a solvent and a mixture of water and LiCl as additive, and applied to remove 1,1,1-trichloroethane (TCA) by VMD [21]. The produced membranes (mean pore radius: 1.62×10^{-7} m; thickness: 0.19 mm; effective porosity ε/Lp: 118 m^{-1}) were packed into a tubular module (number of fibers: 143; total membrane area: 0.159 m^2) and the feed solution was sent into the lumen side of fibers. High TCA fluxes were obtained at high temperatures and low permeate pressure; however, the water vapor flux increased as well and, in order to avoid an excessive permeate dilution, optimal operating conditions were identified: at 50°C, 80–105 mbar, and 10^{-3} m^3/h, a 97% TCA removal was achieved from a feed with a TCA content of 620 ppm.

In the work of Wu et al. [22], the effect of membrane properties on the efficiency of TCA removal by VMD was analyzed. Five modules equipped with the prepared PVDF (blended with LiCl) hollow fibers were tested. The five types of hollow fibers mainly differed in terms of thickness (M1: 188 μm; M2: 155 μm; M3: 150 μm, M4: 177 μm; M5: 172 μm), mean pore radius (M1: 5.4×10^{-8} m; M2: 6.8×10^{-8} m; M3: 3.1×10^{-8} m; M4: 6.4×10^{-8} m; M5: 6.7×10^{-8} m) and effective porosity ε/Lp (M1: 456 m^{-1}; M2: 542 m^{-1}; M3: 1516 m^{-1}; M4: 160 m^{-1}; M5: 71 m^{-1}). The five modules had the same packing density (2000 m^2/m^3), while the membrane areas were M1: 0.159 m^2; M2: 0.0492 m^2; M3: 0.0498 m^2; M4: 0.0482 m^2; M5: 0.0486 m^2. High separation factors, but low TCA permeation fluxes, were obtained using membranes with low effective porosity, while membranes with high porosity and small mean pore size were able to combine high TCA fluxes with high TCA separation factors. By working at 50°C, 10^{-3} m^3/h, and 105 mbar with a feed containing 430 ppm of TCA, the separation factor was 120 and the TCA flux was about 4.3 g/m^2h when using the module equipped with the M3 membranes. The TCA removal obtained in these conditions was about 88%.

Mokhtar et al. [23] prepared PVDF mixed matrix membranes incorporating Cloisite 15 A and studied their performance in DCMD tests on rubber processing effluents. The feed stream was circulated in the shell side at 55.5°C, while the distillate was sent in countercurrent in the lumen of fibers. The module had an effective membrane area of 50 cm^2 and was realized packing 10 fibers with the following properties: pore size, 0.088 μm; porosity, 83.70 %; thickness, 126 μm). A significant reduction in flux with time was observed (from 7.19 kg/m^2h to 5.05 kg/m^2h within the first 30 min of test, reaching an asymptotic value of about 1.5 kg/m^2h after 120 min), which was mainly attributed to fouling. In fact, the stream treated contained complex components, like latex and protein, that interacted with the membrane surface and deposited on it. Nevertheless, high removal efficiencies were registered: >99% for TOC; 98.7% for TDS and conductivity; 97% for color and turbidity; >96% for sulfate.

PVDF hollow fiber membranes blended with ethylene glycol-EG (used as an additive) were prepared by dry-jet wet spinning method and tested in DCMD for the treatment of a dye solution containing 0.5 g/L of reactive black 5 (RB5) [24]. Membranes had a thickness of 155 μm, a porosity of 73.18%, a pore size of 0.15 μm, and a contact angle of 87.30°. Twenty membranes were packed in a tubular module, and the feed stream was recirculated in the lumen side of fibers, while the cold water was sent in countercurrent at the shell side. After a 6 h experiment at 60°C, the transmembrane flux was constant at around 9.71 kg/m^2h with a dye rejection >99.75%. It was found that the flux obtained with the EG (pore former) modified PVDF membrane was 90% higher than that one obtained by the same authors in a previous work. Some fouling was observed, but with negligible impact on flux and rejection. A similar type of membrane was used in the work of Chong et al. [25] to treat a solution containing a reactive blue dye (20–35 ppm) at 55°C. The flux decreased with the dye content in the feed, moving from 9.5 kg/m^2h down to 7.5 kg/m^2h for 20 ppm and 35 ppm, respectively, and was always higher than that of the unmodified membrane (only PVDF) for which the corresponding fluxes were 6.5 kg/m^2h and 4 kg/m^2h, respectively. The rejection also decreased with the feed concentration, moving from 95.7% to 94.6%.

An integrated forward osmosis (FO)-DCMD process was investigated by Ge et al. for the treatment of a model solution containing 50 ppm of acid orange 8 dye [26]. Authors prepared PVDF membranes using NMP, ethylene glycol, and hydrophobic POSS particles on the dope composition, obtaining fibers with a thickness of 162 μm, a pore size of 0.28 μm, a porosity of 78.8%, and a contact angle of 114°. The MD module, equipped with these fibers, was coupled with the FO unit used to treat the model solution, in order to concentrate the dilute draw solution (a solution of poly(acrylic acid) sodium salt: PAA-Na) exiting the FO module. Both FO and MD were able to efficiently reject the PAA-Na solute and the best operating conditions were found to be 66°C and 0.48 g/ml of PAA-Na, at which the water transfer rate was constant in time (around 0.18 L/h) and the acid orange 8 concentration increased from 50 ppm to about 180 ppm, after 120 min of test. DCMD was applied to treat radioactive wastewater by Liu and Wang [27]. The module was assembled with 70 PVDF hollow fibers (effective area, 310 cm^2) produced by State Key Laboratory of Environmental Aquatic Chemistry, Research Center for Eco-Environmental Sciences, Chinese Academy of Sciences, China. The main fiber properties were as follows: pore size, 0.15 μm; thickness, 0.15 mm; porosity, 80%. Three simulated radioactive wastewaters were prepared: S1, containing 20 ppm of CO^{2+}; S2, containing Cs^{2+}, Sr^{2+}, Co^{2+}, each one at 20 ppm; S3, containing Cs^{2+}, Sr^{2+}, Co^{2+}, each one at 100 ppm. Moreover, $NaNO_3$, in the range of 1–200 g/L, was added to the solutions. DCMD was effective in removing Cs^{2+}, Sr^{2+}, Co^{2+}, leading to a distillate that can be discharged or reused. The increase in the $NaNO_3$ content in the feed led to a reduction in the permeate flux, which moved at 47.5°C, from 7.5 kg/m^2h down to 4.75 kg/m^2h for a 0 g/l and 200 g/l $NaNO_3$ feed, respectively.

Table 8.2 Lab-made membranes utilized in MD for wastewater treatment

Application	Liquid Feed Composition	MD Configuration	Membrane Material	Membrane Geometry	Main Membrane Properties	Refs.
Concentration of coking wastewater	COD, 400 ppm	VMD	PVDF	Capillary	Pore size: 0.16 μm Porosity: 85%	Zhao and Sun [17]
Ammonia removal from water	Ammonia, 0.5–3.5 g/L	DCMD	PVDF	Capillary	Pore size: 0.22 μm Porosity: 80% Thickness: 0.1 mm	Qu et al. [18]
Concentration of NMP solutions	NMP, 0.05–40 wt%	VMD	PP	Hollow fiber	PP-1 Porosity: 62% Thickness: 0.050 mm PP-2 Porosity: 53% Thickness: 0.040 mm PP-3 Porosity: 48% Thickness: 0.049 mm	Shao et al. [19]
Removal of water from aqueous lithium bromide solutions	Lithium bromide, 50 wt%	VMD	PVDF	Capillary	Pore size: 0.16 μm Porosity: 85% Thickness: 0.15 mm	Wang et al. [20]
Removal of TCA from water	TCA, 100–750 ppm	VMD	PVDF blended with LiCl	Capillary	Pore radius: 0.16 μm Thickness: 0.19 mm	Wu et al. [21]

Removal of TCA from water	TCA, 430 ppm	VMD	PVDF blended with LiCl	Capillary	M1 Pore radius: 0.054 μm; Thickness: 0.188 mm M2 Pore radius: 0.068 μm; Thickness: 0.155 mm M3 Pore radius: 0.031 μm; Thickness: 0.150 mm M4 Pore radius: 0.064 μm; Thickness: 0.177 mm M5 Pore radius: 0.067 μm; Thickness: 0.172 mm	Wu et al. [22]
Treatment of wastewater from rubber industry	TOC, 6765 ppm; TDS, 18710 ppm; Sulfate, 3339 ppm	DCMD	PVDF –Cloisite 15A	Capillary	Pore size: 0.088 μm Porosity: 83.70% Thickness: 0.126 mm	Mokhtar et al. [23]
Treatment of dye solutions	Reactive black 5, 0.5 g/L	DCMD	PVDF blended with EG	Hollow fiber	Pore size: 0.15 μm Porosity: 73.81% Thickness: 0.155 mm	Mokhtar et al. [24]
Treatment of dye solutions	Reactive blue dye, 20–35 ppm	DCMD	PVDF blended with EG	Hollow fiber	Pore size: 0.45 μm Porosity: 75%	Chong et al. [25]
Treatment of dye solutions	Acid orange 8 dye, 50 ppm	DCMD	PVDF blended with EG and POSS	Capillary	Pore size: 0.28 μm Porosity: 78.8% Thickness: 0.162 mm	Ge et al. [26]
Treatment of radioactive wastewater	Cs^{2+}, Sr^{2+}, Co^{2+}, 20–100 ppm; $NaNO_3$, 1–200 g/L	DCMD	PVDF	Capillary	Pore size: 0.15 μm Porosity: 80% Thickness: 0.15 mm	Liu and Wang [27]

8.4 Seawater Treatment

8.4.1 Commercial

Numerous commercial tubular/capillary/hollow fiber membranes were used to desalinate seawater utilizing various MD configurations. Highly hydrophobic polypropylene (PP) hollow fiber membranes with fluorosilicone coating, by Applied Membrane Technology (AMT) Inc., were used in DCMD with raw Mediterranean seawater as feed [28]. The effects of temperature, hydrodynamic conditions (cross flow and parallel flow), water quality, and rate of biofilm development were investigated. In parallel-flow experiments, biofilm formation did not result in flux loss (3.9 kg/m^{2}h) when operated at 40°C. In cross-flow experiments, a slight flow decline was observed after 180 h of the experiment (from 3.6 to 3.4 kg/m^{2}h) when DCMD was performed at the same temperature (i.e., 40°C). On the contrary, temperature cycling between 70°C and 40°C reduces the tendency of biofilm and the flux loss (practically constant at 3.9 kg/m^{2}h) with respect to the tests run at a constant temperature of 40°C. Krivorot et al. [28] concluded that biofouling can be minimized by utilizing a top brine temperature of 70°C or higher.

Gryta [29] used two different types of capillary PP membranes from PolyMem (Warszawa, Poland) for investigating the effect of talc addition (PP-T) on the morphology and performance of the tested membranes. Both types of PP membranes exhibited properties suitable for the MD process, such as resistance to wetting and almost 100% salt rejection. Furthermore, tests on membrane stability revealed that only the PP-T membranes had the appropriate properties for application to pilot plant MD studies. Additionally, MD tests carried out for 350 h (distillate temperature = 15–22°C; feed temperature = 80°C; linear velocity = 0.55 m/s; feed: 1 g/L NaCl solution; membrane area = 45 cm^2) indicated that the permeate flux, for both membranes, was practically constant, with slightly higher productivity (10.4 L/m^{2}h) for the PP-T membrane (polypropylene with talc) with respect to that of PP-N membrane (net polypropylene) (9.5 L/m^{2}h).

High feed temperatures and PVDF commercial hollow fiber MD membranes were used by Maab et al. [30] for performing Red Seawater desalination via DCMD configuration. Salt selectivity as high as 99.95% and water flux of 31 L/m^{2}h were obtained operating at 80°C feed temperature, 20°C permeate temperature, and utilizing a membrane module with an effective area of 25 cm^2.

Two different hollow fiber PVDF membranes were used by Singh et al. [31], too. In particular, they utilized PVDF hollow fiber from two different sources: (1) Hyflux Filtech (Shanghai), Burnaby BC, Canada, 900 μm ID (indicated as PVDF'H'); (2) Arkema, King of Prussia, PA, 925 μm OD (indicated as PVDF'E'). The performance of the fibers was analyzed over a temperature range of 80–125°C in order to check their potentialities in steam-assisted gravity drainage (SAGD) processes, where the brine obtained as produced water comes out at a temperature as high as 160°C. For these hot streams, reverse osmosis desalination can be utilized only after cooling of such a brine. When tested in DCMD with 1% NaCl solution as feed, the hollow fiber membrane H (area = 28.6 cm^2) showed flux of 27 kg/m^2h at 95.7°C and significant loss of salt due to the presence of defects as large as 5 μm. In the same operative conditions, hollow fiber membranes of type E (area = 37.6 cm^2) achieved water vapor flux of 88.5 kg/m^2h at 124°C.

DCMD tests for the treatment of oilfield-produced water were carried out by Macedonio et al. [32], too. They used two commercial PP membranes: one was acquired from Microdyn-Nadir (MD020CP2N) in the form of an already assembled module (whose principal characteristics, as specified by the manufacturer, are: number of capillaries = 40; membrane pore size = 0.2 μm; effective filtration area of the membrane = 0.1 m^2; LEPw = 140 kPa), whereas a second module was assembled in lab using three polypropylene fibers provided by Membrana (PP Accurel® S6/2 — pore size = 0.2 μm). Transmembrane flux of about 5 L/m^2h and rejection of 99.7% were measured with the lab-made module (feed temperature = 50°C, permeate temperature = 25°C; feed flow rate = 9.6 L/h, permeate flow rate = 6 L/h); rejection of 99.9% but lower flux (~0.8 L/m^2h) were measured with the MD020CP2N module (feed temperature = 50°C, permeate temperature = 25°C; feed flow rate = 50 L/h, permeate flow rate = 50 L/h). Despite the PP fibers of the two different modules having essentially similar properties, the measured flux under the same conditions of temperature was quite different due to the different hydrodynamic conditions. In the case of MD020CP2N, two modules connected in parallel were used and each module consists of 40 fibers, each 45 cm long. In the case of the lab-made PP module, the fibers were shorter (17 cm). On the other hand, the two membrane modules had comparable shell diameter (2.1 cm and 2.0 cm). Therefore, the higher flux registered in the case of the lab-made membrane module was due to the different packing density of the two membrane modules and to the different amount of cold flow available to remove the heat flux transferred from the retentate to the permeate side of the membrane.

The commercial shell-and-tube capillary membrane module MD020CP2N consisting of an array of PP porous hydrophobic capillaries was used in various other works reported in the literature [33–35]. Kim et al. [34] utilized MD020CP2N module for the development of a solar-assisted multistage vacuum membrane distillation (SMVMD) system. The proposed system employed a heat recovery unit (HRU) for the energy recovery from the permeate vapor to the feed seawater. The mean permeate flux was of about 26 kg/m^{2}h for a feed velocity of 1.0 m/s and a feed temperature equal to 65°C. Moreover, for a solar-assisted VMD system with 24 stages, the total water production of SMVMD system with 10 HRUs was found to be 3.37 m^3/day, which is about 34% higher as compared to the system with a single HRU. For a VMD system without solar thermal unit, the overall specific thermal energy consumption (OSTEC) decreases by 20% with an increase in the number of HRUs from 1 to 10. For the different numbers of HRUs in the range of 1–10, the system OSTECs with solar thermal system having 150 m^2 of evacuated-tube collectors and 16 m^3 seawater storage tanks are 28–36% lower compared to those without the solar thermal system.

MD020CP2N was also tested for seawater desalination by Al-Obaidani et al. [35], together with MD080CO2N and MD020TP2N from Enka-Microdyn. MD080CO2N and MD020CP2N are manufactured using the same kind of PP fibers but with different lengths (0.45 m and 1 m, respectively) and number of fibers (40 and 467, respectively). As a consequence, surface area increases from 0.1 m^2 to 2 m^2 passing from MD020CP2N to MD080CO2N. On the contrary, three tubular membranes are contained in MD020TP2N, for a total surface area of 0.036 m^2. The authors performed flux measurements by keeping the temperature at the permeate side constant at 15°C, the mass flow rate at 0.055 kg/s at feed side, and 0.028 kg/s at permeate side for all modules. The highest flux was achieved for the MD020CP2N module (~13 kg/m^{2}h at feed temperature = 70°C and feed concentration = 35 g/L). MD080CO2N module exhibited a flux 28% lower with respect to MD020CP2N module, while the lowest was measured with the MD020TP2N (~4 kg/m^{2}h at feed temperature = 70°C and feed concentration = 35 g/L).

Two commercial hollow fiber membrane modules from Econity [36], differing both in terms of membrane material and membrane characteristics, were used in the framework of the GMVP project [37]. Econity's hollow fiber membrane is made by stretching method with pores on the membrane surface structured in slit and ellipse in order to induce a less-fouling function. The two different hollow fiber membrane modules utilized in the Korean GMVP project were as follows:

1. A small one for the MD prototype: membrane material PVDF, effective membrane area 7.6 m^2, capacity 1 m^3/d (with two modules) and flux 6 LMH when operated in DCMD configuration at feed temperature 80°C and temperature difference ΔT across the membrane of 20°C.
2. A larger one used in the MD pilot (capacity 10 m^3/d) and MD demo pilot (400 m^3/d): area 10 m^2, pore size 0.1 μm, module material ABS, length 400 mm, configuration VMD.

Numerous articles can be found in the literature reporting the modification of commercial hollow fiber membranes. An example can be found in the work of Song et al. [38], who coated commercial PP hollow fiber membrane (330 μm ID, 150 μm wall thickness, membrane porosity from 0.6 to 0.8, and maximum pore size of 0.60 μm) from Membrana (Charlotte, NC) with a thin layer of plasma-polymerized silicone fluoropolymer. The coating was deposited on the fiber outside surface by Applied Membrane Technology, Minnetonka, MN. Either 2448 or 2652 coated fibers were assembled for the preparation of modules with an effective fiber length of 24.1 cm, and 0.61 or 0.66 m^2 as surface area, respectively. The modules were tested in DCMD utilizing as feed either city water, or city water containing salt at the level of 3.5, 6 or 10%, or seawater trucked in from Long Island Sound, CT. The tests were performed at high temperature, with feed at 64–93°C and distillate at 20–54°C in order to achieve flux between 15 and 33 kg/m^{2}h for configurations employing 6–10 modules. The plant operated successfully daily for 3 months with a very limited flux reduction at salt concentrations up to 19.5% from seawater.

Table 8.3 Commercial membranes utilized in MD for desalination

Application	Liquid Feed Composition	MD Configuration	Membrane Material	Membrane Geometry	Main Membrane Properties	Refs.
Costal Mediterranean seawater desalination	Mediterranean Sea from Ashkelon desalination plant	DCMD	PP with fluorosilicone coating	Hollow fiber	Maximum pore size: 0.6 μm, Average membrane porosity ~ 70%.	Krivorot et al. [28]
Desalination	1 g/L NaCl solution	Submerged MD	PP-N (net polypropylene) PP-T membrane (polypropylene with talc)	Capillary	PP-N Porosity: 81%; Pore diameter 0.2 μm; PP-T porosity: 84%; Pore diameter: 0.2 μm	Gryta [29]
Red Sea water desalination	Red Sea water	DCMD	PVDF	Hollow fiber	Mean pore size: 55 nm	Maab et al. [30]
Treatment of salt water	City water, city water containing salt at the level of 3.5, 6, or 10%, or sea water	DCMD	Support membrane type: PP 150/330 Accurel MEMBRANA Coating: Silicone fluoropolymer	Hollow fiber	Wall thickness: 150 μm Maximum pore size: 0.60 mm Membrane porosity ~0.6–0.8	Song et al. [38]
Desalination	Seawater	SMVMD	PP	Capillary	Pore size: 0.2 μm; Porosity: 70%	Mengual et al. [33]

Application	Liquid Feed Composition	MD Configuration	Membrane Material	Membrane Geometry	Main Membrane Properties	Refs.
Hot brine	0.1% NaCl	DCMD	PVDF from Hyflux Filtech (Shanghai), Burnaby BC, Canada	Capillary	Pore size: 0.62 μm; Porosity: 50%; Thickness: 350 μm	Singh et al. [31]
Hot brine	0.1% NaCl	DCMD	PVDF from Arkema, King of Prussia, PA	Hollow fiber	Pore size: 0.2 μm; Porosity: 54%; Thickness 117 μm	Singh et al. [31]
Seawater desalination	35 g/L NaCl	DCMD	PP MD020CP2N from Enka-Microdyn	Capillary	Pore size: 0.2 μm; Thickness: 0.65 mm; Porosity: 70%; Length: 0.45 m	Al-Obaidani et al. [35]
Seawater desalination	35 g/L NaCl	DCMD	PP MD080CO2N from Enka-Microdyn	Capillary	Pore size: 0.2 μm; Thickness: 0.65 mm; Porosity: 70%; Length: 1 m	Al-Obaidani et al. [35]
Seawater desalination	35 g/L NaCl	DCMD	PP MD020TP2N from Enka-Microdyn	Tubular	Pore size: 0.2 μm; Thickness: 1.55 mm; Porosity: 70%; Length: 0.75 m	Al-Obaidani et al. [35]
Oilfield-produced water treatment	Real produced water	DCMD	PP Accurel® S6/2 from Membrana	Capillary	Pore size: 0.2 μm; Thickness: 0.45 mm; Porosity: 73%	Macedonio et al. [32]

8.4.2 Research

Various studies on the manufacture of hollow fiber and capillary membranes for desalination applications using MD technology can be found in the literature.

Considering that the main requirements for the successful MD scaling-up are the achievement of high flow and LEP, and low fouling rate, most research has focused on achieving these goals.

In membrane distillation, the molar flux (N) through a pore is related to the membrane's average pore size and other characteristic parameters by:

$$N \propto \frac{< r^{\alpha} > \cdot \varepsilon}{\tau \cdot \delta} \tag{8.1}$$

where ε is the membrane porosity, τ is the membrane tortuosity, δ is the membrane thickness, $< r^{\alpha} >$ is the average pore size (for Knudsen diffusion $\alpha = 1$ or the average squared pore size for viscous flux $\alpha = 2$).

Therefore, for obtaining high permeability, the surface layer that governs the membrane transport must be as thinnest as possible, the transport path length through the membrane as lowest as possible, and its surface porosity, as well as pore size, must be as highest as possible.

The goal of achieving high permeability must, however, be balanced with the achievement of good LEP values. In fact, the large pores, on one hand, allow to increase the permeability and, on the other hand, reduce the resistance to wetting of the membranes. Therefore, membranologists stressed the importance of optimal pore sizes and considering other factors that can help achieve the preceding characteristics. Desired values of porosity, maximum pore size, and thickness are 80–90% [39], 0.5–0.6 μm [40], and 100–200 μm [41], respectively. An ideal tortuosity value will be close to zero.

Factors that strongly influence the structure and morphology of the resulting membranes are as follows:

- The membrane material and its concentration (using high polymer concentrations in the casting solution produces a relatively dense and thick outer layer that may not have the required porosity for high mass transfer).

- Solvent–nonsolvent affinity (if low, an asymmetric membrane with a dense non-porous top layer is obtained; if high, a porous membrane is achieved).
- Composition of the coagulation bath (e.g., increasing solvent content in the coagulation bath leads, on one side, to the reduction in membrane porosity with a thick and dense outer layer and, on the other side, to lower polymer concentration at the interface thus causing a more porous outer layer and to decrease the tendency toward macrovoid formation in the porous top layer).
- Composition of the casting solution.
- Temperatures of the casting solution and the coagulation bath.

Not to mention the gelation and crystallization behavior of the polymer, the location of the liquid–liquid demixing gap and the evaporation time (before immersion in the coagulation bath) [42].

This section deals with the research activities reported in the literature to achieve the desired characteristics in MD hollow fiber membranes in the saltwater treatment sector. Various strategies implemented to achieve these characteristics are summarized in Table 8.4.

Table 8.4 Desired characteristics of MD membranes and efforts for achieving them

Desired Characteristics	Research Efforts
High permeability	• Increase in porosity and pore size • Reduction in tortuosity and membrane thickness • MMM • Development of dual- and triple-layer hydrophobic–hydrophilic membrane
High LEP	• Production of superhydrophobic, amphiphobic, and omniphobic membranes through surface modification • Incorporation of hydrophobic nanoparticles (inorganic and carbon-based nanoparticles) in MMM

(*Continued*)

Table 8.4 (*Continued*)

Desired Characteristics	Research Efforts
High mechanical strength	• Production of superhydrophobic membranes through surface modification • Production of MMM with proper nanoparticles concentration • Production of multi-bore membrane
Low thermal conductivity	• Production of highly porous membrane • Use of low thermal conductivity materials for composite membranes production • Development of dual- and triple-layer hydrophobic–hydrophilic membrane
Low fouling rate	• Production of amphiphobic and omniphobic membranes • Improvement in the surface hydrophilicity of membranes • Development of hydrophilic–hydrophobic membranes • Incorporation of hydrophilic nanoparticles in MMM

Note: Modified from [43].

Porous hollow fiber membranes were fabricated from fluorinated polyoxadiazole and polytriazole by a dry–wet spinning method by Maab et al. [30]. The membranes (with LEP values of 2.1 and 2.0 bar and mean pore sizes of 110 nm and 140 nm for fluorinated polyoxadiazole and fluorinated polytriazole hollow fibers, respectively) were applied in desalination of Red Seawater by DCMD. Rejections of 99.95% and transmembrane fluxes equal to 35 and 41 L/m^2h were obtained, respectively, for polyoxadiazole and polytriazole hollow fiber membranes, operating at 80°C feed temperature, 20°C permeate, and utilizing a membrane module with an effective area of 25 cm^2.

The influence of polymer concentration on fiber morphology and, as a consequence, on fiber properties was analyzed by Figoli et al. [44] by preparing PVDF hollow fiber membranes for application in VMD using PVDF types with different molecular weights and different crystallinity. In particular, six polymeric dopes were prepared utilizing PVDF Solef® homopolymers, also in a blend. The polymeric dopes were chosen in order to have the same additive composition (water and poly(vinyl pyrrolidone) (PVP K-17) as additives), the same viscosity (about 7000 mPa/s) and the same solvent (N-methyl pyrrolidone), but different types and concentrations of PVDF .

The porosity of the produced PVDF hollow fibers ranged from 78.88 to ~84%, clearly evidencing typical trade-off with mechanical properties. The authors determined that polymer concentration strongly influences membrane morphology, particularly as regards the formation of macrovoids, fibers' mechanical properties, porosity and pore size. Moreover, VMD tests showed that polymer concentration affects also VMD performance, both in terms of flux (in the range from 13.93 to ~21 kg/m^2h) and rejection (between 99.98% and 99.99%).

Simone et al. [45] analyzed the influence of spinning parameters (i.e., composition and flow rate of the inner coagulant-bore fluid) on PVDF hollow fiber morphology. They proved that spinning parameters affect fiber thickness and porosity: the increase in bore fluid flow rate caused the enlargement of fiber inner diameter and the reduction in fiber thickness; the lowering of the liquid/liquid demixing rate during coagulation allowed to reduce fiber porosity, due to the reduction in macrovoids. The fabricated hollow fibers were tested in VMD (T_{feed} = 50°C, P_{vacuum} = 40 mbar; flow rate = 27 L/h), both feeding double distilled water and synthetic seawater (composition: NaCl 27.85 g/L, $MgCl_2$*\6H_2O 12.7 g/L, $CaCl_2*2H_2O$ 1.79 g/L, Na_2SO_4 4.71 g/L, $NaHCO_3$ 0.21 g/L). The fabricated fibers reached average transmembrane water vapor fluxes as high as 19.50 (pure water) and 14.46 kg/m^2h (synthetic seawater), coupled with solute separation factors of about 99.99%.

Extensive studies can be found related to the addition of chemicals (e.g., poly(vinylpyrrolidone) (PVP), calcium carbonate ($CaCO_3$), lithium chloride (LiCl), tri-hydrated lithium perchlorate $LiClO_4·3H_2O$, ethylene glycol) to the casting solution for increasing membrane surface porosity. However, the recourse to these pore-forming additives caused the formation of macrovoids or cavities that reduced the mechanical stability of the membrane.

Another aspect to consider is connected to the often-incomplete removal of pore-forming additives from the polymeric matrix. The presence of hydrophilic domains dispersed onto the membrane surface reduces its hydrophobic character. The use of hydrophilic pore former represents, therefore, a critical factor, especially if highly hydrophobic membrane surfaces are required.

Bonyadi and Chung [46] reported the fabrication of highly porous and macrovoid-free PVDF hollow fiber membranes. They used a two-phase flow, which consists of a solvent and a dope solution in the air gap region of spinning through a nonsolvent-induced phase separation process (NIPS). They achieved, in DCMD, fluxes as high as 67 kg/m^{2}h at 80°C, utilizing PVDF fibers with a macrovoid-free, sponge-like, and open-cell structure (50 μm wall thickness, bulk porosity 65%, average pore size 0.08 μm, saline water with 3.5 wt% sodium chloride (NaCl) concentration as feed).

In order to improve the mechanical stability of the PVDF membranes, a possibility is to use the PVDF-HYFLON AD superhydrophobic membranes. Cui et al. [47] prepared and tested in DCMD and VMD six different Hyflon/PVDF composite membranes. The hollow fibers were fabricated via two different techniques (i.e., NIPS and TIPS). The fabricated NIPS membranes appeared more porous than the TIPS ones. On the contrary, the latter were characterized by larger pore, higher tensile stress, lower LEPw and were thicker and less hydrophobic than those prepared via NIPS. In the case of Hyflon/PVDF membranes prepared via NIPS, AD60-N was the one with the highest flux (2.5 L/m^2 h) and rejection (99.99%) due to its larger pore size (0.127 μm), higher porosity (70.7%), LEP_w (0.425 MPa), and Φ (134.3°). Moreover, the comparison between NIPS and TIPS membranes revealed the slightly better performance of the TIPS membrane with respect to the NIPS one (flux of AD60-T in DCMD test equal to 2.6 L/m^2 h, same rejection R = 99.99%).

Tang et al. [41] suggested a wall thickness of 100–200 μm and pore diameter smaller than 0.2 μm as guidelines for achieving hollow fiber membranes with good mechanical strength. Lu et al. [48] showed that the mechanical strength can be increased with the number of bores. They proved that tri-bore hollow fiber offered better mechanical strength than the conventional single-bore fiber. To improve membrane anti-wetting properties, Teflon® AF 2400 was deposited on the top of the PVDF membrane surface. DCMD experiments showed that a superhydrophobic surface with a contact angle of 151.35° could be achieved by coating the tri-bore membrane with a 0.025 wt% Teflon® AF 2400 solution for 30 s. The resultant membrane exhibited an increment of 109% in LEP and a 21% sacrifice in flux. Long-term performance tests also confirmed that the Teflon® AF 2400 coated membrane had enhanced performance stability with an average flux of 21 kg/m^2 h (feed and permeate temperatures at 60°C and 16°C, respectively) and rejection of 99.99% at 60°C.

On the contrary, mechanical strength can be decreased by some surface modifications or additives. An example can be found in the agglomeration of nanoparticles in mixed matrix membranes (MMMs). Ravi et al. [43] estimated the optimum concentration of nanoparticles to be in the range of 0.5 to 2.0 wt%. The recourse to nanoparticles was at the forefront of nanotechnology applications in MD due to the favorable qualities imparted by nanomaterials to the membranes. In fact, the addition of nanoparticles usually rises the roughness of the surface and lowers the surface energy of the membrane, thus increasing their wetting resistance. An example can be found in the polyvinylidene fluoride (PVDF) composite hydrophobic hollow fiber membranes fabricated by Hou et al. [49] by dispersing $CaCO_3$ nanoparticles in the casting solution and using a mixture of LiCl and polyethylene glycol (PEG) as nonsolvent additive. The prepared membranes were tested in DCMD for the desalination of 3.5 wt% sodium chloride solution. The studies of Hou et al. [49] revealed that the membrane surface roughness and contact angle increased with the $CaCO_3$ nanoparticles addition. Moreover, compared with the pure PVDF membrane, the composite membranes exhibited stronger mechanical properties. During the DCMD desalination process, the maximum permeate flux of about 46.3 kg/m^{2}h was obtained with a mixing ratio for $CaCO_3$/PVDF equal to 20/100, the hot feed and cold distillation temperatures at 80.5°C and 20.0°C, the feed and permeate flow rate at 0.50 and 0.15 m/s, respectively. Furthermore, salt rejection as high as 99.99% was achieved.

SiO_2, TiO_2, ZnO, CuO, Ag, clay, and magnetite particles are other inorganic nanoparticles utilized for the preparation of MD membranes. Zhang et al. [50] incorporated hydrophobic nano-fumed SiO_2 particles into PVDF-PDMS membrane for the fabrication of hydrophobic asymmetric hollow fiber tested in VMD for the treatment of a 35 g/L NaCl solution as feed. They found that membrane hydrophobicity, mechanical property, and thermal stability improved on increasing nanoparticle concentration. Moreover, they obtained a stable flux of 44.27 LMH and a stable high salt rejection of over 99.9% under the optimum operation conditions of PVDF/ PVP/SiO mass ratio = 18/3/1.0, feed temperature = 75°C, relative vacuum = 87 kPa, and feed flow rate = 50 L/h.

Other efforts to increase the wetting resistance and antifouling properties of MD membranes focused on the fabrication of (super) hydrophobic, underwater superoleophobic, and omniphobic membrane surfaces [51]. Methods used in modifying the surfaces include coating, grafting and plasma modification. The anti-wetting and antifouling properties of superhydrophobic membranes are limited in the case of treatment of wastewater with low-surface-tension impurities (such as oil, surfactants, and amphiphilic organic contaminants). A contained kinetic barricade toward low-surface-tension substances miscible with or soluble in water (such as surfactants and alcohols) is offered by the underwater superoleophobic membrane surfaces. On the contrary, the ability of omniphobic membranes to repel water and other low-surface-tension contaminants allows to mitigate any type of foulant. These latter membranes are, therefore, particularly attractive for the treatment of produced water from shale gas production, which contains low-surface-tension species easily depositable on the membrane surface and causes severe fouling or wetting problems. An example can be found in the omniphobic PVDF hollow fiber membrane developed via silica nanoparticle deposition, followed by a Teflon AF 2400 coating, by Lu et al. [52]. The membrane DPVDF-250-T (i.e., the membrane coated with nanoparticles of diameter = 250 nm and with Teflon AF 2400) showed good repellency toward various liquids with different surface tensions and chemistries, including water, NaCl, ethylene glycol (EG), cooking oil, and ethanol. It also exhibited relatively stable flux with an average value of 16.2 kg/m^2h in 7 h VMD tests, and rejection higher than 99.5% over the entire test (with a feed solution containing up to 0.6 mM of sodium dodecyl sulfate (SDS) in a 3.5 wt% NaCl solution). Lu et al. [52] concluded that this type of membranes may provide valuable guidance to molecularly design omniphobic VMD membranes for produced water treatment.

Table 8.5 Lab-made membranes utilized in MD for desalination

Application	Liquid Feed Composition	MD Configuration	Membrane Material	Membrane Geometry	Main Membrane Properties	Refs.
Red Sea water desalination	Red Sea water	DCMD	Fluorinated polyoxadiazole	Hollow fiber	Mean pore size: 110 nm	Maab et al. [30]
Red Sea water desalination	Red Sea water	DCMD	Fluorinated polytriazole	Hollow fiber	Mean pore size: 140 nm	Maab et al. [30]
Desalination	Pure water and synthetic seawater (composition: NaCl 27.85 g/L, $MgCl_2$*\6H_2O 12.7 g/L, $CaCl_2$*2H_2O 1.79 g/L, Na_2SO_4 4.71 g/L, $NaHCO_3$ 0.21 g/L).	VMD	PVDF	Hollow fiber	Membrane code: Met30-14 Thickness: 0.27 ± 0.02 mm; Porosity: ~86%.	Simone et al. [45]
Desalination	3.5 wt% NaCl	DCMD	PVDF	Hollow fiber	Thickness: 50 μm, Bulk porosity: 65%, Average pore size: 0.08 μm	Bonyadi and Chung [46]
Desalination	Solution 0.6M NaCl	DCMD and VMD	Composite Hyflon/ PVDF	Hollow fiber	Pore size: 0.127 μm, Porosity: 70.7%, Thickness: 262.2 μm	Cui et al. [47]
Desalination	Solution 0.6M NaCl	DCMD and VMD	Composite Hyflon/ PVDF	Hollow fiber	Pore size: 0.174 μm, Porosity: 60.95%, Thickness: 332.5 μm	Cui et al. [47]

(Continued)

Table 8.5 *(Continued)*

Application	Liquid Feed Composition	MD Configuration	Membrane Material	Membrane Geometry	Main Membrane Properties	Refs.
Desalination	Saline solution comprising 3.5 wt% NaCl	DCMD	Composite 0.025 wt% Teflon® AF 2400/PVDF	Hollow fiber	Surface porosity: 7.2%, Max pore size: 262 nm	Lu et al. [48]
Desalination	3.5 wt% NaCl	DCMD	PVDF composite with $CaCO_3$ nanoparticles in the PVDF casting solution and using the mixture of LiCl and PEG as nonsolvent additive	Capillary	Mean pore diameter: 0.25 μm, Maximum pore size: 0.56 μm; Porosity: 85.3%	Hou et al. [49]
Desalination	3.5 wt% NaCl	VMD	SiO_2-PDMS-PVDF	Capillary	Thickness: 0.32 mm, LEP: 4.58 bar, Mean pore size: 100 nm, Porosity: 80.89%	Zhang et al. [50]
Desalination	0.6 mM of sodium dodecyl sulfate (SDS) in a 3.5wt% NaCl solution	VMD	PVDF with silica nanoparticle deposition followed by a Teflon AF 2400 coating	Hollow fiber	Overall porosity: 66.3±2.1, Average pore size: 226.7 ± 23.1 nm Maximum pore size: 289.7 ± 44.5 nm	Lu et al. [48]
Desalination	3.5 wt% NaCl	VMD	PVDF/Ultem® / Al_2O_3 particles	Hollow fiber	Mean pore diameter: 0.12 μm, Contact angle: 126.5°, LEP: 2.1 bar	Zuo et al. [55]

Another possibility for enhancing water flux on MD is offered by hydrophilic–hydrophobic hollow fiber membranes because the hydrophilic inner layer may facilitate water transport and condensation [53, 54]. An example can be found in the dual-layer hydrophobic/hydrophilic PVDF/Ultem® (with Al_2O_3 particles) hollow fiber membranes prepared via co-extrusion method and tested in seawater desalination via VMD by Zuo et al. [55]. They stated that, compared to the conventional single-layer MD fibers, the developed VMD hollow fiber containing 15 wt% Al_2O_3 has 4.5-fold tensile strength (i.e., about 350% increment) with a VMD flux of 45.8 kg/m^{2}h at 70°C (feed = 3.5 wt% NaCl, feed flow rate = 84 L/h; pressure of the vacuum pump = 20 mbar).

References

1. Souhaimi, M. K. and Matsuura, T. (2011) *Membrane Distillation*, Elsevier.
2. Zhang, L., Wang, Y., Cheng, L.-H., Xu, X., and Chen, H. (2012) Concentration of lignocellulosic hydrolyzates by solar membrane distillation. *Bioresour. Technol.*, **123**, 382–385.
3. Chen, J., Zhang, Y., Wang, Y., Ji, X., Zhang, L., Mi, X., and Huang, H. (2013) Removal of inhibitors from lignocellulosic hydrolyzates by vacuum membrane distillation. *Bioresour. Technol.*, **144**, 680–683.
4. Ji, Z. (2018) Treatment of heavy-metal wastewater by vacuum membrane distillation: Effect of wastewater properties. *IOP Conf. Ser. Earth Environ. Sci.*, **108**, 042019.
5. Banat, F., Al-Asheh, S., and Qtaishat, M. (2005) Treatment of waters colored with methylene blue dye by vacuum membrane distillation. *Desalination*, **174**(1), 87–96.
6. Criscuoli, A., Zhong, J., Figoli, A., Carnevale, M., Huang, R., and Drioli, E. (2008) Treatment of dye solutions by vacuum membrane distillation. *Water Res.*, **42**(20), 5031–5037.
7. Carnevale, M. C., Gnisci, E., Hilal, J., and Criscuoli, A. (2016) Direct contact and vacuum membrane distillation application for the olive mill wastewater treatment. *Sep. Purif. Technol.*, **169**, 121–127.
8. Tomaszewska, M. (2001) Recovery of hydrochloric acid from metal pickling solutions by membrane distillation. *Sep. Purif. Technol.*, **22–23** (1–2), 591–600.

9. Tomaszewska, M. and Mientka, A. (2009) Separation of HCl from HCl–H_2SO_4 solutions by membrane distillation. *Desalination*, **240**(1–3), 244–250.

10. Liu, R., Qin, Y., Li, X., and Liu, L. (2012) Concentrating aqueous hydrochloric acid by multiple-effect membrane distillation. *Front. Chem. Sci. Eng.*, **6**(3), 311–321.

11. Li, X., Qin, Y., Liu, R., Zhang, Y., and Yao, K. (2012) Study on concentration of aqueous sulfuric acid solution by multiple-effect membrane distillation. *Desalination*, **307**, 34–41.

12. He, J., Zhang, L., Zhang, K., Qin, Y., and Liu, L. (2015) Concentrating aqueous urea solution by using continuous-effect membrane distillation. *Chem. Eng. Res. Des.*, **104**, 589–604.

13. Urtiaga, A. M., Ruiz, G., and Ortiz, I. (2000) Kinetic analysis of the vacuum membrane distillation of chloroform from aqueous solutions. *J. Memb. Sci.*, **165**(1), 99–110.

14. Abejón, R., Saidani, H., Deratani, A., Richard, C., and Sánchez-Marcano, J. (2019) Concentration of 1,3-dimethyl-2-imidazolidinone in aqueous solutions by sweeping gas membrane distillation: From bench to industrial scale. *Membranes (Basel).*, **9**(12), 158.

15. Phattaranawik, J., Fane, A. G., Pasquier, A. C. S., and Bing, W. (2008) A novel membrane bioreactor based on membrane distillation. *Desalination*, **223**(1–3), 386–395.

16. Jia, F., Li, J., Wang, J., and Sun, Y. (2017) Removal of strontium ions from simulated radioactive wastewater by vacuum membrane distillation. *Ann. Nucl. Energy*, **103**, 363–368.

17. Zhao, H. and Sun, H. (2009) Study on the experiment of concentrating coking wastewaters by air-blowing vacuum membrane distillation. *Appl. Phys. Res.*, **1**(2).

18. Qu, D., Sun, D., Wang, H., and Yun, Y. (2013) Experimental study of ammonia removal from water by modified direct contact membrane distillation. *Desalination*, **326**, 135–140.

19. Shao, F., Hao, C., Ni, L., Zhang, Y., Du, R., Meng, J., Liu, Z., and Xiao, C. (2014) Experimental and theoretical research on N-methyl-2-pyrrolidone concentration by vacuum membrane distillation using polypropylene hollow fiber membrane. *J. Memb. Sci.*, **452**, 157–164.

20. Wang, Z., Gu, Z., Feng, S., and Li, Y. (2009) Application of vacuum membrane distillation to lithium bromide absorption refrigeration system. *Int. J. Refrig.*, **32**(7), 1587–1596.

21. Wu, B., Tan, X., Li, K., and Teo, W. K. (2006) Removal of 1,1,1-trichloroethane from water using a polyvinylidene fluoride hollow fiber membrane module: Vacuum membrane distillation operation. *Sep. Purif. Technol.*, **52**(2), 301–309.
22. Wu, B., Li, K., and Teo, W. K. (2007) Preparation and characterization of poly(vinylidene fluoride) hollow fiber membranes for vacuum membrane distillation. *J. Appl. Polym. Sci.*, **106**(3), 1482–1495.
23. Mokhtar, N. M., Lau, W. J., Ismail, A. F., and Veerasamy, D. (2015) Membrane distillation technology for treatment of wastewater from rubber industry in Malaysia. *Procedia CIRP*, **26**, 792–796.
24. Mokhtar, N. M., Lau, W. J., and Ismail, A. F. (2015) Dye wastewater treatment by direct contact membrane distillation using polyvinylidene fluoride hollow fiber membranes. *J. Polym. Eng.*, **35**(5), 471–479.
25. Chong, K. C., Lai, S. O., Thiam, H. S., Lee, S. S., Lau, W. J., and Mokhtar, N. M. (2016) Reactive blue dye removal by membrane distillation using PVDF membrane. *Indian J. Sci. Technol.*, **9**(S1).
26. Ge, Q., Wang, P., Wan, C., and Chung, T.-S. (2012) Polyelectrolyte-promoted forward osmosis–membrane distillation (FO–MD) hybrid process for dye wastewater treatment. *Environ. Sci. Technol.*, **46**(11), 6236–6243.
27. Liu, H. and Wang, J. (2013) Treatment of radioactive wastewater using direct contact membrane distillation. *J. Hazard. Mater.*, **261**, 307–315.
28. Krivorot, M., Kushmaro, A., Oren, Y., and Gilron, J. (2011) Factors affecting biofilm formation and biofouling in membrane distillation of seawater. *J. Memb. Sci.*, **376**(1–2), 15–24.
29. Gryta, M. (2019) The influence of talc addition on the performance of polypropylene membranes formed by TIPS method. *Membranes (Basel).*, **9**(5), 63.
30. Maab, H., Al Saadi, A., Francis, L., Livazovic, S., Ghafour, N., Amy, G. L., and Nunes, S. P. (2013) Polyazole hollow fiber membranes for direct contact membrane distillation. *Ind. Eng. Chem. Res.*, **52**(31), 10425–10429.
31. Singh, D. and Sirkar, K. K. (2017) Performance of PVDF flat membranes and hollow fibers in desalination by direct contact membrane distillation at high temperatures. *Sep. Purif. Technol.*, **187**, 264–273.
32. Macedonio, F., Ali, A., Poerio, T., El-Sayed, E., Drioli, E., and Abdel-Jawad, M. (2014) Direct contact membrane distillation for treatment of oilfield produced water. *Sep. Purif. Technol.*, **126**, 69–81.

33. Mengual, J. I., Khayet, M., and Godino, M. P. (2004) Heat and mass transfer in vacuum membrane distillation. *Int. J. Heat Mass Transf.*, **47**(4), 865–875.

34. Kim, Y.-D., Thu, K., and Choi, S.-H. (2015) Solar-assisted multi-stage vacuum membrane distillation system with heat recovery unit. *Desalination*, **367**, 161–171.

35. Al-Obaidani, S., Curcio, E., Macedonio, F., Di Profio, G., Al-Hinai, H., and Drioli, E. (2008) Potential of membrane distillation in seawater desalination: Thermal efficiency, sensitivity study and cost estimation. *J. Memb. Sci.*, **323**(1), 85–98.

36. Econity Hollow fibers. http://econity.com/en/business/products.html

37. Kim, S.-H. (2017) Seawater desalination and its brine solution. http://en.bwf.kr/data/files/B1_SeungHyun_Kim.pdf

38. Song, L., Ma, Z., Liao, X., Kosaraju, P. B., Irish, J. R., and Sirkar, K. K. (2008) Pilot plant studies of novel membranes and devices for direct contact membrane distillation-based desalination. *J. Memb. Sci.*, **323**(2), 257–270.

39. Eykens, L., De Sitter, K., Dotremont, C., Pinoy, L., and Van der Bruggen, B. (2016) How to optimize the membrane properties for membrane distillation: A review. *Ind. Eng. Chem. Res.*, **55**(35), 9333–9343.

40. Schneider, K., Hölz, W., Wollbeck, R., and Ripperger, S. (1988) Membranes and modules for transmembrane distillation. *J. Memb. Sci.*, **39**(1), 25–42.

41. Tang, Y., Li, N., Liu, A., Ding, S., Yi, C., and Liu, H. (2012) Effect of spinning conditions on the structure and performance of hydrophobic PVDF hollow fiber membranes for membrane distillation. *Desalination*, **287**, 326–339.

42. Macedonio, F., Ali, A., and Drioli, E. (2019) Nanostructured porous membranes for membrane distillation and membrane crystallization, in *Functional Nanostructured Membranes* (eds. Drioli, E., Giorno, L., and Gugliuzza, A.), Pan Stanford Publishing Pte Ltd., pp. 393–428.

43. Ravi, J., Othman, M. H. D., Matsuura, T., Ro'il Bilad, M., El-badawy, T. H., Aziz, F., Ismail, A. F., Rahman, M. A., and Jaafar, J. (2020) Polymeric membranes for desalination using membrane distillation: A review. *Desalination*, **490**, 114530.

44. Figoli, A., Simone, S., Criscuoli, A., AL-Jlil, S. A., Al Shabouna, F. S., Al-Romaih, H. S., Di Nicolò, E., Al-Harbi, O. A., and Drioli, E. (2014) Hollow fibers for seawater desalination from blends of PVDF with different

molecular weights: Morphology, properties and VMD performance. *Polymer (Guildf).*, **55**(6), 1296–1306.

45. Simone, S., Figoli, A., Criscuoli, A., Carnevale, M. C., Alfadul, S. M., Al-Romaih, H. S., Al Shabouna, F. S., Al-Harbi, O. A., and Drioli, E. (2014) Effect of selected spinning parameters on PVDF hollow fiber morphology for potential application in desalination by VMD. *Desalination*, **344**, 28–35.
46. Bonyadi, S. and Chung, T.-S. (2009) Highly porous and macrovoid-free PVDF hollow fiber membranes for membrane distillation by a solvent-dope solution co-extrusion approach. *J. Memb. Sci.*, **331**(1–2), 66–74.
47. Cui, Z., Pan, J., Wang, Z., Frappa, M., Drioli, E., and Macedonio, F. (2020) Hyflon/PVDF membranes prepared by NIPS and TIPS: Comparison in MD performance. *Sep. Purif. Technol.*, **247**, 116992.
48. Lu, K.-J., Zuo, J., and Chung, T.-S. (2016) Tri-bore PVDF hollow fibers with a super-hydrophobic coating for membrane distillation. *J. Memb. Sci.*, **514**, 165–175.
49. Hou, D., Wang, J., Sun, X., Ji, Z., and Luan, Z. (2012) Preparation and properties of PVDF composite hollow fiber membranes for desalination through direct contact membrane distillation. *J. Memb. Sci.*, **405–406**, 185–200.
50. Zhang, H., Li, B., Sun, D., Miao, X., and Gu, Y. (2018) SiO_2-PDMS-PVDF hollow fiber membrane with high flux for vacuum membrane distillation. *Desalination*, **429**, 33–43.
51. Lee, W. J., Ng, Z. C., Hubadillah, S. K., Goh, P. S., Lau, W. J., Othman, M. H. D., Ismail, A. F., and Hilal, N. (2020) Fouling mitigation in forward osmosis and membrane distillation for desalination. *Desalination*, **480**, 114338.
52. Lu, K. J., Zuo, J., Chang, J., Kuan, H. N., and Chung, T.-S. (2018) Omniphobic hollow-fiber membranes for vacuum membrane distillation. *Environ. Sci. Technol.*, **52**(7), 4472–4480.
53. Bonyadi, S. and Chung, T. S. (2007) Flux enhancement in membrane distillation by fabrication of dual layer hydrophilic–hydrophobic hollow fiber membranes. *J. Memb. Sci.*, **306**(1–2), 134–146.
54. Khayet, M., Mengual, J. I., and Matsuura, T. (2005) Porous hydrophobic/hydrophilic composite membranes. *J. Memb. Sci.*, **252**(1–2), 101–113.
55. Zuo, J., Chung, T.-S., O'Brien, G. S., and Kosar, W. (2017) Hydrophobic/hydrophilic PVDF/Ultem® dual-layer hollow fiber membranes with enhanced mechanical properties for vacuum membrane distillation. *J. Memb. Sci.*, **523**, 103–110.

Chapter 9

Hollow Fiber Membranes for Gas Separation

Elisa Esposito, Marcello Monteleone, Alessio Fuoco, Mariagiulia Longo, and Johannes C. Jansen
Institute on Membrane Technology, National Research Council of Italy (ITM-CNR), via P. Bucci, Cubo 17/C Rende (CS), Italy
johannescarolus.jansen@cnr.it

This chapter discusses the application of hollow fiber membranes in the field of gas separation. After an overview of the main gas separation processes, the process and material requirements for these applications, this chapter will focus on the membrane preparation methods that specifically yield dense membranes suitable for gas separation, as well as methods for their subsequent modification. Then, the variety of pure polymers, polymer blends, mixed-matrix membranes, and polymers with various additives will be discussed. Finally, this chapter will discuss how the materials are affected by the separation process conditions, such as gas pressure, temperature, composition, and impurities or humidity in the gas stream, and by phenomena such as physical ageing and plasticization.

Hollow Fibers and Nanofibers in Membrane Science: Preparation, Characterization, and Applications
Edited by Alberto Figoli, Mir Saeed Seyed Dorraji, and Francesco Galiano

ISBN 978-981-4968-03-4 (Hardcover), 978-1-003-25663-2 (eBook)
www.jennystanford.com

This will provide a complete overview of the main developments in this field, with emphasis on the last one to two decades, which have shown particularly strong progress in the development of new materials and emerging separation processes.

9.1 Introduction

Along with desalination, gas separation is one of the main commercial applications of polymeric membranes, which is also expected to make a large contribution to the development of sustainable processes [1]. After the breakthrough in the development of synthetic polymer membranes by the discovery of integrally skinned asymmetric membranes by Loeb and Sourirajan [2], developments in this field have been very fast, with emphasis on the design of efficient modules in the beginning, and on the continuous search for more efficient membrane materials in later years. A timeline of the most important developments is given in Fig. 9.1.

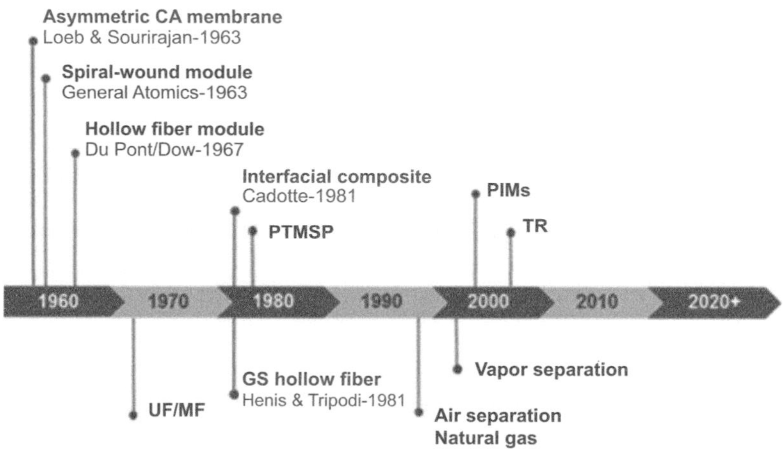

Figure 9.1 Main events marking the historical development of membranes for gas separation. Reprinted from Ref. [3] with permission of Elsevier.

In the field of gas separation, hollow fiber (HF) membranes are usually the preferred configuration because of the low viscosity of the medium, which allows the use of fibers with a very small diameter, and thus with a high surface-to-volume ratio. Compared to flat-sheet membranes, the HF configuration gives a higher productivity per unit volume, which arises from the high packing density and large specific surface. The surface-to-volume ratio is related to the fiber diameter, and a HF module of 0.04 m^3 can host over 500 m^2 of membrane surface area, much higher than the area of flat-sheet membranes in a spiral-wound or plate and frame module [4]. HF membranes furthermore have advantages in terms of low energy consumption, handling simplicity, and low maintenance [5]. More than in any other application, the material properties of membranes are important for gas separation, where the separation principle is based on an intimate contact of the permeating species and the polymer matrix at the molecular level. A good view on the necessary and possible developments in membrane materials is of prime importance, both in terms of materials properties [6, 7] and in terms of process-economical aspects [8]. The introduction of the so-called Robeson diagram in 1991, showing the permeability–selectivity trade-off [9], allowed a good comparison of the performance of novel materials. Since then, Freeman discussed the theoretical background of this trade-off [10] and analyzed the individual contributions of solubility and diffusion to the upper bound [11], while there is a continuous search for new materials to shift the upper bound to higher permeability and higher selectivity [12]. Most of the progress in the last decade is dominated by the polymers of intrinsic microporosity and thermally rearranged polymers, which have significantly improved the upper bounds for important gas pairs such as O_2/N_2 [13], CO_2/N_2, and CO_2/CH_4 [14] for pure gases or for gas mixtures [3]. Although the materials are of fundamental importance, this is not the only aspect, and for practical application, membranes must have low resistance to transport and must be as thin as possible [15]. Thin-film composite (TFC) membranes with an ultrathin and highly selective layer on top and a mechanically resistant porous substrate will play a key role to meet the future demands. This, in turn, requires advanced coating techniques of porous support membranes or, alternatively, coextrusion of the two polymers [16], and this identifies the membrane preparation, after the material selection, as the second most important aspect in the development of the next generation of polymeric HF gas separation membranes. The last aspect that is of fundamental importance for the development of next-generation gas separation membranes is the detailed knowledge of the membrane properties under the operation conditions, as a function of time, temperature, feed gas composition, and in the presence of trace contaminants or humidity.

The scope of this chapter is to give a comprehensive overview of the use of hollow fiber membranes for gas separation, the consolidated and emerging application fields, the required membrane types and preparation methods, the available materials, and their thermal, mechanical, and transport properties.

9.2 Hollow Fibers for Gas Separation Processes

The application of hollow fiber membranes on industrial scale in different gas separation processes is a valid technology to convert the traditional industry into a sustainable industry, which represents one of the most important challenges of our modern society. Hollow fibers (HFs) allow to minimize energy cost, equipment size, waste production, and, in some cases, to convert by-products into new products with innovative value in the light of an increasingly necessary circular economy [17]. Their main applications in gas separation, summarized in Table 9.1, are hydrogen recovery (H_2/CO_2) [18], oxygen-enriched air production (O_2/N_2) [19–21], biogas upgrading [22], natural gas treatment (CO_2/CH_4) [23], and post-combustion carbon capture from flue gas (CO_2/N_2) [24].

Table 9.1 Commercial HF membranes used in industrial gas separation processes

Separation	Application Field	Main Brands and Suppliers	Ref.
O_2/N_2	O_2-enriched Air production Nitrogen generation	SEPURAN® N2 (Evonik) A/G Technology-Permea (Air Products) Generon (Messer) IMS (Praxair) Medal (DuPont, Air Liquide) Aquilo (Parker Hannifin) Ube	[25]
CO_2/CH_4	Biogas upgrading Natural gas treatment	SEPURAN®Green (EVONIK)	[22]
CO_2/N_2	CO_2 from flue gas	SEPURAN® N2 (EVONIK)	

Separation	Application Field	Main Brands and Suppliers	Ref.
He/I	He separation	SEPURAN® He (EVONIK)	[26]
H_2O/Air	Gas/liquid separation Air dehydration	Permea (Air Products) Medal (Air Liquide) Air Products, Ube	[25]
H_2 separation	H_2 recovery	SEPURAN®Noble (EVONIK) Ube industry (Japan) Polysep (UOP Honeywell company) PRISM® (Air production)	[27] [28]
CO_2 separation	Acid gas treating Enhanced oil recovery Landfill gas upgrading	Cynara (Natco) Kvaerner (Grace Membrane System), Air Products, Ube	[29]

9.2.1 Air Separation

HF membranes can be used to separate air for yielding either oxygen- or nitrogen-enriched streams or the pure gases as the final products. Nitrogen with different levels of purity is required in several sectors such as aerospace, energy, chemical, and food industries. Oxygen is used by many industries in gasification processes, for optimizing hydrogen production, by filtration combustion of methane and for the oxy-combustion process, and it is also used in the healthcare sectors. The air separation process is traditionally carried out by means of cryogenic processes, but the incorporation of HF membrane unit into an air separation plant has the potential to reduce the N_2 or O_2 production cost by about 10–15% [30]. In 1987, the A/G Technology was the first company to enter the air separation market with Union Carbide's HF membranes, which were then applied in Johnson & Johnson's industrial plant. The HFs were based on polysulfone (PSf) supports, covered by polydimethylsiloxane (Sylgard®), which present a very low O_2/N_2 selectivity (PO_2 = 500 Barrer; O_2/N_2 = 2). Since then, numerous efforts have been made to increase the membrane gas transport properties to improve the achievable O_2 purity. In large-scale production, the use of membranes can be classified as follows: (1) membranes with selectivities of 3–4, combined with high permeabilities (~600 Barrer) for O_2 production with low purity, i.e., <50%; (2) membranes with selectivity of 8–10 and permeability of 10 Barrer for N_2 production with high purity; (3) membranes with very high selectivity (>30) and modest permeabilities (>50 Barrer) for O_2 production with high purity, i.e., over 90% [31]. However, despite the presence of highly O_2 permeable polymers such as PIM-1 (PO_2 = 370 Barrer; O_2/N_2 = 4) and PIM-7 (PO_2 = 190 Barrer; O_2/N_2 = 4.5), which have defined the new 2008 Robeson upper bound for these gas pairs, they still have not spread to the industrial market due to their difficulty of spinnability and problems related to the ageing phenomena. PIM-1/Matrimid HF membranes at different blend ratios were prepared to combine the high selectivity of Matrimid with the high permeability of PIM-1. The best trade-off was obtained for the combination PIM-1/Matrimid ratio of 10–90 (PO_2 = 50 Barrer; O_2/N_2 = 6), which still requires further improvement to be interesting for industrial uses [32]. Many polymer membranes are suitable for the production of low- to moderate-purity O_2 and for the production of high-purity N_2. In recent years, Evonik Industries introduced SEPURAN® N2, a new module of hollow fiber membranes for efficient separation of nitrogen from air. The HFs are based on polyimides, which are known to have high O_2/N_2 selectivity (PO_2 = 4 Barrer; O_2/N_2 = 6), which was further increased (PO_2 = 18 Barrer; O_2/N_2 = 9) for several other variants of functionalized polyimide [33]. SEPURAN® N2 HF membranes fall in the second class of HFs capable to produce N_2 with high purity, leading to a nitrogen purity up to 99% when the Air/N_2 ratio is about 3.7 and producing 7.8 Nm^3/h (7 bar, 25°C) [34]. Aquilo (Parker Hannifin) produced HF modules for nitrogen and oxygen generator systems based on Parker membranes made with polyphenylene oxide (PPO). Parker HF modules can provide nitrogen gas with a purity ranging from 95% to 99.9%. The Parker Oxy membrane modules can supply an oxygen content ranging from 28% to 40% [35].

9.2.2 Biogas Upgrading

Biogas upgrading is a strongly emerging application field of polymeric gas separation membranes [36, 37] that has already reached a large-scale industrial application for the separation of biogas from organic waste [38], while thousands of smaller units are available worldwide. Traditionally, biogas was used directly for heat or electrical power generation, but there is increased request to purify biogas to grid quality [39–41] or for vehicle fuel applications [42] where a purification step is needed to remove CO_2 and most impurities. The removal of impurities and CO_2 increases the heating value of biogas, which is proportional to the methane concentration. In CO_2/CH_4 separation, both the difference in the diffusion coefficient (related to the kinetic diameters) and the difference in the solubility (related to the critical temperature) of the two gases (CO_2 3.2 Å, T_c 304°C; CH_4 3.18 Å, T_c 190°C) are important and both are in favor of CO_2 permeation. The currently available commercial HF modules for CO_2/CH_4 separation are mostly based on PI and CA due to their good trade-off between permeability and selectivity. Rigid glassy polymers such as PSf, cellulose acetate (CA), and polyimide (PI) present a high CO_2/CH_4 selectivity (22, 30, and 42, respectively) but relatively low CO_2 permeability (5.6, 6.3, and 10.7 Barrer, respectively). On the other hand, rubbery polymers, such as Pebax® and Poly-dimethylsiloxane (PDMS), have a higher polymer chain mobility and are, therefore, more permeable to CO_2 but less CO_2/CH_4 selective [37]. For industrial applications, PI and CA are preferred over the rubbery Pebax® and PDMS due to their thermal stability and to the possibility to spin them in the form of HFs. The first company to commercialize PI-HFs was Ube industry (Japan) in 1978. They were first utilized for hydrogen recovery and only successively for CO_2 separation from different mixtures, including CO_2/CH_4 separation [27]. While CA HFs were first applied by UOP in reverse osmosis, today they are considered among the best industrial HF membranes for the CO_2/CH_4 separation, particularly for natural gas treatment [43]. Nowadays, Evonik's SEPURAN® Green PI membranes are the most used HF membranes in biogas-upgrading processes. They are applied in more than 140 industrial biogas-upgrading plants with small-, medium-, and large-scale installations having a biogas treatment capacity exceeding 1000 Nm^3/h. The SEPURAN® Green system presents a high ratio between membrane area and module volume (>1000 m^2/m^3), which results in compact systems with high productivity and good energy efficiency (0.2 $kWel/Nm^3$) of biogas upgrading [44]. A three-stage separation process in one of Europe's largest industrial biogas-upgrading plants located in Italy (Montello/

Tecno Project plant) has a capacity to treat 15,000 Nm^3/h with a biomethane production capacity of 9000 Nm^3/h and simultaneous recovery of 7000 tonnes of CO_2 per year [38].

9.2.3 Natural Gas Treatment

9.2.3.1 Natural gas sweetening: CO_2/CH_4 and CH_4/H_2S separation

Natural gas, as well as the aforementioned biogas, will have an important role in the 2050 energy strategy [45] based on the massive reduction in fossil fuels. The main difference between the two types is about CO_2 emission, which is about 35% for natural gas, while it does not exceed 1% for biogas. The HFs that dominate the current industrial CO_2 removal from natural gas are asymmetric membranes based on CA. Their H_2S removal capacity has not been systematically studied in detail and only recently, Liu et al. [46] proposed a comprehensive study on asymmetric cellulose triacetate (CTA) membranes for natural gas sweetening under realistic application conditions. Their $H_2S/CO_2/CH_4$ ternary mixture included also other contaminants such as C_2H_6, C_3H_8, and condensable hydrocarbons, e.g., toluene. They demonstrated a controlled plasticization benefit for the CTA HF membrane, with attractive CO_2 and H_2S permeance (>110 GPU) and selectivity (22–28) for CO_2 and H_2S over CH_4 at 35°C and 31 bar [46]. Advanced HF membranes for combined CO_2 and H_2S separation from CH_4 are high priorities for the membrane community and the natural gas processing industry. Recently, Liu et al. [47] crosslinked 6FDA-based polyimides, containing bulky CF_3 groups (PDMC–CF_3), in order to obtain more resistant and rigid HFs for H_2S/CH_4 separation. The crosslinked membranes exhibited H_2S permeance and H_2S/CH_4 permselectivity as high as 32 GPU and 30, respectively, at 31 bar and 35°C. The CO_2 permeance and CO_2/CH_4 permselectivity were also well maintained with values of 35 GPU and 32, respectively [47]. When the natural gas sweetening processes take place at high temperature and pressure, PI-based polymers are preferred due to their good thermal resistance. However, the polyimides can be also affected by the phenomena of plasticization and swelling during natural gas sweetening, similar to other glassy polymers [48]. Crosslinking and thermal treatment may be used to minimize the effect of plasticization and swelling of HFs. Typical HF membranes used in natural gas treatment are reported in Table 9.2.

Table 9.2 Permeance and selectivity of HFs for CO_2/CH_4 and H_2S/CH_4 separation

Polymer	CO_2 [GPU]	CO_2/CH_4 [-]	H_2S [GPU]	H_2S/CH_4 [-]	Ref.
Cellulose triacetate-HF	110	22	140	28	[46]
6FDA-based polyimides with bulky CF_3 groups (PDMC–CF_3)	35	32	32	30	[47]
Crosslinking-modified 6FDA-2,6-DAT-HFs	55	60	—	—	[49]
Copolyimide grafted with β-Cyclodextrin-HFs	130	15	—	—	[50]
PDMC-ester-crosslinkable PI-HFs	120	30*mix 50/50	—	—	[51] [52]

Cao et al. [49] fabricated 6FDA-2,6-DAT asymmetric HFs with increased resistance to plasticization by immersing them into a p-xylenediamine or m-xylenediamine/methanol solution. Koros et al. [51] fabricated ester-crosslinked HFs, in order to improve CO_2 permeance from 50 GPU to 117 GPU with a CO_2/CH_4 selectivity of 37 in testing at 14 bar with 50/50 CO_2/CH_4 feed at 35°C. Moreover, the crosslinked HFs maintain a high CO_2 permeance under highly aggressive feed pressures up to 55 bar and without CO_2 plasticization. Thermal crosslinking was reported as an efficient technique to avoid plasticization by Chen et al. [53], who studied the CO_2/CH_4 separation performance of thermally crosslinked hollow fiber membranes. The separation factor increased after crosslinking, and they exhibited excellent stability under high CO_2 pressure. In contrast, it was recently shown for cellulose triacetate-HF membranes that controlled plasticization may offer benefits in the process of natural gas sweetening [46]. A potential problem of asymmetric PI-HFs used for natural gas treatment is physical ageing, which leads to some loss of permeation productivity. Koros et al. [52] showed that for ester-crosslinked PI-HFs, the physical ageing may be essentially quenched as long as the membranes remain in contact with a high CO_2 partial pressure.

9.2.3.2 Helium separation from natural gas

Helium recovery from natural gas by means of HF membrane separation is a valid alternative to the traditional cryogenic distillation. Helium has a wide range of applications in scientific, medical, and industrial applications. Helium is used in hospitals and analytical labs as a coolant in magnetic resonance imaging and NMR analysis, in analytical and scientific equipment as a carrier gas, in welding as an inert gas, while mixtures of helium and oxygen are used in deep sea for pressurizing and purging of pressure vessels, and it is even used in rocket technology [54]. Häussinger et al. [55] developed an aromatic PI HF system with an effective area of 9000 m^2/m^3, capable to recover helium at 95% by means of several stages of HF membrane separation processes. Shoji and Moriya describe a helium purification method based on glassy polymer (polyolefin/cellulose/silicon) HF membranes. The glass HF walls have pores of 1.5 nm, which were created by the incorporation and the subsequent leaching of alkali metal ions in the wall surface. The selectivity of these HFs for helium over N_2 is about 1800–2000 [56]. Despite the variety of commercial companies producing HF membranes for helium extraction from natural gas, only a few studies in the literature are available for HF membranes for helium recovery. Dibrov et al. [57] presented an extensive study on Asymmetric Torlon® HFs for helium separation from natural gas. The HFs were treated with PDMS to plug the defects in the thin dense skin layer. The module of 200 fibers with a selective layer of 82 nm showed an active area of 0.177 m^2 with a helium permeance of about 0.1 m^3 (STP)/(m^2·h·bar), and an He/CH_4 selectivity of 340 [57]. In an integrated process design, an initial cryogenic distillation is followed by the separation of the crude helium stream by an HF system, and finally helium upgrading can occur in a pressure swing adsorption (PSA). In recent years, Linde Engineering company built in Canada the world's first plant to recover high-purity helium from N_2-rich natural gas without cryogenic stage by using only an integrated system of PSA with Evonik's SEPURAN-Noble membrane technology. SEPURAN® modules upgrade the crude gas to a helium content of about 50%. From the resulting gas mixture, almost 90% pure helium is then obtained by pressure swing adsorption [26].

9.2.4 H_2 Recovery

Hydrogen separation was the first large-scale industrial gas separation process using HF membrane technology. In 1970, Monsanto developed PSf-HFs that were installed in a synthesis gas pilot plant for ammonia production. [6] The PSf-HFs were soon overtaken by the PI asymmetric HFs technology developed by Ube industry in 1980 [27]. Much more recently, UOP Honeywell Company developed the Polysep™ HF system, which can separate H_2 with a purity up to 98% and an H_2 recovery of 95%. Polysep™ HF technology was installed in a Korean refinery for hydrogen recovery from off-gas, and it is also used to correct the H_2/CO ratio in syngas production. [58] One of the latest modules produced for H_2 recovery is the SEPURAN® Noble membrane, manufactured by Evonik®. It is a PI-HFs system with 90% H_2 recovery capacity [26]. Air products produce PRISM® membranes based on PSf-HFs, which can recover 98% of H_2. PRISM® membrane systems are employed in different industrial gas separation processes, such as ammonia purge gas recovery, oil refinery applications, CO purification, methanol purge gas recovery, and petrochemical applications [43]. In all cases, the result is a gaseous mixture containing mainly hydrogen and carbon dioxide. CO_2 removal is necessary in order to increase the hydrogen fuel power. The membranes used for this separation can be divided in two types: H_2-selective membranes and CO_2-selective membranes. Generally, H_2-selective membranes are suitable for achieving high H_2 recovery with moderate H_2 purity; instead, CO_2-selective membranes are favored for high H_2 purity combined with a moderate H_2 recovery.

H_2-selective HF membranes utilize molecular sieving mechanisms to perform efficiently H_2 separation over other bigger gases. Kumbharkar et al. [59] were the first to develop polybenzimidazole (PBI)-based asymmetric HFs for H_2/CO_2 separation at high temperatures. The PBI fibers showed an increment of up to ~8 times in H_2 permeance ($PH_2 = 2.6 \times 10^{-6}$ cm^3(STP)/cm^2·s·cmHg) and an H_2/CO_2 selectivity of about 27 at 400°C compared to its performance at 100°C. The good separation performance at elevated temperatures demonstrates the potential applicability of these HFs for high-temperature H_2 separation. Hydrogen separation from carbon monoxide was also carried out with a PI HF membrane module produced by Peer's group [60]. Permeance for hydrogen and carbon monoxide was about 70–100 GPU and 3–5.5 GPU, respectively. Hosseini et al. [61] adopted a blend HF membrane of PBI and Matrimid, in order to reduce CO_2-induced plasticization phenomena that often occur during H_2/CO_2 separation with pure Matrimid. Results indicated an H_2/CO_2 selectivity of about 11 and an H_2 permeance of 29 GPU with a great resistance to CO_2 plasticization. The H_2/CO_2 selectivity of polymeric membranes is quite low due to the competition between the kinetic factor, favoring fast H_2 diffusion, and thermodynamic factor, favoring the highly soluble CO_2 [43]. Therefore, it is not possible to build a one-stage process for hydrogen recovery from multicomponent gas mixtures, which contain CO_2. On the other hand, polymeric membranes can be very effective for gas pairs such as H_2/CO, H_2/N_2, or H_2/CH_4, having notable differences in kinetic diameters. For this reason, in H_2 separation processes, size-selective glassy polymers such as PSf, CA, and PI are preferred for the development of HFs compared to rubbery polymers [59, 61–64]. PI presents the further advantage of having a high glass transition temperature (T_g), which confers high thermal stability during thermal H_2 recovery process. Favvas et al. [63] produced Matrimid 5218 HFs, with an H_2 permeance from 20 to 52 GPU in the temperature range from 40 to 100°C, with the highest H_2/CO_2 selectivity of 37.8 and the highest H_2/CH_4 selectivity of 137 at 40°C and 60°C, respectively. Since HF-fiber membranes based on neat PBI show a very low H_2 permeability of 2.6 GPU for an H_2/CO_2 selectivity of about 27 at 100°C [59], Hosseini et al. reported HF membranes with improved H_2 separation performance at less drastic conditions (30°C and 10 bar), using a polymer blend based on PBI and Matrimid [61]. The blend HFs showed an H_2 permeance of 29 GPU with an H_2/CO_2 selectivity of 27. In recent years, the PIM-EA(Me_2)-TB polymer of intrinsic microporosity (PIM) containing Tröger's base units in the chain was investigated for the fabrication of highly H_2-selective HFs. In general, PIMs are CO_2-selective, but PIM-EA(Me_2)-TB has a remarkably strong size-sieving character, a high H_2 permeability of 7760 Barrer with interesting selectivity over O_2, CH_4, and N_2 (7, 11.1, 14.8, respectively) [65]. Bernardo et al. [66] reported thin-film composite HF membranes based on the closely related PIM-EA(H_2)-TB, reaching H_2/N_2 selectivities up to 38, depending on the coating conditions. Another strategy to increase the size selectivity of polymeric HFs, and thus the H_2/CO_2 selectivity, is to introduce porous fillers with well-defined pore size, like metal-organic frameworks (MOFs). Zhang et al. [67] synthesized ZIF-7 membranes on polyvinylidene fluoride (PVDF) HF supports, exhibiting high H_2 permeance of about 7000 GPU with an H_2/CO_2 ideal selectivity of 18.43 at 1 bar and 298 K.

CO_2-selective HF membranes, also called reverse-selective HF membranes, are generally based on rubbery polymers with chemical groups having high affinity for the quadrupolar CO_2 and, therefore, with a gas transport dominated by solubility selectivity [68]. Polymers such as poly(ethylene oxide) (PEO), PEO-crosslinked copolymers such as PEO-PI, PDMS, and Pebax® have shown the best performance in terms of trade-off between CO_2 permeability and CO_2/H_2 selectivity. However, they cannot be spun directly, and for this reason, the few available literature studies only concern composite HFs for CO_2/H_2 separation. In 2014, Chen et al. [69] synthetized for the first time composite HFs for CO_2/H_2 separation based on the Pebax/PDMS/polyacrylonitrile (PAN)-HFs. The composite HFs display attractive gas separation performance with a CO_2 permeance of 481.5 GPU and CO_2/H_2 selectivity of 8.1. Later, Hu's group improved the CO_2/H_2 solubility selectivity of Pebax® 1657 composite HFs up to 22, thanks to the loading of SAPO-34-NH_2 and ionic liquid affine to CO_2 [70]. SAPO-34/PDMS HF-MMMs were investigated for the emerging biological hydrogen production process, for which CO_2/H_2 separation does not necessitate high temperatures. The mean values of the permeability coefficients for CO_2 and H_2 were found to be 3285 and 569 Barrer, respectively; reaching a maximum CO_2/H_2 selectivity of 6.1 at 120 kPa of feed pressure [71]. Bio-hydrogen purification by HF membranes can represent a new valid alternative to the conventional method of high-temperature H_2 separation [72].

9.2.5 Flue Gas Purification

Flue gas purification mainly involves CO_2-separation processes treating the waste gas produced by the combustion of fossil fuel. The post-combustion CO_2 capture occurs at relatively low pressure and temperature, from exhaust gas streams containing mostly nitrogen (about 80 vol.%), low CO_2 concentration (about 10–25 vol.%), and small percentage of a number of pollutants, such as particulate matter (like soot), CO, NO_2, and SO_2. CO_2 capture from flue gas by means of HFs can play an important role in addressing current CO_2 emission challenges. HFs studied for flue gas treatment are reported in Table 9.3. P84® copolyimide proved to be one of the most selective glassy polymers. Pure P84® HF membranes have been developed for CO_2/N_2 separation by several authors [73–76]. In all cases, the intrinsic polymer selectivity was achieved only after defect healing by silicone rubber coating. Recently, Barsema et al. [74] fabricated defect-free ultra-thin (~56 nm) P84® asymmetric HF membranes, without the necessity of silicone coating, also demonstrating that the spinning process is reproducible at large scale of 5000 m of fibers. The fibers present the highest CO_2/N_2 selectivity of about 40 and a CO_2 permeance of 28 GPU. On the other hand, PDMS/PAN HF-composite membranes presented the highest CO_2 permeance for CO_2/N_2 separation, equal to 3700 GPU, and a selectivity of about 10. The best compromise between permeability and selectivity is for Pebax® 2533/PSf HF-composite (PCO_2 = 260 GPU, CO_2/N_2 = 32) and PIM-1 composite multilayer-HF (PCO_2 = 483 GPU, CO_2/N_2 = 22). In the separation of CO_2 from N_2, glassy polymers dominate industrial applications, mainly CA and polyimides. Liu et al. [77] successfully validated enhanced low-temperature membrane performance reported by Air Liquide, using Matrimid 5218. They reported asymmetric HFs with a CO_2/N_2 selectivity of 209 and a CO_2 permeance of 227 GPU at –50°C. Air Products commercialized polyvinylamine (PVAm)-based HF modules for CO_2/N_2 separation. These HF modules were applied in a pilot plant for post-combustion CO_2 capture in a cement factory in Norway, showing 70% CO_2 recovery in a single stage process from a flue gas containing a CO_2 concentration of 17 mol% [78]. The same HF membranes were installed in the SINTEF Tiller plant of Trondheim (Norway), for CO_2 recovery from real flue gas, containing 9.5% of CO_2 [79]. However, further efforts must be made to increase the gas transport properties of HFs for CO_2/N_2 separation.

To be interesting from an industrial point of view, and to be applied in industrial processes on a large scale, the overall CO_2 capture system must have performance that allows a CO_2 capture cost lower than $15/ton and this can be achieved only with HFs with a CO_2 permeance of 4000 GPU and a CO_2/N_2 selectivity higher than 20, or CO_2 permeance of 1000 GPU and a CO_2-N_2 of 30 [91].

Table 9.3 HFs for CO_2/N_2 separation

Polymer	CO_2 [GPU]	CO_2/N_2 [-]	T [°C]	Press. [bar]	Ref.
	Rubbery				
PDMS/PAN composite HF membranes	3700	10	25	1	[80]
PDMS/PAN composite HF, PVP grafted	2500	12	—	—	[81]
Pebax2533/PSf HF-composite	260	32	25	7	[82]
Pebax3533/PEI composite HF	10	2	25	1	[83]
poly(ethylene glycol) (PEG) HF-composite	30	50	25	2	[84]
	Glassy				
CA	13	39	25	3	[85]
PSf HF-commercialized by Airrane Co. (Daejeon, Korea)	120	26	25		[86]
Polyethersulfone (PES)					
TPSU (new pes)	85	34	25	1	[87]
Polyimide (PI)					
Matrimid	86	33	25	1	[32]
P84® (BTDA-TDI/MDI) co-polyimide	23	40	35	1	[73]
6FDA-DAM-DABA	520	24	30	2.3	[88]
BTDA-TDI/MDI co-PI (P84)	2.2	45–50	—	20	[74]
	Polymer of intrinsic microporosity (PIMs)				
PIM-1 composite multilayer-HF	483	22			[89]
Blend polymers					
PES/Matrimid	60	39	25	4	[90]
PIM-1/Matrimid	235	25	25	1	[32]

9.2.6 Facilitated Transport Membranes for Olefin/Paraffin Separation

Olefin/paraffin separation is one of the most important processes in the petrochemical industry [92]. Light olefin as ethylene is used as a precursor in the synthesis of polyethylene and several chemicals [93]. The similar dimensions and condensability of olefins and paraffins make their separation through membranes very difficult. Facilitated transport membranes represent a special class of olefin/paraffin separation membranes, usually containing an active metal ion that interacts reversibly with the double bond of the olefin, increasing considerably the solubility and consequently the performance of the separation process [94]. Silver ions are the most used in this field, even if their limited stability over time remains a problem to be solved. HF membranes have been used for olefin/paraffin separation in combination with an absorption process, where a solution with the absorbent metal ion flows through the lumen of the porous HF, acting as a contactor for the gas phase and the liquid phase. The flux and the concentration of adsorbent solution are two important parameters that affect the performance of the separation process [95]. Several systems were developed based on this approach. Aqueous solutions of silver nitrate were used as absorbent for ethane/ethylene separation in various HF membranes of PVDF [95], polyethersulfone (PES) [96], PSf [97], sulfonated poly(ether ether ketone) (SPEEK) supported on polypropylene (PP) [97], and poly(1-(trimethylsilyl)-1-propyne) (PTMSP) coated on PSf [98]. Only in the latter, the membrane is effectively dense, but in this case the thin PTMSP film separates the liquid phase from the gas phase and has a secondary role in the separation of the gases. Solutions of silver salts in ionic liquids have also given interesting results as adsorbents. A 1 M solution of $AgNO_3$ in 1-ethyl-3-methylimidazolium dicyanamide ([EMIM][DCA]) was employed in PSf-HFs where the surface was modified by the deposition of a layer of Protect Guard® (a perfluorinated acrylic copolymer) [99]. Propylene/propane separation was studied with a solution of $AgBF_4$ in the ionic liquid 1-buty-l-3-methylimidazolium tetrafluoroborate [BMIM][BF_4] and LiquiCel® microporous PP HFs Celgard® X50 [100] and PP membranes [101]. Systems, where the ion carrier is dispersed directly in the dense skin of the polymer membrane, have not been developed for HFs but only for flat-sheet membranes, where $AgBF_4$ has shown good compatibility with polymers such as Pebax®2533 [102, 103], Pebax®2533/PEI [104], poly(ethylene oxide/$Al(NO_3)_3$ [105], and poly(2-ethyl-2-oxazoline) [106]. However, these membranes do not have a stable performance over time, so HF gas–liquid membrane contactors that offer the possibility to replace periodically the adsorbent solution offer better perspectives for future applicability.

9.3 General Concepts of HF Membrane Preparation

The morphology plays a crucial role in the properties and performance of the membranes. During the preparation, all parameters that affect the membrane's morphology must be carefully controlled to obtain the desired transport properties. Membranes for gas separation are characterized by the presence of a dense skin layer, which performs the separation process. Generally, HF membranes for gas separation have an asymmetrical structure, where the dense layer can be at the inside or the outside of the fiber [107]. The thickness of the dense layer determines the productivity of the membranes, and the two parameters are inversely proportional. For a given material, the membrane integrity determines the selectivity. For this reason, the goal in the preparation of membranes is to obtain very thin defect-free dense layer. Asymmetric HF membranes can be divided into two main structures: Loeb-Sorurajan HF, in which the porous support layer that provides mechanical strength and the relatively dense surface layer that performs the separation are formed at the same time from the same material [108]. Alternatively, composite HF membranes, which consist of a microporous support layer, are coated with one (dual-layer HF) or more thin layers of a different polymer (multi-layer HF) that perform the separation (Fig. 9.2).

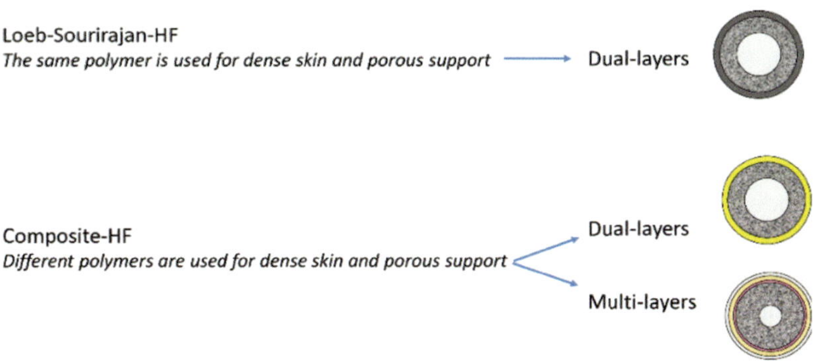

Figure 9.2 Different morphologies of HFs for gas separation.

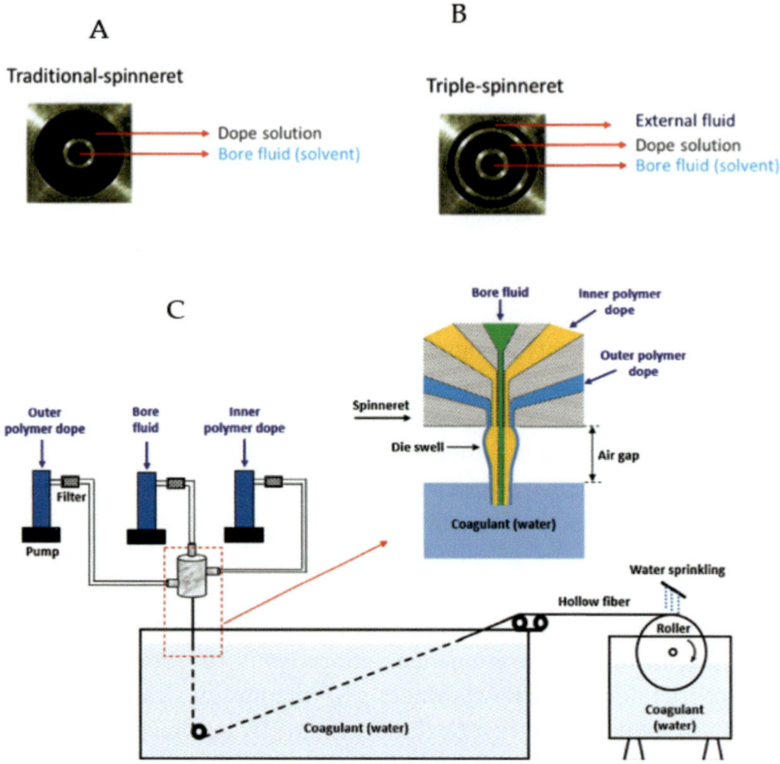

Figure 9.3 Traditional and triple-orifice spinneret with schematic dual-layer HF spinning setup. Adapted from Ref. [15] with permission of Elsevier.

9.3.1 Dual-Layer Asymmetric Hollow Fiber

Asymmetric HF membranes were made using the so-called "dry–wet" solution processing technique, which is illustrated in Fig. 9.3. The dry–wet spinning is the most important technique for the preparation of industrial HF membranes. The polymeric solution is extruded through a spinneret into the air gap ("dry") and then into the nonsolvent bath ("wet") where the polymer solution undergoes phase separation because of the exchange of solvent and nonsolvent, leading to the formation of the HF. The traditional orifice, characterized by two holes (Fig. 9.3a), offers no control of the process in the air gap, in which the HF formation occurs. In 1994, Li et al. [109] reported a new type of spinneret with three openings, which allows the simultaneous preparation of a defect-free top layer with a different material compared to that of porous support membrane. Different from the traditional dual-orifice spinneret, the triple-orifice spinneret allows the coextrusion of the polymer solution, the nonsolvent and an external fluid (Fig. 8.3b), and the formation of dual-layer HFs [15].

The outer fluid, the inner dope, and the bore fluid are fed into the spinneret separately by three syringe pumps. In the case of an outer and inner dope, sometimes these are pre-contacted before exiting the spinneret, to improve their integration. After that, the three fluids are simultaneously co-extruded through the spinneret into the coagulant bath. Finally, the dual-layer HF membrane is collected by a take-up drum (Fig. 9.3b). Naturally, the advantage to prepare TFC-HF by means of triple-orifice is that the deposition of the dense skin layer on top of the porous support takes place in a single step. The thickness of the outer layer can be controlled by adjusting the spinning parameters [110].

9.3.1.1 To create different outside coagulation conditions

The triple-orifice spinneret can be used to spin fibers in which the selective layer and the porous support are made up of two different polymers [61], but the coextrusion of an external fluid with different affinity for the polymer constituting the fiber enables to adjust the final morphology of the HF [111, 112]. Chung's group [61] reported the fabrication of Matrimid/PBI polymer blend dual-layer HF by means of dry-jet wet spinning process using a triple-orifice spinneret. The outer-layer dope was prepared from a 1:1 blend of PBI and Matrimid (weight ratio) at 22 wt.%, while the inner dope was composed of a PBI/Matrimid blend at 25 wt.%. In the resulting HFs, two distinct morphologies were obtained: a thin dense-selective layer at the outermost edge and an inner supporting layer with a cross section occupied by open cell pores and disrupted by finger-like voids. The morphological difference between the inner and outer layers can be explained by taking into account the chemistry of dope components and the coagulation agent [61]. Fang et al. reported the effect of mixed diluents during HFs preparation using a triple-orifice spinneret [111]. Using mixed diluents with different concentrations in the solvent extruded through the outermost layer of the spinneret, controlling the affinity of the solvent for diluents and for polymer, they could tailor the network morphology of the fibers (Fig. 9.4). Coextrusion of solvents that have good compatibility with the solvent in the dope solution reduces the polymer surface concentration and provides the membrane surface a larger pore size and higher porosity. Instead, the coextrusion of solvents with good compatibility with the polymer facilitates the movement of the polymer from the bulk dope solution to the surface to enhance the polymer surface concentration. In this case, the membrane surface pore size and porosity decrease, even to the point to get a dense layer suitable for gas separation.

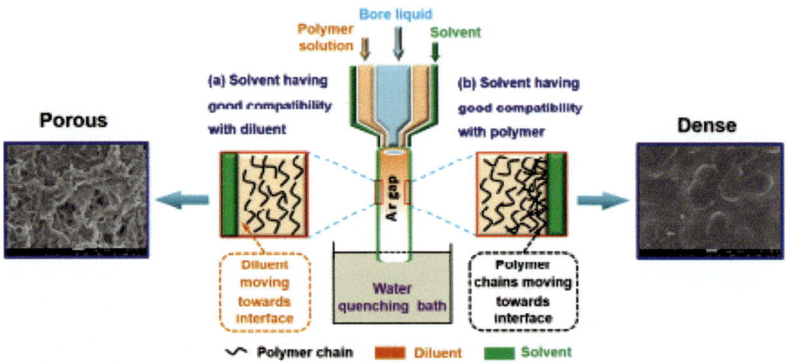

Figure 9.4 Schematic representation of different HF surface structures obtained according to the different affinity of solvents for (a) diluent or (b) polymer chains when solvents are extruded at the contact interface using a triple-orifice spinneret. Reprinted from Ref. [111] with permission of Elsevier.

9.3.1.2 Spinning to prepare TFCs in a single operation

The "dry–wet" spinning process allows the formation of integrally skinned membranes with a thickness of the dense top layer of a few hundred nanometers to a few microns. TFC-HFs are characterized by a thin selective layer with a thickness from 0.1 μm to 1 μm, supported by a porous substrate. Ultrathin composite HF membranes achieve high gas fluxes and ensure the economic viability of membrane gas separation processes. Compared to the integral asymmetric membranes prepared by the Loeb–Sourirajan technique, the principal advantages of composite membranes include fewer limitations on the material's mechanical properties, more flexible processability, and less material needed for the selective layer (0.1–2 g/m^2) [113]. This allows the preparation of high-performance HFs for gas separation using even exotic or expensive materials. Examples of TFC-HFs obtained by means of the dry–wet spinning process are reported in Table 9.4.

Table 9.4 Examples of TFC-HFs fabricated by spinning

Preparation Technique	Top-Layer Thickness [μm]	PCO_2 [GPU]	Selectivity [-]	Ref.
	Single layer			
Torlon®	0.41	0.84	CO_2/N_2 25	[114]
Matrimid® +PIM-1	0.07	243	CO_2/N_2 25	[32]
ODPA–TMPDA/DAT Copolyimide	0.27	39	CO_2/N_2 44	[115]
PES+Ag^+	0.79	38	CO_2/N_2 53	[116]
	Dual layer			
Ultem®1010+PIM-1	0.10	49	CO_2/N_2 21	[117]
TR-PBO	0.19	2500	CO_2/N_2 16	[118]

9.3.1.3 Multi-layer TFC-HF

Multi-layer thin-film composites are built by assembling different materials in a multi-layer structure: the porous support, the intermediate layer, the selective film, and, in some cases, a protective layer (Fig. 9.5). The porous support is generally made of cheap material such as PI, PSf, or PAN. It has the function to ensure mechanical strength without inducing significant resistance to the mass transport across the membrane. Therefore, the optimization of the properties and morphology of the substrate in composite membranes is as important as the formation of defect-free thin selective layer. Besides the requirements of strong mechanical strength and good thermal and chemical stability, the substrate must have a highly porous bulk structure, and the surface where the selective layer will be deposited must be smooth. The surface must have small pores with a high surface porosity and a narrow pore size distribution.

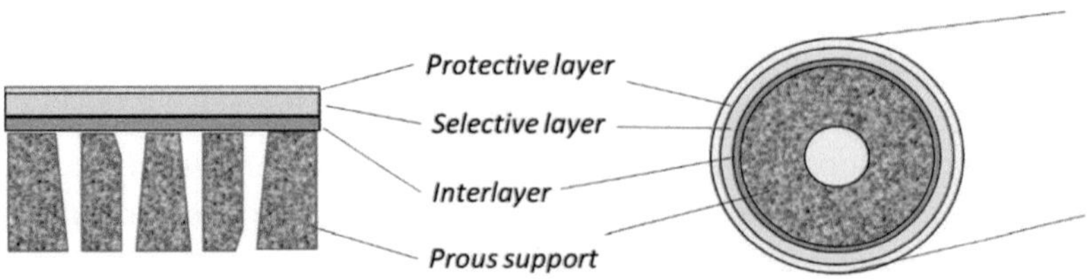

Figure 9.5 Multi-layer thin-film composite HF structure.

The intermediate layer, also called the gutter layer, has the main function of preventing the dilute solution of the selective polymer from penetrating the porous structure and blocking the pores. It may also improve the adhesion between the two surfaces even when they are made of different materials. Moreover, an intermediate layer can guarantee a smoother surface for the coating with the thin selective layer. The materials used for the gutter layer must be highly permeable polymers, such as PDMS and PTMSP. Generally, PDMS is preferred compared to the PTMSP because PDMS, being a rubber, does not lose efficiency in time, whereas PTMSP may be subject to physical ageing [119]. The selective layer is the dense layer in which the actual separation takes place and which defines the membrane efficiency. In the composite HF, it could be located in the inner surface or on the outside of the fibers (Fig. 9.5). In the first case, the dense layer will be deposited in the lumen of the fibers avoiding the negative effects such as sticking or accidental mechanical damages, which instead can occur in the case of external coating [120]. For this reason, in many cases, the external selective layer is coated with a protective layer, which can have two different functions. On the one hand, it protects the selective layer, which must be as thin as possible to guarantee a high flux and is, therefore, very delicate. In this case, the protective layer does not affect the gas transport properties of the selective layer. On the other hand, it fixes small defects that may be present in the selective layer, thus improving the membrane selectivity. TFC-HFs can be distinguished in dual-layer TFC-HFs or multi-layer TFC-HFs. Examples of TFC dual-layer-HFs are the fibers reported in 2017 by Ling et al. [21], which developed an ultrathin PDMS/PAN composite HF membrane with a PDMS thin layer of 0.23 μm having a CO_2 permeance >5000 GPU and a CO_2/N_2 selectivity of about 11, thanks to a novel post-crosslinking method of PDMS [21]. Chung's group reported an example of multi-layer TFC-HFs containing a defect-free PIM layer [89]. The selective top layer was made by the nucleophilic substitution copolymerization between the PIM and beta-cyclodextrin (β-CD), while the gutter and substrate materials were PDMS and PAN, respectively. The introduction of a crosslinked PDMS gutter layer between the PIM-β-CD selective layer and PAN substrates was the key to mitigate the negative effects that can occur during solvent evaporation in the dip-coating procedure, allowing PIM to adhere on support with a thickness ranging from 0.129 μm to 1.3 μm. The fibers have shown a pure O_2 and CO_2 permeance of 69 and 483 GPU, respectively, and O_2/N_2 and CO_2/N_2 selectivities of 3.2 and 22.5, respectively [89].

9.3.1.4 TFCs by dip coating and dynamic coating

The two most common methods for coating a thin dense-selective layer on a porous support are dip coating [113] and cross-flow filtration deposition [121]. In both cases, the film thickness and permeability can be tailored by controlling various experimental parameters, such as the concentration and the viscosity of the polymer solution, the contact time, and the number of successive coating treatments. Dip coating is the simplest method to prepare TFC-HF membranes, and it involves the immersion of the porous support into the polymer solution, lifting at a controlled speed, and subsequent evaporation of the solvent. Examples of TFC-HFs obtained by the dip-coating method are reported in Table 9.5.

Table 9.5 Examples of TFC-HFs fabricated by dip coating

HF	Thickness of the Top Layer [μm]	PCO_2 [GPU]	Selectivity [-]	Ref.
PEG	0.15	30	CO_2/N_2 50	[84]
Pebax®1657/PAN	0.5	59	CO_2/CH_4 12	[121]
Pebax®1074/PTMSP	1	970	CO_2/N_2 33	[122]
Pebax®1657/PTMSP	1	111	CO_2/N_2 92	[122]
Pebax®1657/[EmimBF4]	1	306	CO_2/N_2 36	[123]
PES/silicon	0.04	PO_2 10	O_2/N_2 6	[124]

The two most critical points in the fabrication of TFC-HFs are: (1) avoiding penetration of the selective polymer into the porous support; (2) avoiding or repairing defects in the selective layer. The first point ensures that there is limited resistance for mass transport, while the second guarantees good gas transport properties and selectivity. For this reason, the optimization of porous support morphology is as important as the formation of the thin selective layer [15].

A number of techniques have been developed to minimize or reduce the intrusion of the selective polymer into the porous support. In the pre-wetting method, the porous support is pre-wetted by a nonsolvent or by a solvent that is not miscible with that of the polymer solution, before polymer coating. Chung's group reported the fabrication of PDMS-coated PAN HFs, which demonstrated a CO_2 permeance of 3700 GPU and CO_2/N_2 selectivity of 10 [80]. Partial crosslinking of PDMS and pre-wetting of the PAN support with Fluorinert 72 or deionized water before dip coating prevent polymer intrusion. Alternatively, pre-coating of the porous substrate with a highly permeable intermediate layer, known as gutter layer, can achieve better performance than the pre-wetting procedure. This intermediate layer can effectively prevent the penetration and provide a more flat and smooth surface [125]. Fam et al. [123] reported PVDF/Pebax®1657/[Emim][BF4] thin-film composite HFs with a selective layer thickness of about 1 micron. This low thickness was obtained only when using an intermediate PTMSP gutter layer that provides a smooth coating surface for the selective layer and prevents penetration of the IL into the support pores [123]. TFC Pebax®/IL80 gel membranes did not exhibit any kind of mechanical resistance to the transport, due to the partial polymer introduction into the pores of support, showing a CO_2 permeance of 300 GPU and CO_2/N_2 and CH_4/N_2 selectivities of 36 and 15, respectively. Another technique of coating is the vacuum-assisted dip coating used by Bazzarelli et al. for the fabrication of HF multi-layer poly(styrene-butadiene-styrene) (SBS) composite membranes [126]. However, they demonstrated that standard dip-coating procedure yields lower SBS layers compared to the vacuum-assisted dip coating (thickness ca. 40 μm). Instead, the vacuum-assisted method resulted in thin top layers of about 1 μm in the case of Hylon AD® membranes on a modified poly(ether ether ketone) (PEEK) support [127].

Minimizing the formation of defects in the selective layer requires a careful study of the effect of the polymer concentration, its affinity for the polymer support, and the speed of the solvent evaporation. Furthermore, the surface of the porous support should be smooth to guarantee a homogeneous distribution of the polymer solution during the coating process. A strategy to repair defects is to coat the selective layer with a highly permeable silicone layer. Following this strategy, Li et al. [124] fabricated ultrathin HFs with an extraordinary thickness of 0.04 μm (Fig. 9.6B). They coated unselective PES HFs, which became selective after silicon dip coating and showed an O_2/N_2 selectivity of 6.0 and a permeance of 10.8 GPU (Fig. 9.6). In other cases, the silicone layer is used as a protective coating that avoids sticking or accidental mechanical damage, especially when the selective layer is deposited on the outside of the fibers.

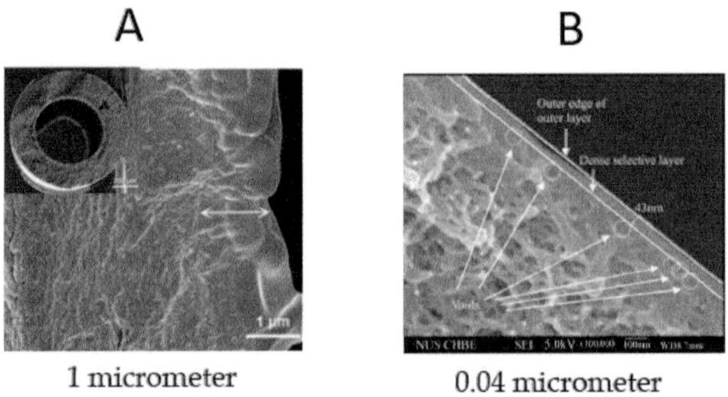

Figure 9.6 (A) Thin selective layer [123]. (B) Ultrathin selective layer, reprinted from Ref. [124] with permission of Elsevier.

In cross-flow filtration deposition, the polymer solution can be fed into the lumen side of the porous support fibers by means of a circulation pump. In this case, the external porous support acts as a protective support for the selective layer. The further advantage of cross-flow coating is that it can be simultaneously applied to a large number of fibers, when they are already mounted in a membrane module. Esposito et al. [121] demonstrated the usefulness of this dynamic coating method for the preparation of thin-composite Pebax®1657/PAN HFs for CO_2/CH_4 separation. They studied how the experimental parameters affect the thickness of the selective layer, and the gas transport properties of the resulting HFs. They reached TFC-HF membranes with a minimum thickness of 0.5 μm, a CO_2/CH_4 selectivity of about 6, and a CO_2 permeance of 48 GPU. Lasseuguette et al. [128] revealed that the thickness of the liquid film coated into a fiber is related to the coating velocity, the physical and chemical properties of the liquid, and the coating geometry. Several regimes are defined: two unstable regimes, in which a thin and regular polymer layer cannot be produced, and two stable regimes—the viscocapillary, which is of interest for the laboratory, and the high-velocity boundary layer regime, which is preferred on industrial scale [128].

9.3.2 Melt Spinning

Melt spinning is the most common process for synthetic textile fibers such as polyester and polyamide, because of the extremely high speed that can be reached, and the small fiber dimensions due to high draw ratios that can be applied. This is more difficult for HFs, which require a cylindrical instead of a spherical opening of the spinneret, with an inner tube for the bore gas. Liu et al. [129] reported the melt spinning of dense PVDF HFs, using a tube-in orifice spinneret with inner and outer diameters of 3.6 and 5.2 mm, and with air as the bore gas. With this method, they produced dense membranes, but in spite of the high drawing rate of 70 m/min, the thickness of the membranes was still 115 μm and the inner diameter was 900 μm. Much thinner membranes of approximately 40 μm are obtained from a spinneret with OD = 0.21–0.25 μm [130], but this thickness is still extremely high compared to the thickness of the dense skin that can be obtained by the nonsolvent induced phase separation (NIPS) process, approximately 0.2 μm [131]. Given the very low intrinsic permeability of PVDF, the melt-spun HFs are, therefore, not likely to be suitable for separation processes. Much thinner poly(4-methyl-1-pentene) fibers with an outer diameter in the order of 40–90 μm and a wall thickness of 8–20 μm were obtained from a smaller spinneret, using very high take-up speeds [132]. Under the appropriate conditions, the fibers maintained a high selectivity, $O_2/N_2 \approx 4$ and $CO_2/N_2 \approx 12$. Nevertheless, this thickness cannot compete with the extremely thin films that can be produced by NIPS, which remains the preferential process for dense HF membrane preparation.

9.3.3 Special Design Features

One of the main advantages of HF membranes for gas separation is their high surface-to-volume ratio compared to flat-sheet membranes, which is critical for the treatment of large fluid volumes. Special attempts have been made to further increase the active surface by using a specially designed spinneret that produces a structured inner or outer surface of a 20/80 PES/Matrimid blend membrane for O_2/N_2 separation [133]. Although this may be a valid reason for ultrafiltration [134] and for membrane distillation [135], where the fluid phase is water with a much higher viscosity than gases. If there is no significant reduction in the skin thickness of the corrugated membrane compared to the traditional cylindrical form [133], the simple reduction in the fiber dimensions with a traditional cylindrical shape is probably equally effective to increase the specific surface area of gas separation membranes.

9.4 Post-Treatment

9.4.1 Crosslinking and Chemical Modification of Membranes

Membranes in industrial separation plants often undergo high physical and chemical stresses (high pressure, presence of aggressive chemical compounds) with the risk of compromising their integrity and their performance over time. A frequent phenomenon is plasticization, which occurs especially in the presence of highly condensable gases (e.g., CO_2) at high pressure and low temperature or in the presence of organic vapors. Plasticization involves an increase in the mobility between the polymer chains, which changes the free volume and the inter-chain space, causing generally faster diffusion of all gases and loss in selectivity [136]. Crosslinking is an efficient strategy to prevent plasticization. There are several possible approaches, including chemical crosslinking (by crosslinking agents like polyamines and diols), thermal crosslinking, or UV irradiation. The formation of new chemical bonds by a crosslinking agent or by intramolecular reactions stabilizes the membranes with changes in morphology, which improves the mechanical properties and increases the selectivity for some gas pairs. Since crosslinking tends to decrease the permeability, the process must be optimized for the best compromise between stability and performance of the membranes. The crosslinking process usually requires one or two additional steps in the spinning process: the first for the treatment of fiber with the crosslinker, and if that is done inline [75], the second for the post-curing process. Polyimides are very suitable for crosslinking processes, due to the presence of imide bonds and other functional groups in their chemical structure. In chemical crosslinking, generally, the HFs are immersed for a certain period in a solution containing the crosslinker agent. P84® copolyimide HFs were crosslinked by an aqueous solution of 1,5-diamino-2-methylpentane (DAMP) directly in the spinning process, before the take-up roll, obtaining an improvement in mechanical strength and a threefold increase in selectivity for H_2/CO_2 gas pair (from 5.3 to 16.1) [75]. Another way of in line crosslinking for P84® was proposed by Kopéc et al. [137] where the chemical crosslinking was performed in the spinning process by poly(ethylene imine) dissolved in the bore fluid. The obtained fibers exhibit a CO_2/N_2 mixed gas selectivity around 12. Chemical crosslinking by the immersion of the HF in an m-xylenediamine/methanol solution has been used for 6FDA-2,6-DAT PI [138] and for PI/PES dual-Layer HFs [139]. In both cases, better resistance against CO_2-induced plasticization was obtained, and a higher CO_2/N_2 and CO_2/CH_4 selectivity. A solution of N-(1-naphthyl)ethylenediamine dihydrochloride (NED) in methanol was used to crosslink commercial PI-HFs, resulting in an increase in the CO_2/CH_4 selectivity from 72 to 82–96, depending on the degree of crosslinking [140]. Thermal treatment is another easy way of crosslinking where the time and temperature of the treatment must be optimized for each type of material. Thermally induced decarboxylation of 6FDA-DAM:DABA (3:2) PI for 2 h at 350°C gives a high resistance to CO_2 plasticization in CO_2/CH_4 separation [141]. Koros et al. [142] proposed the thermal crosslinking of mono-esterified 6FDA-DAM:DABA with 1,3-propanediol at 200°C for 2 h, reducing plasticization and maintaining good CO_2/CH_4 separation performance up to 800 psi. Fibers of Matrimid® for CO_2 removal were thermally annealed by Dong et al. [143], exhibiting the best performance in terms of permeance and resistance to plasticization after treatment at 250°C for 30 min. Propanediol mono-esterified polyimide(PDMC)/Torlon composite HFs were crosslinked at 200°C for 2 h. They showed an improvement in the CO_2/CH_4 selectivity from 28 to 39, and they maintained a good performance also in the presence of up to 1000 ppm of toluene as a contaminant [51]. Crosslinking can also suppress the physical ageing: PDMC PI treated at 200°C for 2 h under vacuum showed just a minimum decrease in CO_2 permeability of only 3% during a continuous permeation test with a CO_2/CH_4 mixture for 400 h [52]. PDMS is usually provided as a viscous liquid or waxy pre-polymer and must be crosslinked with a crosslinking agent or by humidity to form the rubber film. To guarantee a better coating of PAN HF supports, PDMS was crosslinked in solution by thermal treatment before the coating. The resulting high-flux composite HFs presented a permeance of 10,000 and 11,000 GPU for C_3H_8 and C_3H_6, respectively, while the selectivity values for the C_3H_8/N_2 and C_3H_6/N_2 gas pairs were about 21 and 24, respectively [144].

Fluorination was reported to increase significantly the He/CH_4, CO_2/CH_4, and He/N_2 selectivity of Matrimid 5218 HF membranes [145].

9.4.2 Thermal Rearrangement: TR Polymers

Thermally rearranged (TR) polymers form a relatively new class of materials, which have been extensively investigated for the preparation of membranes for gas separation due to their good separation performance [146, 147]. Up to date, their properties are close to the most recent 2019 upper bounds for CO_2 separation [14], but they are not included in the UB definition since they are not solution-processable polymers [12]. TR polymers are generally prepared starting from a PI precursor with an ortho-functional group, which undergoes a thermal rearrangement. Before the thermal treatment, the precursor can be easily processed in the HF structure as normal polymers. High temperatures are usually required for the thermal treatment, e.g., higher than 300°C, which is close to the T_g of the TR-able polymers [148]. At this high temperature, the polymer chains gain in mobility [149], which causes a collapse of the substructure of the HF and may lead to a decrease in the membrane permeance [138, 150]. This effect, called *densification*, has to be seriously taken into account during the scale-up from dense thick films to asymmetric HFs because it may lead to the opposite effect in both membrane types. For instance, Jo et al. [151] prepared dense self-standing membranes of TR- poly(benzoxazole-co-imide) (PBOI), and both the degree of conversion and the permeability increased with increasing thermal rearrangement temperature. Instead, Woo et al. [138] showed that it is not necessarily true for HFs that a higher TR temperature corresponds a higher permeance, because it also leads to a thicker selective layer, which suppresses the gas transport. They found that the best TR temperature that guarantees the highest permeance is just below the temperature at which the polymer chain relaxation takes place, thus just before the densification [138]. For a like-to-like comparison, a TR-PBOI-based HF membrane treated at 375°C has an active layer of 0.2 μm and a CO_2 permeance of 560 GPU, while the HF treated at 450°C has a more than 10 times thicker selective layer and a CO_2 permeance of 97 GPU, even if the latter has a higher intrinsic permeability.

9.4.3 Pyrolysis: Carbon Molecular Sieve Membranes

Carbon molecular sieve (CMS) membranes are an emerging alternative to polymeric membranes. Although they are strictly no polymer membranes, they are briefly treated here because their main application is gas separation, and they are prepared from polymer membrane precursors. This class of membranes has high porosity with slit-like pore structure, which makes them highly permeable with interesting selectivity for industrial applications in gas separation [152]. However, poor mechanical properties and high production costs are the main issues limiting their practical use. The preparation procedure requires the pyrolysis of the polymeric precursor, and the membrane properties can be tailored by the optimization of the process conditions (temperature, time, atmosphere of the pyrolysis). CMS membranes derived from two 6FDA-polymer precursors showed good performance for olefin/paraffin separation, favoring the permeation of the olefins with respect to the paraffin [152]. Koros et al. [153] have evaluated the pyrolysis of the coated porous HFs with 6FDA:BPDA-DAM PI for CO_2/CH_4 separation, highlighting that the increase in the pyrolysis temperature from 550°C to 800°C increases the selectivity from

10 to 87.6, with a final CO_2 permeance of 236 GPU. Another study confirmed the relationship between the pyrolysis temperature and the final selectivity for CMS membranes obtained from Matrimid® and 6FDA/BPDADAM, and further demonstrated a good stability at high pressures (up to 1000 psia) of [154]. P84 copolyimide fibers were used as precursor in the preparation of CMS by varying the time of isothermal treatment and using argon as an inert gas [155]. The resulting fibers were highly size selective and suitable for the separation of He/CH_4, H_2/CH_4, He/N_2, and H_2/N_2 gas pairs, achieving selectivities of 2925, 5500, 350, and 600, respectively. Very interesting results were also obtained with carbon molecular sieve HF membranes prepared from PIM-1 [8]. These membranes were prepared in two steps: (1) coating of PIM-1 on a support made from PVDF and (2) subsequent pyrolysis. The PIM-1–based CMS membranes had a performance above the upper bound for O_2/N_2, CO_2/CH_4, He/N_2, and C_2H_4/C_2H_6. The Robeson upper bounds for CO_2/CH_4 and O_2/N_2 were also surpassed by CMS HFs made from blends of cellulose/1-ethyl-3-methylimidazolium acetate, and their performances were affected both from the spinning conditions and from the pyrolysis procedure [156].

9.5 Membrane Materials and Properties

More than in most other membrane processes, for gas separation the membrane material rather than the membrane morphology is the critical factor that determines its separation performance. The selection criteria for successful gas separation membranes [6, 8] and the systematic comparison of different materials [157, 158] have, therefore, been the topic of various studies. The first choice to make is that between glassy or rubbery polymers, and between temperature-resistant, pressure-resistant, and/or plasticization-resistant membranes, depending on the separation needs and the expected process conditions [159].

9.5.1 Rubbery Polymer Membranes

Rubbery membranes have a glass transition temperature below their operation temperature, and the high chain mobility facilitates gas diffusion. Therefore, they are generally more permeable than glassy polymers, while their separation performance relies to a large extent on solubility selectivity because they lack in the rigidity needed for size-selective transport. A consequence of their flexibility is that rubbery HF membranes are generally not self-supported: They would collapse if pressurized from the lumen side, and they would blow up if pressurized from the bore side. Therefore, rubbery HF membranes of poly(ether block amide) [82], silicones [160], or polymer/ionic liquid blends [161] are used in the form of thin-film composites, with a thin dense layer of the rubber on a suitable mechanically resistant porous support at the lumen or in the bore of the HF [69, 121]. The separation mechanism based on the solution-diffusion mechanism, with a dominant role for solubility selectivity, makes rubbery membranes especially effective for vapor permeation, and for instance applications like removal of VOCs from air [162].

9.5.2 Glassy Polymer Membranes

Glassy polymers are by far the most used polymers in HF gas separation membranes, since they offer a wide range of permeabilities and sufficient mechanical strength to produce robust membranes. The material is responsible for the transport properties, but it also determines the membrane preparation method that can be used, as described in Section 9.2. An enormous number of materials has been studied, although commercial membranes [158] are based on relatively few different polymers. Some of the most common polymers studied as HF gas separation membranes are PI [141, 163], polyetherimide (PEI), PES [87], and PSf [164, 165]. These can all be conveniently produced by the NIPS method because of their good solubility in a number of organic solvents. Polyaramide HF membranes have superior mechanical properties and temperature resistance [166], but the polymer is soluble in very few solvents that can be used in the NIPS process. Also PBI has superior thermal properties, allowing high-temperature separations [59, 167]. Homogeneous poly(4-methyl-1-pentene) membranes were prepared by melt spinning [132].

The relatively new class of PIM has received a lot of attention as gas separation membrane materials since their introduction in 2004

[168]. Due to their high cost and often brittle nature, neat PIM HFs are rarely reported [169] and usually they are studied as TFCs [89] and/or as blends with other polymers [32, 170]. Both approaches have the advantage that much less PIM is needed.

Glassy perfluoropolymers form another class of high-free volume membrane materials with unique properties, such as extraordinary chemical inertness and high resistance to common organic solvents. Due to their high cost and low mechanical resistance, they are best used in the form of thin-film composite membranes. Hyflon® AD was supported on PAN and PEEK HF ultrafiltration membrane supports [127] and Teflon® AF2400 was supported on commercial PP and sulfonated PES membranes [128]. Thanks to their hydrophobic nature, they can be used as a coating to improve the water resistance of composite membranes as well [171].

9.5.3 Polymer Blend Membranes

Polymer blends can be economically profitable and suitable for the preparation of HF, with a superior combination of permeability and thermophysical properties compared to the individual neat polymers. Different studies reported the preparation and characterization of HF gas separation membranes based on polymer blends, both as single-layer [90, 172–174] and as dual-layer [61] HFs.

Chung's group was among the first to fabricate single-layer polymer blend HFs from PBI/PEI blends [172]. They studied the effect of the PBI/PEI blend composition on the gas transport properties for H_2, O_2, and N_2. Unfortunately, the initial HF membranes presented several structural imperfections (Fig, 9.7), with thick and tight substructure characterized by the presence of pores. Since then, numerous efforts have been realized to obtain defect-free blend HF membranes. Kapantaidakis and Koops studied the preparation of HF gas separation membranes based on PES and Matrimid 5218 PI blends, using three different blend compositions (PES/PI: 80/20, 50/50, and 20/80 wt.%) [90]. Macrovoids appeared only at the highest PES concentration (80 wt.%), and pinholes were healed using a silicone rubber coating to yield HFs with a CO_2 permeance varying from 31 to 60 GPU and a CO_2/N_2 selectivity of about 40 [90, 175].

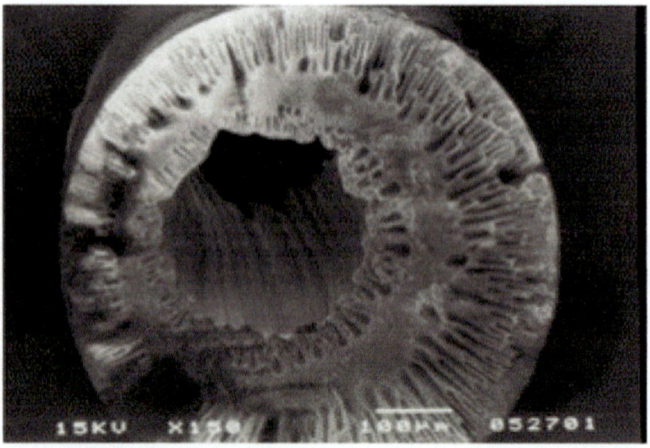

Figure 9.7 Single-layer HF from a PBI/PEI polymer blend. Reprinted from Ref. [172] with permission of Elsevier.

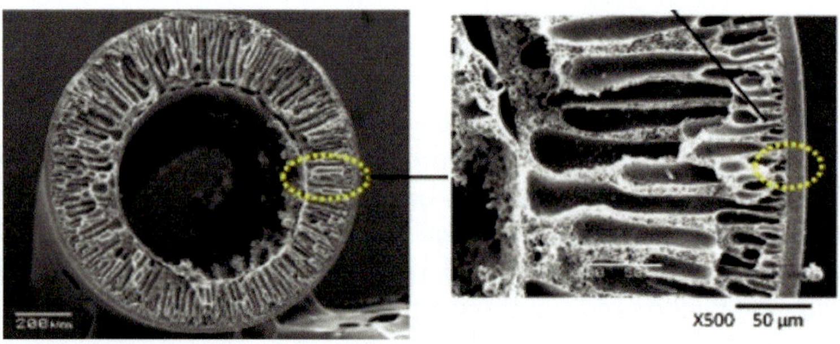

Figure 9.8 Dual-layer HF of PBI/Matrimid polymer blend. Reprinted from Ref. [61] with permission of Elsevier.

HFs based on the PES/PI blend showed a better resistance to ageing compared to neat PI-HFs. This is ascribed to the intrinsic strength of pure PES and is probably also due to specific chemical interactions between the two polymers [176]. One of the first studies on the application of miscible polymer blends for the fabrication of dual-layer HF was conducted by Hosseini et al. [61]. They developed PBI/Matrimid dual-layer HFs (Fig. 9.8) with a CO_2/CH_4 selectivity up to 42 (PCO_2 = 4.81 GPU). The gas separation performance of the HF membranes was found to be highly influenced by the type of spinning (i.e., wet-spun/dry-jet) and to the air-gap distance, with an increase in CO_2/CH_4 selectivity with decreasing air gap. PIM-1/Matrimid blend HFs were reported by Yong et al. [32] with the aim to combine the high permeability of PIM-1 with the high selectivity of Matrimid. The newly developed membranes showed these synergistic characteristics of higher selectivity compared to PIM-1 and higher permeability compared to Matrimid. The incorporation of only 5 and 10 wt.% of PIM-1 in Matrimid increased the CO_2 permeance from 83 GPU to about 200 GPU without compromising the CO_2/CH_4 selectivity very much.

9.5.4 Mixed-Matrix Membranes

Mixed-matrix HF membranes (MMM-HFs) have great potential for energy-efficient gas separation processes [177, 178]. As with the polymer blends above, an appropriate combination of different materials can give novel HFs with improved gas transport properties compared to the neat materials. The incorporation of fillers in the dense-selective layer of HFs can improve the gas transport performance by three different effects. First, the filler can selectively promote the diffusion and, as a consequence, the permeability of smaller gases compared to bigger ones; in this case, the selectivity will be more influenced by diffusivity selectivity. Second, the fillers can have a higher affinity for specific gases, and in this case, the permeability will increase due to enhanced solubility, and the selectivity will be more affected by the solubility selectivity. Finally, the presence of fillers in glassy polymers, prone to ageing phenomena, could influence the local polymer dynamics, mitigating the chain motion and the gradual loss of fractional free volume, which is characteristic for physical ageing. Typical fillers that offer attractive gas transport properties are zeolites, CMS, carbon nanotubes (CNTs), MOFs, covalent organic frameworks (COFs) or organic cages, graphite, and non-porous silica. The pioneers in the fabrication of successful MMM-HFs for gas separation are Husain and Koros [179]. Before them, only Jiang et al. have reported the formation of zeolite–polymer MMM-HFs, which were characterized by Knudsen gas permeation mechanism [180].

If the material selection is inappropriate, the following three undesirable morphologies at the polymer–fillers interface can be formed [182]: sieve in a cage, matrix rigidification, and pore blocking, which have negative effects on the gas transport properties of HF membranes. The sieve-in-a-cage morphology is characterized by the presence of voids at the interface, which provide a preferential path for gas passage, causing much higher permeability compared to that of neat polymer, while the selectivity can be higher, nearly equal, or lower than neat polymer, depending on the void size. The role of the filler material is mostly cancelled by the presence of these voids. In the case of "matrix rigidification," the rigidified polymer layer at the sieve/polymer interface reduces permeation with an insignificant enhancement in the selectivity (Fig. 9.9). During the preparation

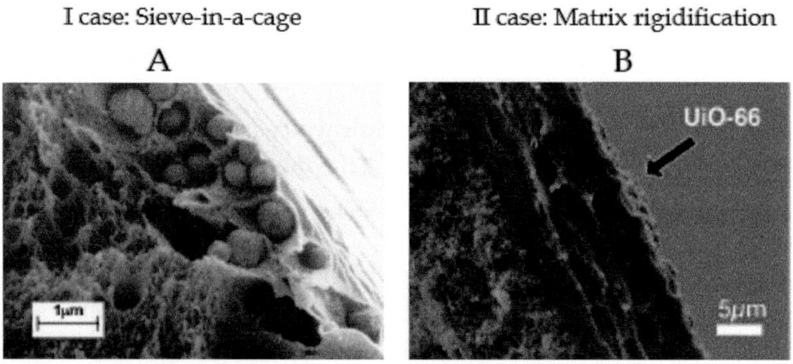

Figure 9.9 Nonideal interface in (A) Ultem/HSSZ-13 MMM-HFs [179] and (B) UiO-66/Pebax MMM-HFs. Reprinted from Ref. [181] with permission of Elsevier.

of asymmetric HF membranes, the solvent or the nonsolvent can be adsorbed in the cage structure of the fillers, which is one of the mechanisms of "pore blocking." Therefore, the filler usually needs to be activated. Attempts to solve this issue were made by Husain and Koros [179]. They modified zeolite surface by means of Grignard reagent, making the cages of zeolite hydrophobic and impermeable to coagulant water. Defect-free MMM-HFs based on HSSZ-13 zeolite/polyetherimide showed an improvement of 17% for CO_2/CH_4 and 10% for O_2/N_2 selectivity compared to the neat PEI HF membranes. A second modification of the zeolite was made by replacing the hydroxyl groups of zeolite surface with methyl groups in order to make the zeolite hydrophobic and to minimize zeolite–solvent/

nonsolvent interaction and avoid the "pore blocking" in Ultem/HSSZ-13 HF-MMMs [179]. The chemical modification of the zeolite surface is used to increase the zeolite compatibility for the polymer phase [183–185]. In this view, a new class of fillers was fabricated, called zeolitic imidazolate frameworks (ZIFs), which are expected to be more compatible with the polymer phase, due to their partially organic nature. Typical MMM-HFs are reported in Table 9.6.

- **Metal-organic frameworks (MOFs)** form a relatively new class of microporous materials, which can overcome the limited compatibility of zeolite-based MMM-HFs. MOFs are formed by the self-assembly of transition metal ions or metal clusters linked by bridging organic ligands. They present many attractive characteristics, such as adjustable chemical functionality, pore shape and connectivity, high thermal stability, low density, ordered structure, high porosity, and tunable pore structure. Moreover, the CO_2 storage capacity in some MOFs is remarkably higher compared to the traditional zeolites and other types of fillers materials.
- **Zeolitic imidazolate frameworks (ZIFs)** are a class of MOFs with tetrahedral networks that resemble the structure of conventional zeolites. Dai et al. provided a good starting point for the growing area of ZIF/polymer hybrid HF membranes. They produced dual-layer asymmetric HF by dispersing ZIF-8 in Ultem® polymer with increased CO_2 permeance (PCO_2 = 18 GPU) and selectivity (CO_2/N_2 = 44) [117]. **Copper-based MOFs** form a class of MOFs in which the metal or the cluster of metals are based on copper, such as $Cu_3(BTC)_2$, also known as HKUST (named after the Hong Kong University of Science and Technology). They offer an exceptional CO_2 uptake due to their high affinity with polar molecules. The presence of unsaturated open metal sites in Cu-based MOFs after activation is reported as an assisted mechanism in CO_2 sorption. Hu et al. [186] reported the dispersion of $Cu_3(BTC)_2$ MOF in PI-HFs. At a loading of 6 wt.% $Cu_3(BTC)_2$, the permeance of H_2 increased by 45%, and the ideal selectivity increased by a factor of 2–3 compared to the corresponding values for pure PI.

Table 9.6 Typical mixed-matrix HFs membranes (MMM-HFs)

ZEOLITE	HF	Separation Task	Ref.
	Ultem®1000-PEI/HSSZ-13	O_2/N_2 He/N_2 CO_2/CH_4 CO_2/CH_4 mixed	[179]
	PES/zeolite 4A	O_2/N_2 CO_2/CH_4	[183]
	PVDF/ZSM-5	CO_2 gas–liquid absorption	[184]
	PTMSP/zeoliteA	CO_2/N_2	[185]
MOF	Zeolitic imidazolate frameworks (ZIFs)		
	Ultem®/ZIF-8	CO_2/N_2	[117]
	PI/ZIF	hydrocarbon	[187]
	6FDA-DAM/ZIF-8	olefin/paraffin	[188]
	Pebax-1657/ZIF-7	CO_2/N_2	[181]
	Copper-based MOF		
	PI/ $Cu_3(BTC)_2$	H_2/CO_2 H_2/CH_4 H_2/N_2 H_2/O_2	[186]
	Material Institute Lavoisier (MILs)		
	Ultem®1000/MIL-53	O_2 permeance CO_2 permeance	[189]
	Ultem®1000/S-MIL-53	CO_2/N_2	[190, 191]
	CA/NH_2-MIL-53(Al)	CO_2/CH_4	[192, 193]
	Zirconium 1,4-Dicarboxybenzene (UiO-66)		
	Pebax/UiO-66	CO_2/N_2	[181]

- **MILs**, named after the Material Institute Lavoisier, form a subfamily of MOFs based on trivalent metals strongly bonded to oxygen-anion-terminated linkers. MIL-53 is an example of a "breathing" MOF, with a one-dimensional diamond-shaped pore diameter of 8.6Å. Zuh et al. [189] were the first to incorporate the flexible MIL-53 into Ultem®1000 to fabricate asymmetric mixed-matrix HF membranes. Ultem® loaded with 5 wt.% MIL-53 exhibited an O_2 and CO_2 permeance of 8.1 GPU and 27.9 GPU, respectively. This corresponds to an increase of 138% and 129%, respectively, compared to the pure Ultem® membrane, with almost unchanged selectivity for the O_2/N_2 and CO_2/CH_4 gas pairs. The affinity between MOF and polymer must be good to obtain defect-free HF-MMMs with improved gas separation performance. One strategy is to functionalize the MOF with chemical groups that increase the interface affinity. Zuh et al. post-functionalized MIL-53 with aminosilane before its dispersion in Ultem®, demonstrating both an increase in permeance, and an enhanced CO_2/N_2 selectivity, which increased from 26 for pure Ultem® HF to 36 for the MMMs-HF loaded with 15 wt.% of the aminosilane-functionalized MIL-53 [190, 191]. This behavior was ascribed to the higher affinity of S atoms for CO_2 and to the presence of grafted alkyl chains, which endowed better particle dispersion in Ultem® matrix due to the better polymer/filler interfacial affinity [190, 191]. Moreover, the S-MIL-53/Ultem® HF showed a long-time operation stability, which is crucial for HF application on industrial scale. In another work, MIL-53 was functionalized with amino group in order to increase CA polymer/filler interfacial affinity and to obtain defect-free MMMs-HF. The NH_2-MIL-53(Al)/CA HF revealed an increased CO_2/CH_4 separation factor until 14.4 by increasing the feed pressure from 3 bar to 15 bar. This phenomenon is ascribed to the breathing capacity of MIL-53, which at high pressure accelerated the CO_2 transport mechanism. It must be noted that this behavior is opposite to the typical intrinsic property of polymeric membranes. For this reason, this type of HF-MMMs are noteworthy because they can play an important role in natural gas and biogas upgrading, where the retentate has to be kept pressurized in these applications [194].

- **Zirconium 1,4-Dicarboxybenzene-based MOFs (UiO-66)** form a class of MOF in which zirconium oxide is linked by 1,4-benzendicarboxylate as a bridging ligand with an average size of 60–80 nm. This kind of MOF was dispersed in HF membranes by Sutrisna et al., and they studied how the different functional groups (amino, carboxylic, or neat-MOF) can influence the surface affinity for the Pebax selective layer [181]. In all cases, the UiO-66 increased both the CO_2 permeance and the selectivity. The highest CO_2/N_2 selectivity, i.e., 57, and a CO_2 permeance of 338 GPU, were obtained for –COOH functionalized UiO-66. Moreover, the addition of UiO-66 rigidifies the Pebax polymer chains via hydrogen bonds due to the good interfacial compatibility between the MOF and Pebax, which in this case enhances the plasticization and compaction resistance of the composite HF-MMMs.

9.5.4.1 Issues on the fabrication of HF-MMMs

For economically feasible fabrication of HF-MMMs, the fillers must not be distributed across the entire thickness of the HF, but only in the selective layer. In single-layer HF-MMMs, the nanofillers often tend to position themselves through the entire thickness of the membrane and preferentially in the middle region [195]. This gives little improvement in the gas transport properties, because of the low filler concentration in the selective skin layer [180]. Instead, in dual-layer and multi-layer HFs, the fillers are embedded mainly in the thin selective layer, allowing better interaction of the gases with the fillers, and exploiting the specific filler properties (Fig. 9.10) [1]. Jiang et al. have studied the distribution of nano-sized zeolite in HF-MMMs, demonstrating the significant advantages of dual-layer HF-MMMs compared to the single-layer HFs [195]. The preparation of thin-film composite HF-MMMs is another strategy to embed the fillers only in the selective layer of the HFs [185–190]. Zhang et al. [196] reported HF-MMMs with improved CO_2 permeance up to 413 GPU and without compromising the CO_2/N_2 selectivity (43.2), thanks to the deposition of the thin selective layer of graphene oxide/Pebax composite layer on a pre-existing porous PVDF support. In this case, it is important to prevent the infiltration of the polymer solution into the pores of the support [121]. Sutrisna et al. fabricated composite multi-layer HF-MMMs based on the coating of a PTMSP gutter layer with a ZIF-8/Pebax-1657 top layer [181]. They demonstrated that the PTMSP gutter layer prevents the penetration of the polymer and the fillers into the pores of the PVDF support, improving the CO_2 permeance from 200 to 350 GPU.

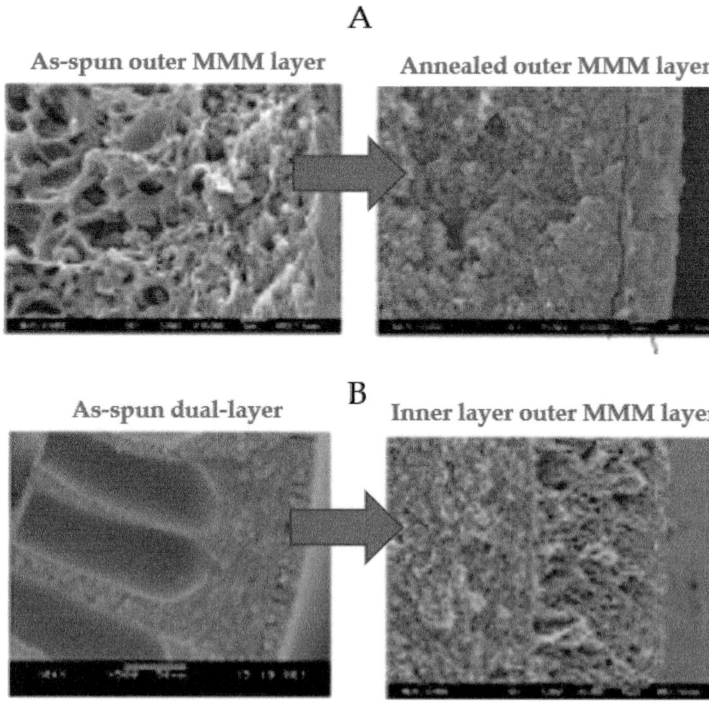

Figure 9.10 Typical distributions of fillers in a dual-layer HF-MMMs (A) and in a multi-layer HF-MMS (B) [1].

9.6 Hollow Fiber Properties and Characterization

9.6.1 Mechanical Properties

Good mechanical stability is one of the key factors needed for the successful industrial application of HF membranes. For gas separation applications, membranes are used in the pressure range of 10–15 bar for biogas separation [22] and up to 50 bar and higher pressures for other separation of industrial interest such as O_2/N_2 separation. The mechanical properties of the polymer are also important for the gas separation performance, where the rigidity of the polymer strongly affects the transport resistance. In a study involving a family of polyimides, Hirayama et al. [197]

show a clear correlation between the diffusion coefficient and the storage modulus of the polymer, which represents an index of the segmental mobility. A higher modulus leads to lower diffusion coefficients, which is further correlated with the cohesive energy density [198]. A similar trend was recently observed for various light gases in a series of PIMs [199], in which the high size selectivity is ascribed to their high rigidity. Glassy polymers often show a further increase in the rigidity in time [200, 201]. This phenomenon, known as *physical ageing*, is a consequence of the thermodynamic nonequilibrium nature of the glassy polymer. It is associated with a gradual decrease in the fractional free volume of the polymer, which leads to a reduction in the chain mobility, a rearrangement of the molecular structure. The mechanical strength of a polymer membrane strongly depends on the type of polymer, but also on the membrane morphology. A sponge-like morphology without macrovoids is needed for membranes that have to withstand high pressures [202] because macrovoids represent weak spots, where the damage of the thin skin layer is more likely to happen under compression. Controlling the macrovoids is useful to develop mechanically more stable membranes [203]. In the case of MMMs, the mechanical properties are influenced by the presence of one or more fillers in the polymer matrix. The addition of nanoparticles into a polymer matrix often leads to a decrease in the elongation at break point, an enhancement in terms of membrane rigidity [204], and in extreme cases to highly brittle materials. Ionic liquids (ILs) are also used to modulate the transport parameters, but at the same time, they also affect the membranes' mechanical behavior. The IL leads to a decrease in the interactions between the polymer chains and reduces the number of entanglements, resulting in a reduction in its rigidity and a consequent increase in the maximum elongation [205, 206]. The mechanical stability is one of the main issues to solve also for a full exploitation of microporous materials such as the TR polymers, which are usually very brittle and require careful handling. TR polymers exhibit a low elongation at break, and this further decreases with increasing temperature of the thermal rearrangement [207]. Spiro-TR-PBO-based polymers are among the most mechanically stable micro/nanoporous polymer-based materials [208]. They show remarkably high elongation at break even after the thermal rearrangement, even higher than that of PIM-

1, which does not undergo thermal treatment. The lower mechanical stability of PIM-1 is probably due to the presence of the rigid dioxane unit that hinders the entanglement between the polymer chains, showing that specific functional groups could lead to the preparation of mechanically robust HFs with microporous polymers.

9.6.2 Swelling/Plasticization

Glassy polymers may undergo strong changes in the transport properties in the presence of highly absorbing gases or vapors, due to the relaxation of polymer chains, known as plasticization [209]. Plasticization by CO_2 may be a serious problem for the separation efficiency of polymer membranes involved in the upgrading of natural gas, and therefore Adewole and Ahmad have compared the performance of 29 different polymers, establishing new criteria to define their performance in CO_2 removal [159]. Plasticization is generally associated with an increase in the permeability of all penetrating gases, and with a consequent loss of selectivity. For a single gas that has no plasticization effect, the permeability generally decreases with increasing feed pressure, while an increase in the permeability is observed with a plasticizing gas. In the case of gas mixtures that do not plasticize the polymer, the competitive sorption between the two components often lowers the absorption of the least soluble gas, which likely leads to a decrease in permeability. When a self-plasticizing gas is present in the mixture, the complexity of the transport increases considerably since both an increase or a decrease in permeability can occur. This depends on the balance between the reduction caused by competitive absorption and the increase due to plasticization. Visser et al. [209] evaluated the plasticization effect in an asymmetric HF blend of Matrimid and copolyimide P84 during the permeation of a propane/propylene mixture. The study was performed also with a ternary mixture adding methane. The membrane undergoes strong plasticization with the binary mixture, reducing the selectivity with increasing partial pressure of propylene in the feed, because the latter is responsible for the plasticization. A weaker effect of plasticization was observed with HFs of neat Matrimid. When CH_4 is added to the mixture, significant changes in the permeability occur with the Matrimid/P84 membranes, but the blending does not stabilize the membranes sufficiently and other routes should be investigated, such as crosslinking. The effect of traces of condensable vapors such as toluene and n-heptane on the CO_2/CH_4 separation was investigated for annealed and non-annealed Matrimid® asymmetric HFs [210]. Both contaminants, alone or together, decrease the CO_2 permeability and the selectivity in non-annealed fibers. Lower hysteresis of permanence and selectivity shows that the effect is weaker in annealed fibers.

Plasticization is often associated with swelling of the polymer. Swelling causes physical and structural changes in the polymer membranes, where it penetrates the free volume of the polymer and into the spaces between the polymer chains where it then increases the free volume. This expansion of the free volume causes an increase in the permeability and a decrease in the selectivity [48]. Humidity in feed gas is also a frequent cause of swelling, especially at high pressures. A study by Koros et al. [211] on the conditioning of polycarbonate asymmetric HFs by CO_2 highlighted the changes in morphology after exposure to CO_2. It was proposed that the selective dense skin was formed by several layers, defined as nodules. At high pressure, the CO_2 causes the elongation or the disconnection of the nodules, resulting in a decrease in the effective thickness of the dense skin.

9.6.3 Thermal Properties

Some gas separation processes operate at high, low, or variable temperature and this requires a careful choice of the materials that maintain integrity and performance under the given operation conditions. When the temperature increases near the glass transition temperature (T_g) of the polymer, there is an increase in the polymer chain mobility, which may cause compaction of the porous asymmetric membrane structure, and loss in selectivity and of permeation occurs either by a change in the intrinsic polymer properties or by the damage of the dense skin. Consequently, for separation plants working at high temperatures, the development of materials with a high T_g is required. A practical example of an industrial plant where separation occurs at high temperature (maximum 200°C) is an integrated gasification combined cycle power plant (IGCC), with pre-combustion H_2 recovery and CO_2 capture from the synthesis gas. For this purpose, Singh et al. [212]

developed PBI-based HFs membranes with a $T_g \approx 450°C$, through an accurate spinning procedure with particular care to control the thickness of dense skin layer (between 160 to 2180 nm). They obtained membranes with high performances for H_2/CO_2 separation (selectivity > 20) and H_2 permeance in the range of 200–400 GPU at 350°C. PBI was also used for the preparation of composite HF on a stainless steel support, for the H_2 separation from syngas [167]. These membranes are very selective for H_2/CO_2 separation (α = 47), with an H_2 permeance of 4.67 GPU at 250°C. Importantly, they also proved to be stable in the presence of H_2S for 330 days in dry simulated syngas conditions. Polyimides, such as P84®, Matrimid®, and 6FDA-6FpDA, were employed for the preparation of ceramic-supported thin-film composite HFs [213]. Optimization of the spinning procedure yielded fibers with a defect-free skin layer of less than 1 μm. Their transport properties as a function of temperature in the range 30–80°C show an increase in the permeance, in particular for H_2, and thus an increase in the H_2/CO_2 selectivity for all three membranes. Prediction of the gas transport properties at high temperatures (up to 300°C) suggests that the selectivity increases with decreasing fractional free volume: P84® > Matrimid® > 6FDA-6FpDA [213]. The obtained results show the possibility to use these membranes in gas separation processes at high temperatures. In addition to the high-temperature processes, a cold membrane process for post-combustion CO_2 capture was developed by Air Liquide in order to reduce the effective costs. Asymmetric Matrimid 5218 HFs were prepared by Liu et al. [77] for a sub-ambient process. These membranes were coated with PDMS to seal defects and then tested with a mixture of CO_2/N_2 20/80 mol%. The permeation tests show that there is an improvement in the performance at a temperature below –20°C, and increasing feed pressure.

9.6.4 Physical Ageing

Physical ageing is a characteristic phenomenon of glassy polymers, and especially the high-free volume polymers, due to their nonequilibrium state. It is a consequence of the relaxation of the polymeric chains over time, which induces changes in their physical properties, including the reduction in the fractional free volume, with a consequent decrease in permeability. Consequently, this phenomenon must be addressed during the development of HFs to avoid non-negligible changes in the productivity of the gas separation system. This effect is particularly important for high-free volume materials such as PIMs and TR polymers. A tailored selection of monomers and supports is needed to overcome this issue. For this purpose, Lee et al. studied a thermally rearranged poly(benzoxazole-co-imide) (XTR-PBOI) HF membrane module that proved to be stable for more than 10 days with different feed conditions, including changes in the feed pressure and humidity [214]. Chung et al. [200] investigated the ageing of PES HFs, considering transport and mechanical properties, for 4 months. The permeance rapidly decreases in the first 40 days, while the O_2/N_2 selectivity shows a slight decrease over time. The flux drop is higher for the fibers spun with high shear rates. Young's modulus and the tensile strength increase slightly with the ageing, while there is no clear correlation between the mechanical properties and the spinning shear rates. Asymmetric hollow fibers of the polyamide-imide Torlon® show both a decrease in helium permeance and selectivity during He/CH_4 separation at high feed pressure (≈80 bar) [215]. The study by resistance model reveals that the dense skin layer and the support were affected by ageing. In detail, the thickness of the selective layer increased by 17% while the resistance of the support versus the helium increased 2.8 times. The ageing of Matrimid® membranes of different thicknesses was evaluated with pure gases and with CO_2/CH_4 and CO_2/N_2 gas mixtures [153] and showed that the ageing was more rapid for the thinner membranes. Asymmetric HFs of Matrimid® prepared with different protocols [216] were investigated over a long time to check their stability. All the membranes showed rapid ageing in the first six months, mainly due to the densification of the selective layer. Also in this case, faster ageing was observed for the thin membranes, while crosslinked membranes were found to be more stable over time.

References

1. Peng, N., Widjojo, N., Sukitpaneenit, P., Teoh, M. M., Lipscomb, G. G., Chung, T.-S., and Lai, J.-Y. (2012) Evolution of polymeric hollow fibers as sustainable technologies: Past, present, and future. *Prog. Polym. Sci.*, **37**(10), 1401–1424.

2. Loeb, S. and Sourirajan, S. (1963) Sea water demineralization by means of an osmotic membrane. *Saline Water Conversion-II, Adv. Chem. Ser.*, **38**, 117–132.

3. Wang, Y., Ma, X., Ghanem, B. S., Alghunaimi, F., Pinnau, I., and Han, Y. (2018) Polymers of intrinsic microporosity for energy-intensive membrane-based gas separations. *Mater. Today Nano*, **3**, 69–95.

4. Moch, I. (2005) Membranes, hollow-fiber. *Kirk-Othmer Encycl. Chem. Technol.*, (4), 1–31.

5. Mubashir, M., Fong, Y. Y., Leng, C. T., and Keong, L. K. (2018) Issues and current trends of hollow-fiber mixed-matrix membranes for CO_2 separation from N_2 and CH_4. *Chem. Eng. Technol.*, **41**(2), 235–252.

6. Baker, R. W. and Low, B. T. (2014) Gas separation membrane materials: A perspective. *Macromolecules*, **47**(20), 6999–7013.

7. Nunes, S. P., Culfaz-Emecen, P. Z., Ramon, G. Z., Visser, T., Koops, G. H., Jin, W., and Ulbricht, M. (2020) Thinking the future of membranes: Perspectives for advanced and new membrane materials and manufacturing processes. *J. Memb. Sci.*, **598** (October 2019), 117761.

8. Ding, Y. (2020) Perspective on gas separation membrane materials from process economics point of view. *Ind. Eng. Chem. Res.*, **59**(2), 556–568.

9. Robeson, L. M. (1991) Correlation of separation factor versus permeability for polymeric membranes. *J. Memb. Sci.*, **62**(2), 165–185.

10. Freeman, B. D. (1999) Basis of permeability/selectivity tradeoff relations in polymeric gas separation membranes. *Macromolecules*, **32**(2), 375–380.

11. Robeson, L. M., Smith, Z. P., Freeman, B. D., and Paul, D. R. (2014) Contributions of diffusion and solubility selectivity to the upper bound analysis for glassy gas separation membranes. *J. Memb. Sci.*, **453**, 71–83.

12. Robeson, L. M. (2008) The upper bound revisited. *J. Memb. Sci.*, **320**(1–2), 390–400.

13. Swaidan, R., Ghanem, B., and Pinnau, I. (2015) Fine-tuned intrinsically ultramicroporous polymers redefine the permeability/selectivity upper bounds of membrane-based air and hydrogen separations. *ACS Macro Lett.*, **4**(9), 947–951.

14. Comesana-Gandara, B., Chen, J., Bezzu, C. G., Carta, M., Rose, I., Ferrari, M.-C., Esposito, E., Fuoco, A., McKeown, N. B., and Jansen, J. C. (2019) Redefining the Robeson upper bounds for CO_2/CH_4 and CO_2/N_2 separations using a series of ultrapermeable benzotriptycene-based

polymers of intrinsic microporosity. *Energy Environ. Sci.*, **12**(9), 2733–2740.

15. Liang, C. Z., Chung, T.-S., and Lai, J.-Y. (2019) A review of polymeric composite membranes for gas separation and energy production. *Prog. Polym. Sci.*, **97**, 101141.
16. Ullah Khan, I., Othman, M. H. D., Ismail, A. F., Matsuura, T., Hashim, H., Nordin, N. A. H. M., Rahman, M. A., Jaafar, J., and Jilani, A. (2018) Status and improvement of dual-layer hollow fiber membranes via co-extrusion process for gas separation: A review. *J. Nat. Gas Sci. Eng.*, **52** (February), 215–234.
17. Drioli, E., Brunetti, A., Di Profio, G., and Barbieri, G. (2012) Process intensification strategies and membrane engineering. *Green Chem.*, **14**(6), 1561–1572.
18. Li, P., Wang, Z., Qiao, Z., Liu, Y., Cao, X., Li, W., Wang, J., and Wang, S. (2015) Recent developments in membranes for efficient hydrogen purification. *J. Memb. Sci.*, **495**, 130–168.
19. Cacho-Bailo, F., Caro, G., Etxeberría-Benavides, M., Karvan, O., Téllez, C., and Coronas, J. (2016) MOF-polymer enhanced compatibility: Post-annealed zeolite imidazolate framework membranes inside polyimide hollow fibers. *RSC Adv.*, **6**(7), 5881–5889.
20. Hosseini, S. S., Najari, S., Kundu, P. K., Tan, N. R., and Roodashti, S. M. (2015) Simulation and sensitivity analysis of transport in asymmetric hollow fiber membrane permeators for air separation. *RSC Adv.*, **5**(105), 86359–86370.
21. Liang, C. Z., Yong, W. F., and Chung, T. S. (2017) High-performance composite hollow fiber membrane for flue gas and air separations. *J. Memb. Sci.*, **541** (July), 367–377.
22. Esposito, E., Dellamuzia, L., Moretti, U., Fuoco, A., Giorno, L., and Jansen, J. C. (2019) Simultaneous production of biomethane and food grade CO_2 from biogas: An industrial case study. *Energy Environ. Sci.*, **12**(1), 281–289.
23. Scholes, C. A., Stevens, G. W., and Kentish, S. E. (2012) Membrane gas separation applications in natural gas processing. *Fuel*, **96**, 15–28.
24. Kárászová, M., Zach, B., Petrusová, Z., Červenka, V., Bobák, M., Šyc, M., and Izák, P. (2020) Post-combustion carbon capture by membrane separation, Review. *Sep. Purif. Technol.*, **238** (August 2019).
25. Baker, R. W. (2002) Future directions of membrane gas separation technology. *Ind. Eng. Chem. Res.*, **41**(6), 1393–1411.

26. Evonik (2011) SEPURAN® Noble Membrane technology for helium and hydrogen generation. 1–6.

27. Ohya, H., Kudryavsev, V. V., and Semenova, S. I. (1997) *Polyimide Membranes: Applications, Fabrications and Properties*. CRC Press, p. 328.

28. Visser, T., Ungerank, U., Balster, J., and Führer, C. (2015) Method for producing polyimide membranes (12) Patent Application Publication (10) Pub. No.: US 2006/0222585 A1 Figure 1. Evonik Fibres GmbH, Method for producing polymide membranes patent No: US 2016/0144323 A1, issued 2015.

29. Liu, G. (2012) *Greenhouse Gases: Capturing, Utilization and Reduction*. IntechOpen .

30. Minhas, B. S. (1988) Membrane separation systems. *Chem. Eng. News*, **66**(32), 60.

31. Prasad, R., Notaro, F., and Thompson, D. R. (1994) Evolution of membranes in commercial air separation. *J. Memb. Sci.*, **94**, 225.

32. Yong, W. F., Li, F. Y., Xiao, Y. C., Chung, T. S., and Tong, Y. W. (2013) High performance PIM-1/Matrimid hollow fiber membranes for CO_2/CH_4, O_2/N_2 and CO_2/N_2 separation. *J. Memb. Sci.*, **443**, 156–169.

33. Kim, Y. H., Kim, H. S., and Kwon, S. K. (2005) Synthesis and characterization of highly soluble and oxygen permeable new polyimides based on twisted biphenyl dianhydride and spirobifluorene diamine. *Macromolecules*, **38**(19), 7950–7956.

34. Evonik (2004) SEPURAN N2 Technology for efficient nitrogen generation. 1–11.

35. Parker, T. (2016) Membrane modules for nitrogen and oxygen generator systems Parker modules the heart of OEM tailor-made nitrogen generators. 1–7.

36. Chen, X. Y., Vinh-Thang, H., Ramirez, A. A., Rodrigue, D., and Kaliaguine, S. (2015) Membrane gas separation technologies for biogas upgrading. *RSC Adv.*, **5**(31), 24399–24448.

37. Jeon, Y. W. and Lee, D. H. (2015) Gas membranes for CO_2/CH_4 (biogas) separation: A review. *Environ. Eng. Sci.*, **32**(2), 71–85.

38. Esposito, E., Dellamuzia, L., Moretti, U., Fuoco, A., Giorno, L., and Jansen, J. C. (2019) Simultaneous production of biomethane and food grade CO_2 from biogas: An industrial case study. *Energy Environ. Sci.*, **12**(1), 281–289.

39. Hengeveld, E. J., van Gemert, W. J. T., Bekkering, J., and Broekhuis, A. A. (2014) When does decentralized production of biogas and centralized

upgrading and injection into the natural gas grid make sense? *Biomass and Bioenergy*, **67**, 363–371.

40. Pasini, G., Baccioli, A., Ferrari, L., Antonelli, M., Frigo, S., and Desideri, U. (2019) Biomethane grid injection or biomethane liquefaction: A technical-economic analysis. *Biomass and Bioenergy*, **127**, 105264.
41. Cavana, M. and Leone, P. (2019) Biogas blending into the gas grid of a small municipality for the decarbonization of the heating sector. *Biomass and Bioenergy*, **127**, 105295.
42. Haider, S., Lindbråthen, A., Lie, J. A., Carstensen, P. V., Johannessen, T., and Hägg, M.-B. (2018) Vehicle fuel from biogas with carbon membranes: A comparison between simulation predictions and actual field demonstration. *Green Energy Environ.*, **3**(3), 266–276.
43. Iulianelli, A. and Drioli, E. (2020) Membrane engineering: Latest advancements in gas separation and pre-treatment processes, petrochemical industry and refinery, and future perspectives in emerging applications. *Fuel Process. Technol.*, **206** (March), 106464.
44. Dabbah, M. A., Goatman, K., and Poole, I. (2016) Medical image data processing apparatus and method for determining the presence of an abnormality. Evonik Fibres GmbH, Method for producing polymide membranes patent No: US 2016/0144323 A1, issued 2016.
45. Mathieu, P. (2006) The IPCC special report on carbon dioxide capture and storage. *ECOS 2006: Proc. 19th Int. Conf. Effic. Cost, Optim. Simul. Environ. Impact Energy Syst.*, (October).
46. Liu, Y., Liu, Z., Morisato, A., Bhuwania, N., Chinn, D., and Koros, W. J. (2020) Natural gas sweetening using a cellulose triacetate hollow fiber membrane illustrating controlled plasticization benefits. *J. Memb. Sci.*, **601**, 117910.
47. Liu, Z., Liu, Y., Liu, G., Qiu, W., and Koros, W. J. (2020) Cross-linkable semi-rigid 6FDA-based polyimide hollow fiber membranes for sour natural gas purification. *Ind. Eng. Chem. Res.*, **59**(12), 5333–5339.
48. Suleman, M. S., Lau, K. K., and Yeong, Y. F. (2016) Plasticization and swelling in polymeric membranes in CO_2 removal from natural gas. *Chem. Eng. Technol.*, **39**(9), 1604–1616.
49. Cao, C., Chung, T. S., Liu, Y., Wang, R., and Pramoda, K. P. (2003) Chemical cross-linking modification of 6FDA-2,6-DAT hollow fiber membranes for natural gas separation. *J. Memb. Sci.*, **216**(1–2), 257–268.
50. Askari, M., Yang, T., and Chung, T. S. (2012) Natural gas purification and olefin/paraffin separation using cross-linkable dual-layer hollow fiber membranes comprising B-Cyclodextrin. *J. Memb. Sci.*, **423–424**, 392–403.

51. Ma, C. and Koros, W. J. (2013) Ester-cross-linkable composite hollow fiber membranes for CO_2 removal from natural gas. *Ind. Eng. Chem. Res.*, **52**(31), 10495–10505.

52. Ma, C. and Koros, W. J. (2018) Physical aging of ester-cross-linked hollow fiber membranes for natural gas separations and mitigation thereof. *J. Memb. Sci.*, **551** (November 2017), 214–221.

53. Chen, C. C., Miller, S. J., and Koros, W. J. (2013) Characterization of thermally cross-linkable hollow fiber membranes for natural gas separation. *Ind. Eng. Chem. Res.*, **52**(3), 1015–1022.

54. Scholes, C. A. and Ghosh, U. K. (2017) Review of membranes for helium separation and purification. *Membranes (Basel).*, **7**(1), 1–13.

55. Ussinger, P. H. (2006) Nitro and nitroso dyes. *Ullmann's Encyclopedia of Industrial Chemistry*. Weinheim: Wiley-VCH., 149–174.

56. Shoji, K. and Moriya, A. (2003) Method of producing high-purity helium. US20030221448A1, issued 2003.

57. Dibrov, G., Ivanov, M., Semyashkin, M., Sudin, V., and Kagramanov, G. (2018) High-pressure aging of asymmetric Torlon® hollow fibers for helium separation from natural gas. *Fibers*, **6**(4).

58. (2000) Gas processes 2000: Hydrogen (Polysep membrane). *Hydrocarb. Process.*, **79**(4), 73.

59. Kumbharkar, S. C., Liu, Y., and Li, K. (2011) High performance polybenzimidazole based asymmetric hollow fibre membranes for H_2/CO_2 separation. *J. Memb. Sci.*, **375**(1–2), 231–240.

60. Peer, M., Mehdi Kamali, S., Mahdeyarfar, M., and Mohammadi, T. (2007) Separation of hydrogen from carbon monoxide using a hollow fiber polyimide membrane: Experimental and simulation. *Chem. Eng. Technol.*, **30**(10), 1418–1425.

61. Hosseini, S. S., Peng, N., and Chung, T. S. (2010) Gas separation membranes developed through integration of polymer blending and dual-layer hollow fiber spinning process for hydrogen and natural gas enrichments. *J. Memb. Sci.*, **349**(1), 156–166.

62. Favvas, E. P., Kouvelos, E. P., Romanos, G. E., Pilatos, G. I., Mitropoulos, A. C., and Kanellopoulos, N. K. (2008) Characterization of highly selective microporous carbon hollow fiber membranes prepared from a commercial co-polyimide precursor. *J. Porous Mater.*, **15**(6), 625–633.

63. Favvas, E. P., Kapantaidakis, G. C., Nolan, J. W., Mitropoulos, A. C., and Kanellopoulos, N. K. (2007) Preparation, characterization and gas permeation properties of carbon hollow fiber membranes based on Matrimid® 5218 precursor. *J. Mater. Process. Technol.*, **186**(1–3), 102–110.

64. David, O. C., Gorri, D., Nijmeijer, K., Ortiz, I., and Urtiaga, A. (2012) Hydrogen separation from multicomponent gas mixtures containing CO, N_2 and CO_2 using Matrimid® asymmetric hollow fiber membranes. *J. Memb. Sci.*, **419–420**, 49–56.

65. Carta, M., Malpass-Evans, R., Croad, M., Rogan, Y., Jansen, J. C., Bernardo, P., Bazzarelli, F., and McKeown, N. B. (2013) An efficient polymer molecular sieve for membrane gas separations. *Science*, **339**(6117), 303–307.

66. Bernardo, P., Scorzafave, V., Clarizia, G., Tocci, E., Jansen, J. C., Borgogno, A., Malpass-Evans, R., McKeown, N. B., Carta, M., and Tasselli, F. (2019) Thin film composite membranes based on a polymer of intrinsic microporosity derived from Tröger's base: A combined experimental and computational investigation of the role of residual casting solvent. *J. Memb. Sci.*, **569**, 17–31.

67. Zhang, Y. and Wang, R. (2013) Fabrication of novel polyetherimide-fluorinated silica organic-inorganic composite hollow fiber membranes intended for membrane contactor application. *J. Memb. Sci.*, **443**, 170–180.

68. Barillas, M. K., Enick, R. M., O'Brien, M., Perry, R., Luebke, D. R., and Morreale, B. D. (2011) The CO_2 permeability and mixed gas CO_2/H_2 selectivity of membranes composed of CO_2-philic polymers. *J. Memb. Sci.*, **372**(1), 29–39.

69. Chen, H. Z., Thong, Z., Li, P., and Chung, T. S. (2014) High performance composite hollow fiber membranes for CO_2/H_2 and CO_2/N_2 separation. *Int. J. Hydrogen Energy*, **39**(10), 5043–5053.

70. Hu, L., Cheng, J., Li, Y., Liu, J., Zhang, L., Zhou, J., and Cen, K. (2017) Composites of ionic liquid and amine-modified SAPO 34 improve CO_2 separation of CO_2-selective polymer membranes. *Appl. Surf. Sci.*, **410**, 249–258.

71. Ramírez-Morales, J. E., Tapia-Venegas, E., Nemestóthy, N., Bakonyi, P., Bélafi-Bakó, K., and Ruiz-Filippi, G. (2013) Evaluation of two gas membrane modules for fermentative hydrogen separation. *Int. J. Hydrogen Energy*, **38**(32), 14042–14052.

72. Bakonyi, P., Nemestóthy, N., and Bélafi-Bakó, K. (2013) Biohydrogen purification by membranes: An overview on the operational conditions affecting the performance of non-porous, polymeric and ionic liquid based gas separation membranes. *Int. J. Hydrogen Energy*, **38**(23), 9673–9687.

73. Etxeberria-Benavides, M., Karvan, O., Kapteijn, F., Gascon, J., and David, O. (2020) Fabrication of defect-free p84® polyimide hollow

fiber for gas separation: Pathway to formation of optimized structure. *Membranes (Basel).*, **10**(1).

74. Barsema, J. N., Kapantaidakis, G. C., Van Der Vegt, N. F. A., Koops, G. H., and Wessling, M. (2003) Preparation and characterization of highly selective dense and hollow fiber asymmetric membranes based on BTDA-TDI/MDI co-polyimide. *J. Memb. Sci.*, **216**(1–2), 195–205.

75. Choi, S.-H., Jansen, J. C., Tasselli, F., Barbieri, G., and Drioli, E. (2010) In-line formation of chemically cross-linked P84® co-polyimide hollow fibre membranes for H_2/CO_2 separation. *Sep. Purif. Technol.*, **76**(2), 132–139.

76. Favvas, E. P., Stefanopoulos, K. L., Nolan, J. W., Papageorgiou, S. K., Mitropoulos, A. C., and Lairez, D. (2014) Mixed matrix hollow fiber membranes with enhanced gas permeation properties. *Sep. Purif. Technol.*, **132**, 336–345.

77. Liu, L., Sanders, E. S., Kulkarni, S. S., Hasse, D. J., and Koros, W. J. (2014) Sub-ambient temperature flue gas carbon dioxide capture via Matrimid® hollow fiber membranes. *J. Memb. Sci.*, **465**, 49–55.

78. Hägg, M. B., Lindbråthen, A., He, X., Nodeland, S. G., and Cantero, T. (2017) Pilot demonstration-reporting on CO_2 capture from a cement plant using hollow fiber process. *Energy Procedia*, **114**(1876), 6150–6165.

79. Lokhandwala, K. A., Pinnau, I., He, Z., Amo, K. D., DaCosta, A. R., Wijmans, J. G., and Baker, R. W. (2010) Membrane separation of nitrogen from natural gas: A case study from membrane synthesis to commercial deployment. *J. Memb. Sci.*, **346**(2), 270–279.

80. Li, P., Chen, H. Z., and Chung, T. S. (2013) The effects of substrate characteristics and pre-wetting agents on PAN-PDMS composite hollow fiber membranes for CO_2/N_2 and O_2/N_2 separation. *J. Memb. Sci.*, **434**, 18–25.

81. Hu, L. Q., Cheng, J., Wang, Y. L., Liu, J. Z., Zhou, J. H., and Cen, K. F. (2019) Improvement on surface hydrophily of hollow fiber-supported PDMS gas separation membrane by PVP modification. *Zhejiang Daxue Xuebao (Gongxue Ban)/Journal of Zhejiang University (Engineering Science)*, **53**(2), 228–233.

82. Liu, L., Chakma, A., and Feng, X. (2004) Preparation of hollow fiber poly(ether block amide)/polysulfone composite membranes for separation of carbon dioxide from nitrogen. *Chem. Eng. J.*, **105**(1–2), 43–51.

83. Kim, K., Ingole, P. G., Kim, J., and Lee, H. (2013) Separation performance of PEBAX/PEI hollow fiber composite membrane for $SO_2/CO_2/N_2$ mixed gas. *Chem. Eng. J.*, **233**, 242–250.
84. Chen, H. Z., Xiao, Y. C., and Chung, T. S. (2011) Multi-layer composite hollow fiber membranes derived from poly(ethylene glycol) (PEG) containing hybrid materials for CO_2/N_2 separation. *J. Memb. Sci.*, **381**(1–2), 211–220.
85. Pak, S. H., Jeon, Y. W., Shin, M. S., and Koh, H. C. (2016) Preparation of cellulose acetate hollow-fiber membranes for CO_2/CH_4 separation. *Environ. Eng. Sci.*, **33**(1), 17–24.
86. Hwang, H. Y., Nam, S. Y., Koh, H. C., Ha, S. Y., Barbieri, G., and Drioli, E. (2012) The effect of operating conditions on the performance of hollow fiber membrane modules for CO_2/N_2 separation. *J. Ind. Eng. Chem.*, **18**(1), 205–211.
87. Yong, W. F., Chung, T.-S., Weber, M., and Maletzko, C. (2018) New polyethersulfone (PESU) hollow fiber membranes for CO_2 capture. *J. Memb. Sci.*, **552** (December 2017), 305–314.
88. Lively, R. P., Dose, M. E., Xu, L., Vaughn, J. T., Johnson, J. R., Thompson, J. A., Zhang, K., Lydon, M. E., Lee, J. S., Liu, L., Hu, Z., Karvan, O., Realff, M. J., and Koros, W. J. (2012) A high-flux polyimide hollow fiber membrane to minimize footprint and energy penalty for CO_2 recovery from flue gas. *J. Memb. Sci.*, **423–424**, 302–313.
89. Liang, C. Z., Liu, J. T., Lai, J.-Y., and Chung, T.-S. (2018) High-performance multiple-layer PIM composite hollow fiber membranes for gas separation. *J. Memb. Sci.*, **563** (December 2017), 93–106.
90. Kapantaidakis, G. C. and Koops, G. H. (2002) High flux polyethersulfone–polyimide blend hollow fiber membranes for gas separation. *J. Memb. Sci.*, **204**(1–2), 153–171.
91. Merkel, T. C., Lin, H., Wei, X., and Baker, R. (2010) Power plant post-combustion carbon dioxide capture: An opportunity for membranes. *J. Memb. Sci.*, **359**(1–2), 126–139.
92. Faiz, R. and Li, K. (2012) Olefin/paraffin separation using membrane based facilitated transport/chemical absorption techniques. *Chem. Eng. Sci.*, **73**, 261–284.
93. Rungta, M., Zhang, C., Koros, W. J., and Xu, L. (2013) Membrane-based ethylene/ethane separation: The upper bound and beyond. *AIChE J.*, **59**, 3475–3489.
94. Azhin, M., Kaghazchi, T., and Rahmani, M. (2008) A review on olefin/paraffin separation using reversible chemical complexation technology. *J. Ind. Eng. Chem.*, **14**(5), 622–638.

95. Ghasem, N., Al-Marzouqi, M., and Ismail, Z. (2014) Gas-liquid membrane contactor for ethylene/ethane separation by aqueous silver nitrate solution. *Sep. Purif. Technol.*, **127**, 140–148.

96. Teramoto, M., Shimizu, S., Matsuyama, H., and Matsumiya, N. (2005) Ethylene/ethane separation and concentration by hollow fiber facilitated transport membrane module with permeation of silver nitrate solution. *Sep. Purif. Technol.*, **44**(1), 19–29.

97. Tsou, D. T., Blachman, M. W., and Davis, J. C. (1994) Silver-facilitated olefin/paraffin separation in a liquid membrane contactor system. *Ind. Eng. Chem. Res.*, **33**(12), 3209–3216.

98. Malakhov, A. O., Bazhenov, S. D., Vasilevsky, V. P., Borisov, I. L., Ovcharova, A. A., Bildyukevich, A. V., Volkov, V. V, Giorno, L., and Volkov, A. V (2019) Thin-film composite hollow fiber membranes for ethylene/ethane separation in gas-liquid membrane contactor. *Sep. Purif. Technol.*, **219**, 64–73.

99. Kostyanaya, M., Vasilevsky, V., Borisov, I., Plisko, T., and Bazhenov, S. (2019) Surface modified polysulfone hollow fiber membranes for ethane/ethylene separation using gas-liquid membrane contactors with ionic liquid-based absorbent. *Fibers*, **7**(1), 4.

100. Fallanza, M., Ortiz, A., Gorri, D., and Ortiz, I. (2011) Improving the mass transfer rate in G-L membrane contactors with ionic liquids as absorption medium. Recovery of propylene. *J. Memb. Sci.*, **385–386**(1), 217–225.

101. Fallanza, M., Ortiz, A., Gorri, D., and Ortiz, I. (2011) Effect of liquid flow on the separation of propylene/propane mixtures with a gas/liquid membrane contactor using Ag+-RTIL solutions. *Desalin. Water Treat.*, **27**(1–3), 123–129.

102. Merkel, T. C., Blanc, R., Ciobanu, I., Firat, B., Suwarlim, A., and Zeid, J. (2013) Silver salt facilitated transport membranes for olefin/paraffin separations: Carrier instability and a novel regeneration method. *J. Memb. Sci.*, **447**, 177–189.

103. Morisato, A., He, Z., Pinnau, I., and Merkel, T. C. (2002) Transport properties of PA12-PTMO/$AgBF_4$ solid polymer electrolyte membranes for olefin/paraffin separation. **145**, 347–351.

104. Wang, Y., Ren, J., and Deng, M. (2011) Ultrathin solid polymer electrolyte PEI/Pebax2533/$AgBF_4$ composite membrane for propylene/propane separation. *Sep. Purif. Technol.*, **77**(1), 46–52.

105. Song, D., Kang, Y. S., and Kang, S. W. (2015) Highly permeable and stabilized olefin transport membranes based on a poly(ethylene oxide) matrix and $Al(NO_3)_3$. *J. Memb. Sci.*, **474**, 273–276.

106. Kang, S. W., Kim, J. H., Won, J., and Kang, Y. S. (2013) Suppression of silver ion reduction by $Al(NO_3)_3$ complex and its application to highly stabilized olefin transport membranes. *J. Memb. Sci.*, **445**(3), 156–159.

107. Feng, C. Y., Khulbe, K. C., Matsuura, T., and Ismail, A. F. (2013) Recent progresses in polymeric hollow fiber membrane preparation, characterization and applications. *Sep. Purif. Technol.*, **111**, 43–71.

108. Baker, R. W. and Lokhandwala, K. (2008) Natural gas processing with membranes: An overview. *Ind. Eng. Chem. Res.*, **47**, 2109–2121.

109. Li, S. G., Koops, G. H., Mulder, M. H. V., Van Den Boomgaard, T., and Smolders, C. A. (1994) Wet spinning of integrally skinned hollow fiber membranes by a modified dual-bath coagulation method using a triple orifice spinneret. *J. Memb. Sci.*, **94**(1), 329–340.

110. Xia, Q.-C., Liu, M.-L., Cao, X.-L., Wang, Y., Xing, W., and Sun, S.-P. (2018) Structure design and applications of dual-layer polymeric membranes. *J. Memb. Sci.*, **562**, 85–111.

111. Fang, C., Rajabzadeh, S., Liu, W., Wu, H. C., Kato, N., Sun, Y., Jeon, S., and Matsuyama, H. (2020) Effect of mixed diluents during thermally induced phase separation process on structures and performances of hollow fiber membranes prepared using triple-orifice spinneret. *J. Memb. Sci.*, **596** (December 2019), 117715.

112. Jung, J. T., Wang, H. H., Kim, J. F., Lee, J., Kim, J. S., Drioli, E., and Lee, Y. M. (2018) Tailoring nonsolvent-thermally induced phase separation (N-TIPS) effect using triple spinneret to fabricate high performance PVDF hollow fiber membranes. *J. Memb. Sci.*, **559** (March), 117–126.

113. Dai, Z., Ansaloni, L., and Deng, L. (2016) Recent advances in multi-layer composite polymeric membranes for CO_2 separation: A review. *Green Energy Environ.*, **1**(2), 102–128.

114. Kosuri, M. R. and Koros, W. J. (2008) Defect-free asymmetric hollow fiber membranes from Torlon®, a polyamide-imide polymer, for high-pressure CO2 separations. *J. Memb. Sci.*, **320**(1–2), 65–72.

115. Jie, X., Duan, C., Wang, L., Jiang, C., Zheng, H., Liu, J., Liu, D., Cao, Y., and Yuan, Q. (2014) Fabrication of an asymmetric 4,4′-oxydiphthalic anhydride-2,4,6-trimethyl-1,3-phenylenediamine/2,6-diaminotoluene copolyimide hollow fiber membrane and its performance for CO_2 separation. *Ind. Eng. Chem. Res.*, **53**(11), 4442–4452.

116. Li, Y. and Chung, T. S. (2010) Silver ionic modification in dual-layer hollow fiber membranes with significant enhancement in CO_2/CH_4 and O_2/N_2 separation. *J. Memb. Sci.*, **350**(1–2), 226–231.

117. Dai, Y., Johnson, J. R., Karvan, O., Sholl, D. S., and Koros, W. J. (2012) Ultem®/ZIF-8 mixed matrix hollow fiber membranes for CO_2/N_2 separations. *J. Memb. Sci.*, **401–402**, 76–82.

118. Woo, K. T., Lee, J., Dong, G., Kim, J. S., Do, Y. S., Hung, W. S., Lee, K. R., Barbieri, G., Drioli, E., and Lee, Y. M. (2015) Fabrication of thermally rearranged (TR) polybenzoxazole hollow fiber membranes with superior CO_2/N_2 separation performance. *J. Memb. Sci.*, **490**, 129–138.

119. Olivieri, L., Meneguzzo, S., Ligi, S., Saccani, A., Giorgini, L., Orsini, A., Pettinau, A., and De Angelis, M. G. (2018) Reducing ageing of thin PTMSP films by incorporating graphene and graphene oxide: Effect of thickness, gas type and temperature. *J. Memb. Sci.*, **555**, 258–267.

120. Esposito, E., Clarizia, G., Bernardo, P., Jansen, J. C., Sedláková, Z., Izák, P., Curcio, S., Cindio, B. de, and Tasselli, F. (2015) Pebax®/PAN hollow fiber membranes for CO_2/CH_4 separation. *Chem. Eng. Process. Process Intensif.*, **94**, 53–61.

121. Esposito, E., Clarizia, G., Bernardo, P., Jansen, J. C., Sedláková, Z., Izák, P., Curcio, S., Cindio, B. de, and Tasselli, F. (2015) Pebax®/PAN hollow fiber membranes for CO_2/CH_4 separation. *Chem. Eng. Process. Process Intensif.*, **94**, 53–61.

122. Wang, Y., Hu, T., Li, H., Dong, G., Wong, W., and Chen, V. (2014) Enhancing membrane permeability for CO_2 capture through blending commodity polymers with selected PEO and PEO-PDMS copolymers and composite hollow fibres. *Energy Procedia*, **63**, 202–209.

123. Fam, W., Mansouri, J., Li, H., and Chen, V. (2017) Improving CO_2 separation performance of thin film composite hollow fiber with Pebax®1657/ionic liquid gel membranes. *J. Memb. Sci.*, **537**, 54–68.

124. Li, Y., Cao, C., Chung, T. S., and Pramoda, K. P. (2004) Fabrication of dual-layer polyethersulfone (PES) hollow fiber membranes with an ultrathin dense-selective layer for gas separation. *J. Memb. Sci.*, **245** (1–2), 53–60.

125. Xie, K., Fu, Q., Qiao, G. G., and Webley, P. A. (2019) Recent progress on fabrication methods of polymeric thin film gas separation membranes for CO2 capture. *J. Memb. Sci.*, **572**, 38–60.

126. Bazzarelli, F., Bernardo, P., Tasselli, F., Clarizia, G., Dzyubenko, V. G., Vdovin, P., and Jansen, J. C. (2011) Multilayer composite SBS membranes for pervaporation and gas separation. *Sep. Purif. Technol.*, **80**(3), 635–642.

127. Jansen, J. C., Tasselli, F., Tocci, E., and Drioli, E. (2006) High-flux composite perfluorinated gas separation membranes of Hyflon® AD

on a hollow fibre ultrafiltration membrane support. *Desalination*, **192**(1–3), 207–213.

128. Lasseuguette, E., Rouch, J.-C., and Remigy, J.-C. (2013) Hollow-fiber coating: Application to preparation of composite hollow-fiber membrane for gas separation. *Ind. Eng. Chem. Res.*, **52**(36), 13146–13158.
129. Liu, J., Lu, X., and Wu, C. (2013) Effect of preparation methods on crystallization behavior and tensile strength of poly(vinylidene fluoride) membranes. *Membranes (Basel).*, **3**(4), 389–405.
130. Li, X. and Lu, X. (2007) Effect of annealing on the structure and properties of polyvinylidene fluoride hollow fiber by melt-spinning. *J. Appl. Polym. Sci.*, **103**(2), 935–941.
131. Choi, S.-H., Tasselli, F., Jansen, J. C., Barbieri, G., and Drioli, E. (2010) Effect of the preparation conditions on the formation of asymmetric poly(vinylidene fluoride) hollow fibre membranes with a dense skin. *Eur. Polym. J.*, **46**(8), 1713–1725.
132. Twarowska-Schmidt, K. and Włochowicz, A. (1997) Melt-spun asymmetric poly(4-methyl-1-pentene) hollow fibre membranes. *J. Memb. Sci.*, **137**(1), 55–61.
133. Nijdam, W., de Jong, J., van Rijn, C. J. M., Visser, T., Versteeg, L., Kapantaidakis, G., Koops, G.-H., and Wessling, M. (2005) High performance micro-engineered hollow fiber membranes by smart spinneret design. *J. Memb. Sci.*, **256**(1), 209–215.
134. Çulfaz, P. Z., Rolevink, E., van Rijn, C., Lammertink, R. G. H., and Wessling, M. (2010) Microstructured hollow fibers for ultrafiltration. *J. Memb. Sci.*, **347**(1), 32–41.
135. García-Fernández, L., García-Payo, C., and Khayet, M. (2017) Hollow fiber membranes with different external corrugated surfaces for desalination by membrane distillation. *Appl. Surf. Sci.*, **416**, 932–946.
136. Zhang, M., Deng, L., Xiang, D., Cao, B., Hosseini, S. S., and Li, P. (2019) Approaches to suppress CO_2-induced plasticization of polyimide membranes in gas separation applications. *Processes*, **7**(1).
137. Kopeć, K. K., Dutczak, S. M., Wessling, M., and Stamatialis, D. F. (2011) Chemistry in a spinneret: On the interplay of crosslinking and phase inversion during spinning of novel hollow fiber membranes. *J. Memb. Sci.*, **369**(1–2), 308–318.
138. Woo, K. T., Lee, J., Dong, G., Kim, J. S., Do, Y. S., Jo, H. J., and Lee, Y. M. (2016) Thermally rearranged poly(benzoxazole-co-imide) hollow fiber membranes for CO_2 capture. *J. Memb. Sci.*, **498**, 125–134.

139. Liu, Y., Chung, T. S., Wang, R., Li, D. F., and Chng, M. L. (2003) Chemical cross-linking modification of polyimide/poly(ether sulfone) dual-layer hollow-fiber membranes for gas separation. *Ind. Eng. Chem. Res.*, **42**(6), 1190–1195.

140. Nisola, G. M., Beltran, A. B., Africa, V. J. L., Cho, E., Han, M., and Chung, W. J. (2011) Carbon dioxide and methane gas permeations in thermally annealed and chemically cross-linked commercial polyimide hollow fiber membrane. *Fibers Polym.*, **12**(5), 572–579.

141. Chen, C. C., Qiu, W., Miller, S. J., and Koros, W. J. (2011) Plasticization-resistant hollow fiber membranes for CO_2/CH_4 separation based on a thermally crosslinkable polyimide. *J. Memb. Sci.*, **382**(1–2), 212–221.

142. Ma, C. and Koros, W. J. (2013) High-performance ester-crosslinked hollow fiber membranes for natural gas separations. *J. Memb. Sci.*, **428**, 251–259.

143. Dong, G., Li, H., and Chen, V. (2011) Plasticization mechanisms and effects of thermal annealing of Matrimid hollow fiber membranes for CO_2 removal. *J. Memb. Sci.*, **369**(1–2), 206–220.

144. Liang, C. Z. and Chung, T. S. (2018) Ultrahigh flux composite hollow fiber membrane via highly crosslinked PDMS for recovery of hydrocarbons: Propane and propene. *Macromol. Rapid Commun.*, **39**(5), 1–7.

145. Syrtsova, D. A., Kharitonov, A. P., Teplyakov, V. V., and Koops, G. H. (2004) Improving gas separation properties of polymeric membranes based on glassy polymers by gas phase fluorination. *Desalination*, **163**(1–3), 273–279.

146. Kim, S. and Lee, Y. M. (2015) Rigid and microporous polymers for gas separation membranes. *Prog. Polym. Sci.*, **43**, 1–32.

147. Park, H. B., Jung, C. H., Lee, Y. M., Hill, A. J., Pas, S. J., Mudie, S. T., Van Wagner, E., Freeman, B. D., and Cookson, D. J. (2007) Polymers with cavities tuned for fast selective transport of small molecules and ions. *Science (80).*, **318**(5848), 254–258.

148. Lee, J., Kim, J. S., Kim, J. F., Jo, H. J., Park, H., Seong, J. G., and Lee, Y. M. (2019) Densification-induced hollow fiber membranes using crosslinked thermally rearranged (XTR) polymer for CO_2 capture. *J. Memb. Sci.*, **573** (December 2018), 393–402.

149. Xu, L., Rungta, M., and Koros, W. J. (2011) Matrimid® derived carbon molecular sieve hollow fiber membranes for ethylene/ethane separation. *J. Memb. Sci.*, **380**(1), 138–147.

150. Kim, S., Han, S. H., and Lee, Y. M. (2012) Thermally rearranged (TR) polybenzoxazole hollow fiber membranes for CO_2 capture. *J. Memb. Sci.*, **403–404**, 169–178.

151. Jo, H. J., Soo, C. Y., Dong, G., Do, Y. S., Wang, H. H., Lee, M. J., Quay, J. R., Murphy, M. K., and Lee, Y. M. (2015) Thermally rearranged poly(benzoxazole-co-imide) membranes with superior mechanical strength for gas separation obtained by tuning chain rigidity. *Macromolecules*, **48**(7), 2194–2202.

152. Xu, L., Rungta, M., Brayden, M. K., Martinez, M. V., Stears, B. A., Barbay, G. A., and Koros, W. J. (2012) Olefins-selective asymmetric carbon molecular sieve hollow fiber membranes for hybrid membrane-distillation processes for olefin/paraffin separations. *J. Memb. Sci.*, **423–424**, 314–323.

153. Cao, Y., Zhang, K., Sanyal, O., and Koros, W. J. (2019) Carbon molecular sieve membrane preparation by economical coating and pyrolysis of porous polymer hollow fibers. *Angew. Chemie Int. Ed.*, **58**(35), 12149–12153.

154. Vu, D. Q., Koros, W. J., and Miller, S. J. (2002) High pressure CO_2/CH_4 separation using carbon molecular sieve hollow fiber membranes. *Ind. Eng. Chem. Res.*, **41**(3), 367–380.

155. Favvas, E. P., Heliopoulos, N. S., Papageorgiou, S. K., Mitropoulos, A. C., Kapantaidakis, G. C., and Kanellopoulos, N. K. (2015) Helium and hydrogen selective carbon hollow fiber membranes: The effect of pyrolysis isothermal time. *Sep. Purif. Technol.*, **142**, 176–181.

156. Lei, L., Lindbråthen, A., Hillestad, M., Sandru, M., Favvas, E. P., and He, X. (2019) Screening cellulose spinning parameters for fabrication of novel carbon hollow fiber membranes for gas separation. *Ind. Eng. Chem. Res.*, **58**(29), 13330–13339.

157. Chen, X. Y., Kaliaguine, S., and Rodrigue, D. (2019) Polymer hollow fiber membranes for gas separation: A comparison between three commercial resins. *AIP Conf. Proc.*, **2139**, 1–15.

158. Chen, X., Kaliaguine, S., and Rodrigue, D. (2017) A comparison between several commercial polymer hollow fiber membranes for gas separation. *J. Membr. Sep. Technol.*, **6**(1), 1–15.

159. Adewole, J. K. and Ahmad, A. L. (2017) Polymeric membrane materials selection for high-pressure CO_2 removal from natural gas. *J. Polym. Res.*, **24**(5).

160. Hashemifard, S. A., Ismail, A. F., Matsuura, T., and DashtArzhandi, M. R. (2015) Performance of silicon rubber coated polyetherimide hollow fibers for CO_2 removal via a membrane contactor. *RSC Adv.*, **5**(60), 48442–48455.

161. Fam, W., Mansouri, J., Li, H., and Chen, V. (2017) Improving CO_2 separation performance of thin film composite hollow fiber with Pebax®1657/ionic liquid gel membranes. *J. Memb. Sci.*, **537**, 54–68.

162. Deng, S., Tremblay, A., and Matsuura, T. (1998) Preparation of hollow fibers for the removal of volatile organic compounds from air. *J. Appl. Polym. Sci.*, **69**(2), 371–379.

163. Clausi, D. T., and Koros, W. J. (2000) Formation of defect-free polyimide hollow fiber membranes for gas separations. *J. Memb. Sci.*, **167**(1), 79–89.

164. Dibrov, G., Ivanov, M., Semyashkin, M., Sudin, V., Fateev, N., and Kagramanov, G. (2019) Elaboration of high permeable macrovoid free polysulfone hollow fiber membranes for air separation. *Fibers*, **7**(5).

165. Ismail, A. F., Dunkin, I. R., Gallivan, S. L., and Shilton, S. J. (1999) Production of super selective polysulfone hollow fiber membranes for gas separation. *Polymer (Guildf).*, **40**(23), 6499–6506.

166. Ekiner, O. M. and Vassilatos, G. (1990) Polyaramide hollow fibers for hydrogen/methane separation: Spinning and properties. *J. Memb. Sci.*, **53**(3), 259–273.

167. Berchtold, K. A., Singh, R. P., Young, J. S., and Dudeck, K. W. (2012) Polybenzimidazole composite membranes for high temperature synthesis gas separations. *J. Memb. Sci.*, **415–416**, 265–270.

168. Budd, P. M., Ghanem, B. S., Makhseed, S., McKeown, N. B., Msayib, K. J., and Tattershall, C. E. (2004) Polymers of intrinsic microporosity (PIMs): Robust, solution-processable, organic nanoporous materials. *Chem. Commun.*, **4**(2), 230–231.

169. Jue, M. L., Breedveld, V., and Lively, R. P. (2017) Defect-free PIM-1 hollow fiber membranes. *J. Memb. Sci.*, **530**, 33–41.

170. Hao, L., Zuo, J., and Chung, T.-S. (2014) Formation of defect-free polyetherimide/PIM-1 hollow fiber membranes for gas separation. *AIChE J.*, **60**(11), 3848–3858.

171. Scholes, C. (2018) Water resistant composite membranes for carbon dioxide separation from methane. *Appl. Sci.*, **8**(5), 829.

172. Chung, T. S. and Xu, Z. L. (1998) Asymmetric hollow fiber membranes prepared from miscible polybenzimidazole and polyetherimide blends. *J. Memb. Sci.*, **147**(1), 35–47.

173. Kapantaidakis, G. C., Koops, G. H., and Wessling, M. (2002) Effect of spinning conditions on the structure and the gas permeation properties of high flux polyethersulfone-polyimide blend hollow fibers. *Desalination*, **144**(1–3), 121–125.

174. Hosseini, S. S., Teoh, M. M., and Chung, T. S. (2008) Hydrogen separation and purification in membranes of miscible polymer blends with interpenetration networks. *Polymer (Guildf).*, **49**(6), 1594–1603.

175. Kapantaidakis, G. C., Koops, G. H., and Wessling, M. (2002) Preparation and characterization of gas separation hollow fiber membranes based on polyethersulfone-polyimide miscible blends. *Desalination*, **145**(1), 353–357.

176. Kapantaidakis, G. C., Koops, G. H., and Wessling, M. (2003) CO_2 plasticization of polyethersulfone/polyimide gas-separation membranes. **49**(7).

177. Hamid, M. R. A. and Jeong, H. K. (2018) Recent advances on mixed-matrix membranes for gas separation: Opportunities and engineering challenges. *Korean J. Chem. Eng.*, **35**(8), 1577–1600.

178. Dong, G., Li, H., and Chen, V. (2013) Challenges and opportunities for mixed-matrix membranes for gas separation. *J. Mater. Chem. A*, **1**(15), 4610–4630.

179. Husain, S. and Koros, W. J. (2007) Mixed matrix hollow fiber membranes made with modified HSSZ-13 zeolite in polyetherimide polymer matrix for gas separation. *J. Memb. Sci.*, **288**(1–2), 195–207.

180. Jiang, L. Y., Chung, T. S., and Kulprathipanja, S. (2006) An investigation to revitalize the separation performance of hollow fibers with a thin mixed matrix composite skin for gas separation. *J. Memb. Sci.*, **276**(1–2), 113–125.

181. Sutrisna, P. D., Hou, J., Zulkifli, M. Y., Li, H., Zhang, Y., Liang, W., D'Alessandro, D. M., and Chen, V. (2018) Surface functionalized UiO-66/Pebax-based ultrathin composite hollow fiber gas separation membranes. *J. Mater. Chem. A*, **6**(3), 918–931.

182. Moore, T. T. and Koros, W. J. (2005) Non-ideal effects in organic-inorganic materials for gas separation membranes. *J. Mol. Struct.*, **739**(1–3), 87–98.

183. Ismail, A. F., Kusworo, T. D., and Mustafa, A. (2008) Enhanced gas permeation performance of polyethersulfone mixed matrix hollow fiber membranes using novel Dynasylan Ameo silane agent. *J. Memb. Sci.*, **319**(1–2), 306–312.

184. Shirzadeh-Gharacheh, A. and Rahbari-Sisakht, M. (2016) Polyvinylidene fluoride hollow fiber mixed matrix membrane contactor incorporating modified ZSM-5 zeolite for carbon dioxide absorption. *RSC Adv.*, **6**(82), 78865–78874.

185. Fernández-Barquín, A., Casado-Coterillo, C., Etxeberria-Benavides, M., Zuñiga, J., and Irabien, A. (2017) Comparison of flat and hollow-fiber mixed-matrix composite membranes for CO_2 separation with temperature. *Chem. Eng. Technol.*, **40**(5), 997–1007.

186. Hu, J., Cai, H., Ren, H., Wei, Y., Xu, Z., Liu, H., and Hu, Y. (2010) Mixed-matrix membrane hollow fibers of $Cu_3(BTC)_2$ MOF and polyimide for gas separation and adsorption. *Ind. Eng. Chem. Res.*, **49**(24), 12605–12612.

187. Zhang, C. and Koros, W. J. (2013) ZIF/polyimide asymmetric mixed-matrix hollow fiber membranes with significantly enhanced hydrocarbon separation performance. *Sep. Div. 2013 - Core Program. Area 2013 AIChE Annu. Meet. Glob. Challenges Eng. a Sustain. Futur.*, 891–892.

188. Zhang, C., Zhang, K., Xu, L., Labreche, Y., Kraftschik, B., and Koros, W. J. (2014) Highly scalable ZIF-based mixed-matrix hollow fiber membranes for advanced hydrocarbon separations. *AIChE J.*, **60**(7), 2625–2635.

189. Zhu, H., Jie, X., Wang, L., Kang, G., Liu, D., and Cao, Y. (2016) Effect of MIL-53 on phase inversion and gas separation performance of mixed matrix hollow fiber membranes. *RSC Adv.*, **6**(73), 69124–69134.

190. Zhu, H., Jie, X., Wang, L., Kang, G., Liu, D., and Cao, Y. (2018) Enhanced gas separation performance of mixed matrix hollow fiber membranes containing post-functionalized S-MIL-53. *J. Energy Chem.*, **27**(3), 781–790.

191. Zhu, H., Jie, X., and Cao, Y. (2017) Fabrication of functionalized MOFs incorporated mixed matrix hollow fiber membrane for gas separation. *J. Chem.*, **2017**.

192. Mubashir, M., Yin fong, Y., Leng, C. T., Keong, L. K., and Jusoh, N. (2020) Study on the effect of process parameters on CO_2/CH_4 binary gas separation performance over NH2-MIL-53(Al)/cellulose acetate hollow fiber mixed matrix membrane. *Polym. Test.*, **81**.

193. Mubashir, M., Yeong, Y. F., Chew, T. L., and Lau, K. K. (2019) Optimization of spinning parameters on the fabrication of NH2-MIL-53(Al)/cellulose acetate (CA) hollow fiber mixed matrix membrane for CO2 separation. *Sep. Purif. Technol.*, **215** (September 2018), 32–43.

194. Esposito, E., Jansen, J. C., Dellamuzia, L., Moretti, U., Giorno, L. (2017) CO_2 recovery from biogas and successful upgrading to food-grade quality: A case study. *World Acad. Sci. Eng. Technol. Int. Sci. Index, Chem. Mol. Eng.*, **4**(11), 2153–2153.

195. Jiang, L. Y., Chung, T. S., Cao, C., Huang, Z., and Kulprathipanja, S. (2005) Fundamental understanding of nano-sized zeolite distribution in the formation of the mixed matrix single- and dual-layer asymmetric hollow fiber membranes. *J. Memb. Sci.*, **252**(1), 89–100.

196. Zhang, Y., Shen, Q., Hou, J., Sutrisna, P. D., and Chen, V. (2017) Shear-aligned graphene oxide laminate/Pebax ultrathin composite hollow fiber membranes using a facile dip-coating approach. *J. Mater. Chem. A*, **5**(17), 7732–7737.

197. Hirayama, Y., Yoshinaga, T., Kusuki, Y., Ninomiya, K., Sakakibara, T., and Tamari, T. (1996) Relation of gas permeability with structure of aromatic polyimides I. *J. Memb. Sci.*, **111**(2), 169–182.

198. Hirayama, Y., Yoshinaga, T., Kusuki, Y., Ninomiya, K., Sakakibara, T., and Tamari, T. (1996) Relation of gas permeability with structure of aromatic polyimides II. *J. Memb. Sci.*, **111**(2), 183–192.

199. Longo, M., De Santo, M. P., Esposito, E., Fuoco, A., Monteleone, M., Giorno, L., Comesaña-Gándara, B., Chen, J., Bezzu, C. G., Carta, M., Rose, I., McKeown, N. B., and Jansen, J. C. (2020) Correlating gas permeability and Young's modulus during the physical aging of polymers of intrinsic microporosity using atomic force microscopy. *Ind. Eng. Chem. Res.*, **59**(12), 5381–5391.

200. Chung, T. S. and Khean Teoh, S. (1999) The ageing phenomenon of polyethersulphone hollow fibre membranes for gas separation and their characteristics. *J. Memb. Sci.*, **152**(2), 175–188.

201. Fuoco, A., Rizzuto, C., Tocci, E., Monteleone, M., Esposito, E., Budd, P. M., Carta, M., Comesaña-Gándara, B., McKeown, N.B., and Jansen, J. C. (2019) The origin of size-selective gas transport through polymers of intrinsic microporosity. *J. Mater. Chem. A*, **7**(35), 20121–20126.

202. Fontananova, E., Jansen, J. C., Cristiano, A., Curcio, E., and Drioli, E. (2006) Effect of additives in the casting solution on the formation of PVDF membranes. *Desalination*, **192**(1–3), 190–197.

203. DashtArzhandi, M. R., Ismail, A. F., Matsuura, T., Ng, B. C., and Abdullah, M. S. (2015) Fabrication and characterization of porous polyetherimide/montmorillonite hollow fiber mixed matrix membranes for CO_2 absorption via membrane contactor. *Chem. Eng. J.*, **269**, 51–59.

204. Amirabedi, P., Yegani, R., and Hesaraki, A. H. (2017) Hydrophobicity optimization of polypropylene hollow fiber membrane by sol–gel process for CO_2 absorption in gas–liquid membrane contactor using response surface methodology. *Iran. Polym. J. (English Ed.)*, **26**(6), 431–443.

205. Wickramanayake, S., Hopkinson, D., Myers, C., Sui, L., and Luebke, D. (2013) Investigation of transport and mechanical properties of hollow fiber membranes containing ionic liquids for pre-combustion carbon dioxide capture. *J. Memb. Sci.*, **439**(0), 58–67.

206. Wickramanayake, S., Hopkinson, D., Myers, C., Hong, L., Feng, J., Seol, Y., Plasynski, D., Zeh, M., and Luebke, D. (2014) Mechanically robust hollow fiber supported ionic liquid membranes for CO_2 separation applications. *J. Memb. Sci.*, **470**, 52–59.

207. Lee, J., Kim, J. S., Moon, S., Park, C. Y., Kim, J. F., and Lee, Y. M. (2020) Dimensionally-controlled densification in crosslinked thermally rearranged (XTR) hollow fiber membranes for CO_2 capture. *J. Memb. Sci.*, **595** (April 2019), 117535.

208. Li, S., Jo, H. J., Han, S. H., Park, C. H., Kim, S., Budd, P. M., and Lee, Y. M. (2013) Mechanically robust thermally rearranged (TR) polymer membranes with spirobisindane for gas separation. *J. Memb. Sci.*, **434**, 137–147.

209. Visser, T. and Wessling, M. (2008) Auto and mutual plasticization in single and mixed gas C3 transport through Matrimid-based hollow fiber membranes. *J. Memb. Sci.*, **312**(1), 84–96.

210. Lee, J. S., Madden, W., and Koros, W. J. (2010) Antiplasticization and plasticization of Matrimid® asymmetric hollow fiber membranes—Part A. Experimental. *J. Memb. Sci.*, **350**(1), 232–241.

211. Jordan, S. M., Henson, M. A., and Koros, W. J. (1990) The effects of carbon dioxide conditioning on the permeation behavior of hollow fiber asymmetric membranes. *J. Memb. Sci.*, **54**(1–2), 103–118.

212. Singh, R. P., Dahe, G. J., Dudeck, K. W., Welch, C. F., and Berchtold, K. A. (2014) High temperature polybenzimidazole hollow fiber membranes for hydrogen separation and carbon dioxide capture from synthesis gas. *Energy Procedia*, **63**, 153–159.

213. Escorihuela, S., Tena, A., Shishatskiy, S., Escolástico, S., Brinkmann, T., Serra, J. M., and Abetz, V. (2018) Gas separation properties of polyimide thin films on ceramic supports for high temperature applications. *Membranes (Basel).*, **8**(1).

214. Lee, J. H., Lee, J., Jo, H. J., Seong, J. G., Kim, J. S., Lee, W. H., Moon, J., Lee, D., Oh, W. J., Yeo, J. G., and Lee, Y. M. (2017) Wet CO_2/N_2 permeation through a crosslinked thermally rearranged poly(benzoxazole-co-imide) (XTR-PBOI) hollow fiber membrane module for CO_2 capture. *J. Memb. Sci.*, **539** (June), 412–420.

215. Weigelt, F., Escorihuela, S., Descalzo, A., Tena, A., Escolástico, S., Shishatskiy, S., Serra, J. M., and Brinkmann, T. (2019) Novel polymeric thin-film composite membranes for high-temperature gas separations. *Membranes (Basel).*, **9**(4), 51.

216. Clarizia, G., Tasselli, F., and Bernardo, P. (2020) Effect of physical aging on gas transport in asymmetric polyimide hollow fibers prepared by triple-orifice spinneret. *Polymers (Basel).*, **12**(2), 441.

Chapter 10

Hollow Fibers for Pervaporation Separations

Roberto Castro-Muñoz
Tecnologico de Monterrey, Campus Toluca,
Avenida Eduardo Monroy Cárdenas 2000 San Antonio Buenavista,
50110 Toluca de Lerdo, Mexico
food.biotechnology88@gmail.com, castromr@tec.mx

Pervaporation, as the latent alternative to distillation technology, has been successfully applied in separating different types of azeotropic solvent mixtures. Nowadays, "membranologists" are looking for membranes that can be useful in large-scale applications. In this regard, according to their less fouling and larger effective surface-area-to-volume ratios, hollow fiber membranes are likely the most feasible configurations compared to flat-sheet or tubular modules. This chapter provides a literature review of various studies in which different types of hollow fiber membranes have been developed and applied in several PV applications. Special attention has been paid to such studies that provide relevant insights into the field.

Hollow Fibers and Nanofibers in Membrane Science: Preparation, Characterization, and Applications
Edited by Alberto Figoli, Mir Saeed Seyed Dorraji, and Francesco Galiano

ISBN 978-981-4968-03-4 (Hardcover), 978-1-003-25663-2 (eBook)
www.jennystanford.com

For a better understanding of the readers, a brief theoretical basis of PV technology is provided. Based on the current ongoing works, concluding remarks are also given.

10.1 Introduction

Pervaporation (PV) technology, as a potential alternative to conventional distillation, was introduced for the first time by Kober in 1917, who reported the selective permeation of water from albumin and toluene aqueous solutions using films [1]. Since that period, PV has been explored until reaching industrial maturity nowadays due to its implementation by membrane manufactures and operation of PV-hybrid units. It is well known that PV displays the ability to break the azeotropic point of different types of solvent mixtures. PV has been, for instance, widely investigated for the dehydration of different water–solvent mixtures such as water–ethanol [2, 3], water–isopropanol [4], water–acetone [5], water–butanol, water–acetic acid [6], water–N,N-dimethylformamide, water–N,N-dimethylsulfoxide, water–N,N-dimethylacetamide, water–hydrogen peroxide [7], water–ethylene glycol [8], water–N-methyl-2-pyrrolidone [9], and water–tetrahydrofuran [10]. In addition to the dehydration of organics, PV is also able to separate organic–water mixtures, including butanol–water [11], furfural–water [12], pyridine–water [13], and ethylene dichloride–water [14], while among organic–organic mixture separations, benzene–cyclohexane [15], dimethylcarbonate–methanol [16], methanol–methyl tert-butyl ether [17–19], and acetone–butanol [20] have been successfully reported. This technology has been currently proposed for other specific purposes, such as the recovery of aroma compounds [21], the assistance of chemical and biochemical reactions [22], and for the de-alcoholization of alcoholic drinks [23].

Thanks to its versatility, PV has been industrially established for specific purposes, e.g., solvent recovery (such as dehydration of isopropanol–water and ethanol–water) [24] and different integrated PV production processes [25, 26]. To date, it is likely that most of the research toward the application of PV membranes has been focused on flat-sheet configuration membranes; however, other feasible configurations, like hollow fibers, present numerous advantages in comparison to their flat-sheet counterparts. At large or industrial scale, hollow fiber membranes are the preferred choice if compared to flat-sheet and tubular membranes, thanks to their lower fouling tendency (which provides them a longer life time) and their bigger surface-area-to-volume ratio (which provides them high packing densities) [27]. The current chapter focuses on a literature survey of PV applications using hollow fiber membranes, describing the different types of membrane materials used. For the new readers in the field, brief theoretical basis of PV technology is initially given. Highlighted studies in the field are then addressed showing the main concepts on hollow fiber membranes, as well as their relevant insights in the separation of azeotropic solvent mixtures. According to the current ongoing development works, general remarks in the field are also provided.

10.2 Theoretical Basis of PV Technology

PV has been well recognized for its ability on separating binary or multi-component azeotropic mixtures by selective partial vaporization using a permselective membrane. Such a membrane must present a dense non-porous structure [28, 29]. Basically, the liquid azeotropic mixture is placed in direct contact with the selective layer of the membrane, while the permeate, which is found at the other side of the membrane, is collected in a vapor phase. In theory, the permeate stream mostly contains the permeating species, which display a higher affinity toward the hydrophobic/hydrophilic nature of the membrane.

By using non-porous or well-known dense membranes, the transport mechanism has been widely described by the so-called solution-diffusion mechanism [28]. Herein, the mass transfer across the PV membrane can be conceptualized as depicted in Fig. 10.1. First, (1) an adsorption of the target compounds from the feed mixture to the membranes according to its chemical affinity takes place; after this, (2) a diffusion of the permeating compounds across the membrane occurs as a product of the concentration gradient; and finally, (3) a desorption of the permeating compounds at the permeate side of the membrane takes place [30, 31].

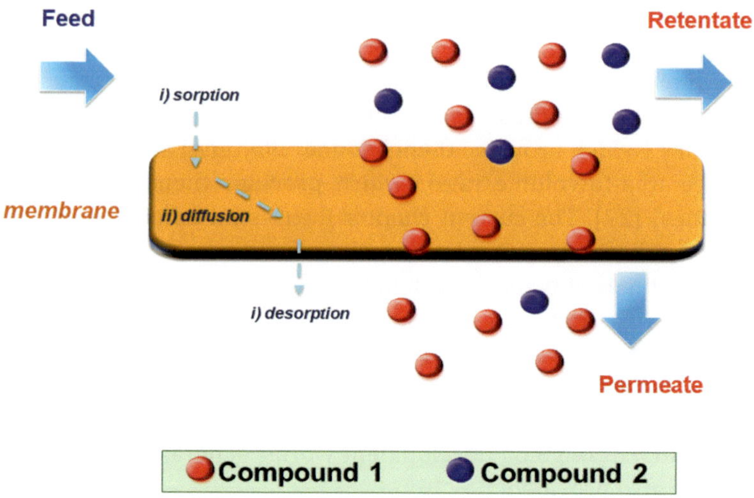

Figure 10.1 Solution-diffusion mechanism for the transport of permeating compounds through a dense PV membrane.

The permeability (P) is a parameter that depends on diffusivity (D) and solubility (S) of the permeating compounds [28], as denoted by Eq. 10.1. Particularly, S is a thermodynamic parameter that gives information about the amount of the penetrant adsorbed by the membrane at equilibrium conditions, whereas D is a kinetic parameter that gives information about the permeating compound transport rate across the membrane [32]. Here, both parameters are thus represented by the diffusion (D) and sorption (S) coefficients:

$$P = D \cdot S \tag{10.1}$$

Concerning the membrane performance, this is mainly expressed, by most of the research community, in terms of permeate flux (J) and separation factor (β) [33]. The permeate flux (J_A) of compound A, which is identified for the faster permeating compound, can be calculated by Eq. 10.2.

$$J_A = \frac{m_A}{A_m \cdot t} \tag{10.2}$$

where m_A is the mass of the compound A transported through the specific membrane area (A_m) to the permeate side and collected in a certain time (t) [34, 35]. On the other hand, the separation factor (β) is the relation between the concentrations of compounds A

and B in the permeate and retentate sides, respectively, as denoted by Eq. 10.3.

$$\beta_{AB} = \frac{\left(C_A / C_B\right)_{\text{PERMEATE}}}{\left(C_A / C_B\right)_{\text{FEED}}} \tag{10.3}$$

In addition to those parameters, the "pervaporation separation index" (PSI) is another relevant parameter used to evaluate the performance of a PV process [36]. The PSI evaluates the trade-off relationship between J and β. Its calculation is the product of multiplying both parameters, as shown in Eq. 10.4, and it is commonly used to determine the overall performance of a membrane.

$$PSI = J \cdot \beta \tag{10.4}$$

10.3 Starting Developments in Hollow Fibers for PV Technology

Promising results on using hollow fiber membranes in PV started to be provided by the research community during the 1990s. Table 10.1 lists some studies developed in this decade.

The following paragraphs focus on the use of hollow fibers in different PV separations such as water–organic separation (mainly dehydration of solvents), organic–water separations, and water desalination.

10.4 Dehydration of Organics

Maeda et al. [37] prepared a hydrophilic polyacrylonitrile (PAN) hollow fiber membrane, which was partially hydrolyzed to incorporate a carboxylic group, and then converted to a polyion complex. These membranes were able to provide a permeation rate up to 0.42 kg/m^{2}h, with a separation factor of 5000 (at 60°C, 95 wt.% ethanol in feed). Previously, Seok et al. [38] also used two types of hydrophilic membranes to dehydrate ethanol and isopropanol (IPA). Here, cellulose acetate and silicon rubber membranes were

Table 10.1 First studies on the use of hollow fibers in PV during the 1990s

Membrane Material	Membrane Nature	Type of Separation	Separation	Reported Performance	Reference
Hydrolyzed PAN	Hydrophilic	Water–organic	Water–ethanol	J: 0.42 kg/m^2h β: 5000	[37]
Cellulose acetate Silicon rubber	Hydrophilic	Water–organic	Water–ethanol Water–IPA	β: <10	[38]
PSF	Hydrophobic	Water–organic	Water–acetic acid	J: 0.2–0.5 kg/m^2h β: 47–63	[39]
Ion-exchange based on sulfonated polyethylene	Hydrophilic	Water desalination	Brine solution	J: 0.8–3.3 kg/m^2h	[40]
PVDF	Hydrophobic	Organic–water	Benzene–water	J_{water}: 0.068 kg/m^2h $J_{benzene}$: 0.008 kg/m^2h β: 1086	[41]

employed for such purposes, displaying low separation factors (β: <10). The dehydration of acetic acid (80 wt.% in feed, 70°C) was carried out by means of a hydrophobic polysulfone (PSF) hollow fibers. Such membranes showed relatively low permeation flux (in the range 0.2–0.5 kg/m^{2}h) and good selectivity (47–63) [39]. From that time on, just a few articles on the use of hollow fibers in PV saw the light.

It was in the 2000s when PV by hollow fibers was again resumed (see Table 10.2). PSF hollow fiber membranes prepared with diethylene glycol dimethyl ether (DGDE) as an additive were produced via the wet spinning process, and later tested for the dehydration of ethanol [42]. The addition of DGDE in dope solution decreased the exchange rate of solvent–nonsolvent in the membrane formation process, resulting in the suppression of macrovoid formation. Regarding the separation performance of those membranes, the permeate flux and separation factor toward water (in a 90 wt.% aqueous ethanol solution) of the asymmetric PSF membranes (with an NMP: DGDE 8:2 ratio) were about 0.172 kg/m^{2}h and 23.9, respectively. Later, employing hydrophilic chitosan (CS)/cellulose acetate hollow fiber membranes, the same group dehydrated an ethanol solution (90 wt.% ethanol in feed, 25°C) obtaining similar permeate fluxes (around 0.169 kg/m^{2}h) in comparison to their previous study [43]. Higher permeation rates, up to 0.186 kg/m^{2}h (90 wt.% ethanol in feed, 120°C), were obtained by Tsai et al. [44], which performed the dehydration of an IPA solution through heat-treated PAN hollow fiber membranes, which also displayed high separation factor (~1,116). In such a study, the authors reported that the permeate stream presented a water concentration up to 99.2 wt.%. However, the membranes presented high degree of swelling (50–60%) in the range of 25–45°C of feed temperature. In a different approach focusing on IPA dehydration, Tsai et al. carried out the preparation of composite fibers that were formed by the CS layer onto the outer surface of NaOH-hydrolyzed PAN hollow fibers [45]. During the preparation of these membranes, γ-glycidoxypropyltrimethoxysilane (GPTMS) was employed as a crosslinking agent for CS. The PV performance values were about 0.145 kg/m^{2}h and 99.7 wt.% for permeation flux and water content (90 wt.% IPA, 25°C), respectively. Additionally, the long-term PV stability for such CS/PAN composite membrane was also analyzed.

Table 10.2 Applications of hollow fiber membranes in PV during the 2000s

Membrane Material	Membrane Nature	Type of Separation	Separation	Reported Performance	Reference
PSF	Hydrophobic	Water–organic	Water–ethanol	J: 0.172 kg/m^2h β: 23.9	[42]
CS-cellulose acetate	Hydrophilic	Water–organic	Water–ethanol	J: 0.169 kg/m^2h	[43]
PAN	Hydrophilic	Water–organic	Water–IPA	J: 0.186 kg/m^2h β: 1116	[44]
CS/PAN	Hydrophilic	Water–organic	Water–IPA	J: 0.145 kg/m^2h	[45]
PAN/polyethylene glycol	Hydrophilic	Water–organic	Water–ethanol	J: 0.125–0.175 kg/m^2h	[46]
PAN	Hydrophilic	Water–organic	Water–NMP	J: 0.130–0.230 kg/m^2h	[47]
Polyamide/PAN	Hydrophilic	Water–organic	Water–ethanol	J: 0.342 kg/m^2h	[48]
PVA/PSF	Hydrophilic	Water–organic	Water–ethanol	J: 0.030 kg/m^2h β: 185	[49]
P84 copolyimide	Hydrophilic	Water–organic	Water–IPA	J: 0.883 kg/m^2h β: 10,585	[50]
PVA/sodium alginate blend on PSF	Hydrophilic	Water–organic	Water–IPA Water–ethanol Water–n-butanol Water–*ter*-butanol	J: 0.10–1.0 kg/m^2h β: >100	[51]

Membrane Material	Membrane Nature	Type of Separation	Separation	Reported Performance	Reference
Poly(acrylic acid)	Hydrophilic	Water–organic	Water–ethanol	J: 0.200 kg/m^2h β: >4	[52]
Crosslinked P84/PES dual layer	Hydrophilic	Water–organic	Water–IPA	J: 0.454 kg/m^2h β: 953	[53]
Thermal and crosslinked asymmetric Matrimid fibers	Hydrophilic	Water–organic	Water–IPA	J: 1.8 kg/m^2h β: 132	[54]
PBI/P84 dual layer	Hydrophilic	Water–organic	Water–tetrafluoropropanol	J: 0.332 kg/m^2h β: 1990	[55]
Crosslinked Torlon/P84 copolyamide–imide blend	Hydrophilic	Water–organic	Water–IPA	J: 1.0 kg/m^2h β: 185	[56]
CS/PSF-PSSA	Hydrophilic	Water–organic	Water–IPA	J: 0.128 kg/m^2h β: 78	[57]

Herein, the PV efficiency of 70 wt.% aqueous IPA solution at 25°C remained constant after 330 days of operation for the crosslinked membrane (i.e., CS–GPTMS/PAN), whereas the GPTMS-free CS/PAN membranes lost its stability after 8 days of operation [45].

More recently, Tsai's group also performed the preparation of different hollow fiber membranes via heat treating [46, 47]. For instance, heat-treated blend PAN/polyethylene glycol hollow fibers were fabricated to investigate the PV performance during ethanol dehydration, which exhibited long-term stability over 250 days (a permeate flux between 0.125 and 0.175 kg/m^{2}h). In the same line, the dehydration of NMP solutions was proposed by the authors [47], where heat-treated PAN hollow fiber membranes were used. In this process, a decrease in the permeation rate, from 0.230 to 0.130 kg/m^{2}h, was observed with the increase in feed NMP concentration in the range of 30–90 wt.%. Furthermore, at a specific NMP concentration (30 wt.% NMP in feed), the effect of the operating temperature was analyzed, observing that such parameter strongly influenced only the mass transport of water molecules. Interestingly, polyamide/PAN thin-film composite (TFC) hollow fiber membranes were prepared to dehydrate ethanol mixtures. Particularly, the fibers were fabricated by interfacial polymerization. During the interfacial polymerization step on PAN hollow fiber membrane, the surface pores of PAN fibers served as the reservoirs of amine monomers, which diffused toward the organic phase to react with trimesoyl chloride to form the nascent polyamide layer [48], as showed in Fig. 10.2.

In general, the PV performances for ethanol dehydration demonstrated that the permeation rate decreased while water content in the permeate increased with the increasing number of amine groups. Such composite membranes provided permeate flux and water content of 0.342 kg/m^{2}h and ~97 wt.% (at 25°C), respectively [48]. Even though the authors did not provide any separation factor value, it is clear that such membranes exhibited high separation efficiency toward water. Concurrently, several authors hardly worked on the development of different types of home-made hydrophilic membranes, including P84 copolyimide [50], poly(vinyl alcohol) (PVA), sodium alginate [51], Matrimid® [58], poly(acrylic acid) [52], aiming at the dehydration of several organic solvents. In particular, Dong et al. [51] developed PVA-sodium alginate (SA) blend supported on PSF hollow fibers, which were able to provide permeation flux up to 1 kg/m^{2}h and separation factor over 100 for the removal of water from different alcohols (e.g., IPA, ethanol, n-butanol, *ter*-butanol, at 80 wt.% alcohol in feed).

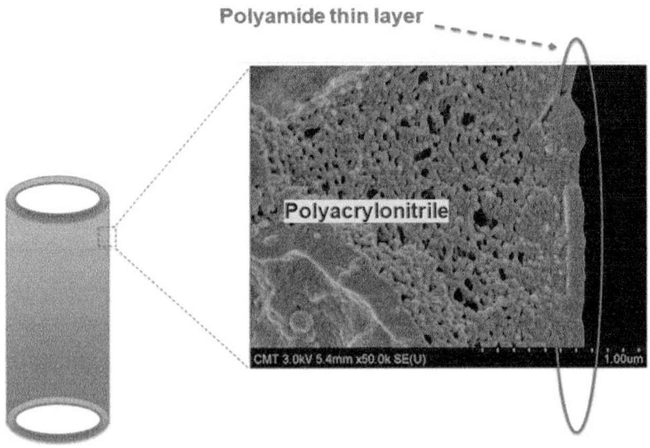

Figure 10.2 Polyamide/PAN thin-film fibers prepared via interfacial polymerization. Reprinted from Ref. [48] with permission of Elsevier.

Unlike Dong's study, Matrimid® membranes, which are commonly used for standard gas separation [59], were implemented to dehydrate acetic acid, providing higher permeation fluxes (up to 2.5 kg/m^{2}h and a separation factor of ~15) [58]. Kujawski and Krajewski [34] initiated to test commercial hollow fiber membranes (Pervasiv Ltd., Israel) based on polyethylene for the dehydration of C_2–C_4 alcohols (i.e., ethanol, propanol, IPA, n-butanol, and isobutanol) and esters (i.e., methyl acetate and ethyl acetate), identifying good water-selective properties. It is important to mention that with the evidence provided by Tsai et al. [45] and Dong et al. [51], the use of biopolymers, such as sodium alginate and CS, started to be proposed as alternatives to the chemically synthesized polymers, above all for the preparation of hydrophilic PV membranes [33, 60].

Nowadays, polymer blending is one of the main approaches aiming at the enhancement of physicochemical and separation properties of polymeric membranes. In this sense, fiber membranes based on blended polymers have been also developed. Teoh et al. [56] explored the potential of Torlon/P84 copolyamide–imide blend fibers and their chemical crosslinking implementation for the dehydration of IPA, in which the fibers have shown enhanced separation preperformance by using *p*-xylenediamine as a crosslinking agent. It has been documented that the role of the p-xylenediamine was to induce a greater chain packing and free volume reduction. Consequently, a separation factor and permeation flux of 185 and 1.0 kg/m^{2}h were, respectively, obtained, during the dehydration of feed solutions containing 85 wt.% of IPA. Similarly, Liu et al. [57] evaluated the effect of surface modifications on CS-PSF composite hollow fiber membranes, which were later tested for IPA dehydration. The authors stated that the presence of poly(styrene sulfonic acid) (PSSA) chains during the grafting of PSF contributed to the chemical bonding with CS chains promoting a good adhesion between the CS layer and the PSF. This successfully improved the stability of the composite fiber membranes. The thickness of the CS layer and the separation properties of the composite membranes were dependent on the concentrations of the CS coating solutions. For example, the membranes, containing 1.5 wt.% of CS/PSF-PSSA showed a permeate flux of 0.128 kg/m^{2}h and a separation factor of 78, dehydrating 70 wt.% IPA aqueous solution at 25°C. It is important to mention that such membranes were stable during 30 days testing, as a proof of their stability in long-term operation [57].

BTDA-TDI/MDI (P84)/polyethersulfone (PES) dual-layer hollow fibers were prepared for the PV dehydration of IPA [53]. In the study, the authors assessed the effect of several preparation conditions, such as air-gap distances and outer- and inner-layer dope flow rates on membrane formation, morphology, and separation performance. They found out that as a consequence of the different phase-inversion rates in the dual layers, the effects of elongational stresses in the air-gap region on the separation performance of dual-layer fibers were more complicated than that of single-layer hollow fibers. Using a crosslinker like *p*-xylenediamine in the P84 outer layer, the best separation factor/selectivity was reached at 2 h crosslinking, which was 0.454 kg/m^{2}h and 953 for the permeate flux and separation factor, respectively. Such results translated to permeance and selectivity values, which were around 32 g/m^{2}h·kPa and 324, respectively. This work demonstrated that a meaningful material cost saving can be reached without compromising the separation performance by selecting the dual-layer hollow fiber approach with the aid of *p*-xylenediamine crosslinking modification [53]. The evaluation of fabrication procedure conditions into other polyimides was also done by Ying et al. [54]. They analyzed the effectiveness of thermal annealing at high temperatures and chemical crosslinking using 1,3-propane diamine (PDA) on the dehydration of alcohols. In general, the authors concluded that the increase in the degree of crosslinking provoked an increase in separation factor accompanied by a subsequent flux decrease. On the other hand, thermal annealing itself did not show any enhancement on hollow fiber performance due to the cracks produced by the inhomogeneous shrinkage in the heating process. For this reason, the authors decided to combine both procedures resulting in the implementation of thermal annealing before the crosslinking procedure. This allowed the manufacture of fibers with optimal separation performance. Comparing the performance of pristine and treated membranes, pristine fibers displayed high permeation rate and low separation factor (around 6.2 kg/m^{2}h and 7.9, respectively), while, the membranes prepared with the thermal procedure followed by crosslinking, showed considerable changes in selective properties, e.g., a flux and separation factor of 1.8 kg/m^{2}h and 132, respectively, during IPA dehydration (at 80°C, 84 wt.% IPA in feed, 0.5 kPa). According to the authors, the formation of charge-transfer complexes within the polymer matrix during heat treatment not only contributed to rigidification and polymeric chain packing, but also facilitated the better action of the crosslinker, which resulted in higher size and shape discrimination in PV dehydration of IPA [54].

Wang et al. [55] also proposed the dehydration of non-typical organic solvents like tetrafluoropropanol (TFP) using polybenzimidazole (PBI)/BTDA-TDI/MDI copolyimide (P84) dual-layer hollow fiber membranes. The interest on this solvent lies in its use in medicine, dyestuff, agrichemicals, as an intermediate of fine chemicals, color agent of photography, textile finishing agent, fluoropolymer assistant ingredient, and detergent for electronic products. During the design of the proposed membranes, PBI was selected as an outer selective layer due to its high hydrophilic nature, as well as excellent solvent resistance and thermal stability. Experimentally, the PBI/P84 dual-layer hollow fiber demonstrated a permeate flux of 0.332 kg/m^{2}h and a separation factor of 1990 (at 60°C, 85 wt.% TFP in feed, 100–200 Pa) [55]. These membranes showed preferred water sorption, and the significant diffusivity difference between TFP and water was the primary cause of the highly selective properties in terms of separation factor. Additionally, by calculating the activation energy, the preferential transport of the PBI membrane toward water was confirmed, thanks to its stronger affinity and its smaller molecular size. By using the same polymer PBI, Yan et al. [61] removed water from ethylene glycol. The importance of such chemical is reflected in its different uses in chemical industries, including antifreeze agents and coolants. Typically, the industries are looking for the possibility of recycling this material. With such purpose, Yan et al. [61] fabricated dual-layer PBI/polyetherimide hollow fiber membranes, which were also processed by thermal annealing. The treated membranes at 75°C showed a separation factor value of 4524 with a total flux of 0.186 kg/m^{2}h, with a permeate stream containing 99.96 wt.% of water. The authors compared their results with previous studies aiming at the dehydration of ethylene glycol, finding out that the thermally treated membranes could provide a higher separation factor than most of other polymeric membranes.

Over the last decade, relevant advances have been done in developing hollow fiber membranes for PV applications. Table 10.3 reports the most recent works on preparing fibers over the last years in solvent dehydration.

Prof. Koros's group started to use commercial polyimides (e.g., Matrimid® and Torlon®) for the fabrication of asymmetric hollow fiber membranes [62]. At the beginning, the as-spun membranes displayed low performance in terms of separation factors for ethanol/water mixtures. In order to enhance their separation performance, thermal annealing of the fibers was carried out. Thermally annealed Matrimid® fibers exhibited a significantly increased separation performance, which depended on the annealing conditions, such as annealing temperature and time. At the end, the observed separation factors for the dehydration of ethanol (containing 15 wt.% and 95 wt.% in feed) were 40 and 240, respectively. The Torlon® fibers provided a suitable separation factor of 130 at 85 wt.% ethanol. The authors also measured long-term stability runs that were done on the annealed Matrimid® fibers at low and high ethanol feed concentrations. Herein, the annealed Matrimid® fibers exhibited stable separations as a function of time at low ethanol concentrations (i.e., 15 wt.% ethanol). However, when the concentration of ethanol was increased (between 85 and 95 wt.% ethanol), such membranes started to be swollen and plasticized resulting in a decrease in separation performance. Therefore, the crosslinking procedure seems to be needed to stabilize the structures for possible long-term applications [62].

Wang et al. [63] concurrently prepared polyamide–imide/polyetherimide dual-layer hollow fiber membranes for the dehydration of C_1–C_4 alcohols. Generally, the hollow fibers annealed at 75°C presented the highest PV performance in terms of separation factors. For water–t-butanol and water–isobutanol binary solutions, for instance, the values were higher than 50,000 with a total flux more than 0.700 kg/m^{2}h. Such impressive separation factors were supported by the high water concentration in permeate measured (99.99 wt.%). Comparing the high separation factor values reported by Wang et al. [63], Zhengbao et al. [64] also obtained impressive separation factor (up to 12,500) during the ethanol dehydration. Nevertheless, in this study, the researchers used zeolite LTA deposited by dip coating-wiping seed deposition, on ceramic hollow fibers, which showed even higher permeation rates (~6.2 kg/m^{2}h) than the ones reported by Wang et al. [63]. High permeate fluxes (5.4 kg/m^{2}h) were also observed for the removal of water from ethanol solutions (95 wt.% ethanol). In this case, polycrystalline zeolite MFI (silicalite-1) membranes were employed by depositing such material on α-Al_2O_3 hollow fibers using a secondary growth method [67]. Unlike Yan Wang and Zhengbao's studies, Shan et al. did not reach outstanding separation factor values (about 54). However, it seems to be that NaA zeolite membranes supported on four-channel α-Al_2O_3 hollow fibers tended to offer even better separation performance for ethanol dehydration. For instance, Liu et al. [74] stated that such membranes, which present a porosity of 55%, are able to provide a notably high water permeation flux of 12.8 kg/m^{2}h and a separation factor over 10,000 (10 wt.% water in feed, at 75°C). These membranes also displayed a water permeation enhancement as a function of temperature, which was attributed to the adsorption–diffusion mechanism, which dominated the water permeation through the four-channel α-Al_2O_3-supported NaA zeolite membrane. Moreover, by analyzing the Arrhenius data, the authors

Table 10.3 Applications of hollow fiber membranes in PV during the last decade

Membrane Material	Membrane Nature	Type of Separation	Separation	Reported Performance	Reference
Thermally annealed Matrimid fibers	Hydrophilic	Water–organic	Water–ethanol	β: 240	[62]
Thermally annealed Torlon fibers	Hydrophilic	Water–organic	Water–ethanol	β: 130	[62]
Thermally annealed polyamide–imide/polyetherimide dual layer	Hydrophilic	Water–organic	Water–t-butanol Water–isobutanol	J: >0.700 kg/m^{2}h β: >50,000	[63]
Zeolite LTA on ceramic hollow fibers	Hydrophilic	Water–organic	Water–ethanol	J: 6.2 kg/m^{2}h β: 12,500	[64]
Thermally annealed PBI/PEI dual-layer fibers	Hydrophilic	Water–organic	Water–ethylene glycol	J: 0.186 kg/m^{2}h β: 4524	[61]
PDM/PSF composite fibers	Hydrophobic	Organic–water	Ethanol–water	J: 0.265 kg/m^{2}h β: 6.4	[65]
6FDA-ODA-NDA/Ultem® **1010 dual-layer hollow fiber**	Hydrophilic	Water–organic	Water–IPA	J: 0.470 kg/m^{2}h β: 3658	[66]
Polycrystalline zeolite MFI on alumina fibers	Hydrophilic	Water–organic	Water–ethanol	J: 5.4 kg/m^{2}h β: 54	[67]
PVA/PAN membranes	Hydrophilic	Water–organic	Water–ethyl acetate	J: 0.34 kg/m^{2}h β: 7270	[68]
Dual PBI/P84 hollow fiber	Hydrophilic	Water–organic	Water–acetone	J: 0.284 kg/m^{2}h β: 1187	[69]

Membrane Material	Membrane Nature	Type of Separation	Separation	Reported Performance	Reference
Thin-film composite polyamide fibers	Hydrophilic	Water–organic	Water–IPA	J: 1.282 kg/m^2h β: 624	[70]
sPPSU/PPSU dual-layer fibers	Hydrophilic	Water–organic	Water–IPA	J: 3.47 kg/m^2h β: 156	[71]
Carbon hollow fiber membranes	Hydrophilic	Water–organic	Water–2-propanol	J: 1.4 kg/m^2h β: 99,900	[72]
PBI/PEI dual-layer hollow fiber	Hydrophilic	Water–organic	Water–ethyl acetate	J: 0.82 kg/m^2h β: 2478	[73]
NaA zeolite supported on α-$Al_{20}3$	Hydrophilic	Water–organic	Water–ethanol	J: 12.8 kg/m^2h β: 10,000	[74]
PDMS composite hollow fiber	Hydrophobic	Organic–water	Butanol–water	J: 1.28 kg/m^2h β: 42.9	[75]
PDMS composite hollow fiber	Hydrophobic	Organic–water	Butanol–water	J: 1.0 kg/m^2h $\beta_{ethanol}$: 6.4 $\beta_{butanol}$: 22.2 $\beta_{acetone}$: 28.6	[76]
T-type zeolite hollow fibers	Hydrophilic	Water–organic	Water–ethanol	J: 0.78 kg/m^2h β: >90	[77]
Multilayer assembled CS-PSS fibers	Hydrophilic	Water–organic	Water–ethanol	J: 0.495 kg/m^2h β: 904	[78]

found out that the ethanol molecules are more difficult to diffuse across NaA zeolitic pores than the water molecules, and the water transport through NaA zeolite membrane was fully independent from the ethanol concentration in the feed.

It is widely known that the swelling phenomenon in membranes can be one of the main drawbacks. In the case of hydrophilic PV membranes, for instance, such phenomenon occurs as a consequence of the solubility of water and highly polar molecules with the membrane material, through hydrogen bonding interactions. Here, the swelling takes place, which is well described as the reduction in the interaction between segments of the polymeric chains caused by polar compounds. Consequently, an increase in segmental mobility and free volume is observed. Of course, such interactions influence the membrane's structure and performance [79]. Typically, the swelling of polymeric materials results in an enhancement in the total flux with a decrease in selectivity [80]. Based on such effect, several authors have started to develop swelling-resistant membranes; Widjojo and Chung [66], for instance, fabricated copoly(4,4′-diphenyleneoxide/1,5-napthalene-2,2′-bis (3,4-dicarboxyl phenyl) hexa fluoro propane diimide) 6FDA-ODA-NDA/Ultem® 1010 dual-layer hollow fiber membranes via the co-extrusion process. These membranes, in fact, presented an enhanced separation performance for the dehydration of C_2–C_4 alcohols (e.g., over 99 wt.% water in permeate side for the dehydration of IPA, n-butanol, and t-butanol at 85 wt.%) and a great swelling resistance. According to the authors' conclusion, the right selection of outer- and inner-layer materials in hollow fiber membranes is fundamental to obtain superior performing membranes with enhanced structural features. The authors also pointed out that the outstanding performance of 6FDA-ODA-NDA/Ultem® 1010 dual-layer hollow fiber membranes in dehydrating alcohols can be related to (1) a precise composition and ratio of the outer and inner layers, (2) the size of d-space in the outer layer for the selective size-exclusion mechanism in the alcohol dehydration systems, and (3) a low water uptake conducting to a less swelling property of the Ultem® 1010 polyimide as a support [66].

The dehydration of specific organics is currently carried out by means of hollow fiber membranes (such as the dehydration of ethyl acetate, which is an important solvent within the chemical industry).

This solvent is generally used in the manufacture of perfumes, varnishes, plasticizers, and synthetic resins among other chemical products [81]. The production of ethyl acetate is conventionally based on an esterification reaction between acetic acid and ethanol [82, 83]. During the esterification reaction, there is the production of a by-product (water), which is forming binary (ethyl acetate and water) and ternary azeotropes (ethyl acetate, water, and residual ethanol). Importantly, this water represents a critical issue, and its removal is needed for two purposes: (1) the purification of the final product and (2) to hinder the reversibility of the reaction caused by the hydrolysis of the ester. In this regard, PV can be considered a valid solution for the removal of the by-product water using hydrophilic membranes in esterification reactions. Yuan et al. [68], for instance, used PVA as coating material and tartaric acid and maleic anhydride as crosslinking agents for the preparation of PAN hollow fiber ultrafiltration membranes. PVA/PAN membranes were manufactured by dip-coating protocol, and later tested for the PV dehydration of ethyl acetate–water solutions. The authors explored the separation performance of these membranes for ethyl acetate–water solution as a function of feed temperature and feed water concentration. In summary, the authors found out that at 2 wt.% water/ethyl acetate solution (at 40°C), the membranes exhibited high separation factor (~7270) but low permeation flux (~0.34 kg/m^{2}h). This study also investigated the PV performance of such membranes for the ternary solution (i.e., ethyl acetate, water, and ethanol), where the membranes still displayed a preferential transport of water in the range of 30–40°C. Such a phenomenon was confirmed by calculating the apparent activation energy of total permeation flux and partial fluxes of water, ethanol, and ethyl acetate whose values were about 9.08, 7.81, 24.47, and 51.72 kJ/mol, respectively. Therefore, the permeation priority was the following: water > ethanol > ethyl acetate. However, such water preference was lost in the range of temperature between 50 and 60°C [68].

The dehydration of ethyl acetate via polybenzimidazole/poly(ether imide) (PBI/PEI) dual-layer hollow fiber membranes was also studied by Yan Wang [73]. The best permeation of 0.82 kg/m^{2}h with a separation factor of 2478 was obtained for the PV dehydration of a solution containing 98 wt.% of ethyl acetate (at 60°C).

One of the most challenging solvent separations is the acetone dehydration since acetone and water molecules have close molecular sizes (0.296 and 0.469 nm, respectively), and acetone also has higher vapor pressure than water. Moreover, acetone is a powerful solvent, which can dissolve or swell most of the polymers. Considering these aspects, Shi et al. developed hollow fibers based on polybenzimidazole/BTDA-TDI/MDI (PBI/P84) polymers [69]. The dual-layer hollow fibers were thermally and chemically crosslinked in order to investigate their effectiveness to overcome acetone-induced swelling. The membranes treated at 300°C were found to have a high water–acetone separation factor (about 1187) and acceptable permeation rates (0.284 kg/m^{2}h) when dehydrating a feed containing 85 wt.% of acetone at 60°C. This study indeed demonstrated that the combination of crosslinking (using p-xylene dichloride) and thermal treatment (higher than 250°C) is suitable for stabilizing the membrane structure and significantly enhancing its separation performance.

High-performance hollow fiber membranes formed by a thin-film composite polyamide layer and a porous Torlon® polyamide–imide substrate were prepared by Zuo et al. [70]. The thin-film layer was fabricated via interfacial polymerization. This technique is typically a type of condensation polymerization in which polymerization occurs at an interface between an aqueous solution enriched in one monomer and an organic solution containing a second monomer [84]. This technique can produce polymerized selective layers as thin as 20 nm [85]. The right selection of the thickness, pore size, porosity, monomers, and polymerization conditions of the selective layer offers the possibility to obtain high performing membranes in terms of selectivity and permeation properties. The membranes prepared by Zuo et al., for instance, showed interesting separation factor and flux of 624 and 1.282 kg/m^{2}h, respectively, in dehydrating IPA (85 wt.% IPA in feed, 50°C) [70] with a permeate stream showing a water concentration of 99.1 wt.%. After a post-methanol treatment, the flux was even more enhanced to 1.920 kg/m^{2}h, slightly compromising the permeate water concentration that decreased to 98.4 wt.%. It is important to highlight that the interfacial polymerization technique also offers the possibility of using small amounts of material to fabricate membranes, representing a potential alternative to optimize the membrane manufacture costs.

Finally, according to the authors, the free volume of the membranes, at a swollen state and measured by positron annihilation spectroscopy (PALS), was able to provide relevant information about the real state and properties of the PV membrane. Tang et al. [71] continued the study of dehydrating IPA using nanometric thin (in the order of 15–90 nm) sulfonated polyphenylsulfone (sPPSU)/ polyphenylsulfone (PPSU) dual-layer hollow fiber membranes. The resulting membranes displayed both high flux (~3.47 kg/m^{2}h) and separation factor (~156) in the dehydration of an IPA aqueous solution (85 wt.%, at 50°C). After a post-thermal treatment (150°C for 2 h) of those membranes, the separation factor was impressively enhanced up to 687, whereas the permeation flux decreased to 2.30 kg/m^{2}h. The authors, therefore, concluded that both types of membranes exhibited superior PV performance compared to most of the polymeric hollow fiber membranes reported in the literature, which was attributed to (1) the anti-swelling resistance of the PPSU and sPPSU polymers, and (2) synergistic combination of the two polymers by means of molecular engineering [71].

Different types of materials have been also used in the preparation of fiber membranes. Yoshimune et al. [72], for instance, prepared carbon hollow fiber membranes derived from sulfonated poly(phenylene oxide)(SPPO) via pyrolysis at 600°C. Interestingly, carbon-based membranes have become more attractive for PV and gas separation applications due to their thermal and chemical resistance while maintaining excellent shape selectivity. To date, carbon-based fiber membranes have been prepared using several polymers, including cellulose-based derivatives [86], polyimides [87], PAN [88], polyetherimide [89], poly(phenylene oxide) [90], among others. Regarding Yoshimune's membranes, they showed high permeate fluxes (about 1.4 kg/m^{2}h) and impressive separation factor (~99,900) in the dehydration of 2-propanol (90 wt.% in feed, 70°C) [72].

The latest reports on the dehydration of organics deal with the removal of water from ethanolic solutions. For instance, Wang et al. [77] evaluated the performance of hollow fiber T-type zeolite membrane modules in the dehydration of ethanol solutions (90 wt.% ethanol in feed, at 70°C) revealing an average water permeation flux in the order of 0.78 kg/m^{2}h and a separation factor over 90. Additionally, the packing density could reach as high as 600 m^2/m^3, which means that hollow fiber membrane modules could reduce the fabrication cost of PV setups. Very recently, for the testing of the same azeotropic separation, multilayer assembled CS-poly(4-styrenesulfonic acid) (PSS) on ceramic hollow fiber membranes was investigated in dehydrating ethanol (10 wt.% water in feed, 70°C) [78]. The authors observed that the separation factor increased with the increase in CS concentration up to a concentration of 3.0 wt.%. Thereby, composite membranes with 3.0 wt.% of CS were selected to analyze the effect of PSS loading. The resulting assembled composite membranes, containing 0.75 wt.% PSS solution, provided the highest PV separation performance, i.e., a total flux of 0.495 kg/m^2h and a separation factor of 904.

10.5 Organic–Water Separations

The separation of organic compounds from aqueous solution is carried out by means of hydrophobic membranes. In this regard, asymmetric hollow polyvinylidene difluoride (PVDF) fiber membranes were prepared, for instance, by Jian and Pintauro [41], and applied for the separation of benzene–water azeotropic solution. At specific conditions (120 ppm benzene in water, vacuum pressure 0.025 atm, 25°C), the membrane displayed a benzene separation factor of 1086. Moreover, these membranes were also effective in separating toluene, chloroform, and styrene from water.

Another typical hydrophobic membrane material is polydimethylsiloxane (PDMS). This polymer, made of a flexible (Si–O) backbone and a repeating ($Si(CH_3)_2O$) unit [91], is usually recognized by its several advantages, including flexibility, nontoxicity, high hydrophobicity, chemical resistance, gas permeability, low cost, and good molding capability [92]. Therefore, according to its high hydrophobicity, Guo et al. [65] successfully fabricated membranes by dynamically coating PDMS on the inner surface of PSF hollow fibers. A graphical representation of such a preparation procedure is illustrated in Fig. 10.3. For the preparation of such membranes, the authors analyzed the effect of polymer concentration, coating time, and crosslink agent concentration on alcohol permselective properties. The best conditions were found at coating time of 60 min; PDMS concentration of 22.8 wt.%; and TEOS concentration of 1.5 wt.%. During the removal of ethanol 8 wt.% in the feed (50°C), the membrane showed a separation factor of 6.4 with a permeate flux of 0.265 kg/m^{2}h. To provide a better understanding of organophilic membranes, a preliminary analysis of the mass transport and prediction of the permeation properties was proposed. Sukitpaneenit et al. [93], for instance, applied a pore-flow model to analyze the mass transport in the PV of ethanol–water solutions through PVDF hollow fiber membranes with a sub-micron pore size.

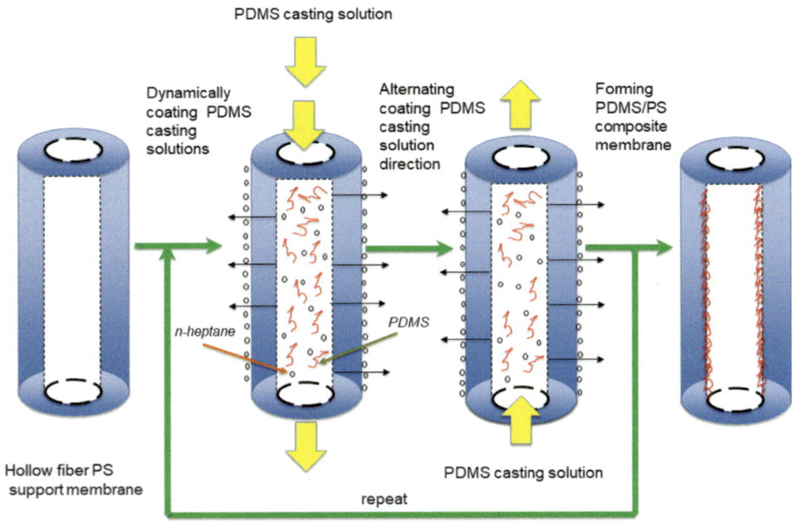

Figure 10.3 Inner skin PDMS/PSF hollow fiber composite membranes prepared via dynamic coating method. Adapted from Ref. [65].

More recently, Wanqin Jin's group explored the ability to recover bio-butanol using PDMS composite hollow fiber membranes [75, 76]. The membranes, which were prepared by dip coating the PDMS layer on the surface of microporous ceramic hollow fiber support, presented a defect-free PDMS layer, showing a total flux of 1.282 kg/m^{2}h and a separation factor of 42.9 for the separation of 1 wt.% n-butanol–water mixtures (at 40°C). Such a performance was exhibited constantly over 200 h of operating time, which gave an idea of the good stability of the membranes. Lately, the authors optimized a module filled with seven bundles of hollow fiber membranes at a high packing density of 560 m^2/m^3, exhibiting high and certainly stable performance in the real acetone–butanol–ethanol (ABE) fermentation broth during 120 h of continuous operation at 40°C. In general, the total flux was 1.0 kg/m^{2}h with separation factors of 6.4, 22.2, and 28.6 measured for ethanol, butanol, and acetone, respectively [76]. As a concluding remark, the authors stated that this approach can be useful for the production of bio-butanol by means of PV.

10.6 Water Desalination

When dealing with the application of PV technology in water desalination, hollow fiber membranes have been also proposed. For example, Korin et al. [40] manufactured hydrophilic ion-exchange membranes on sulfonated polyethylene hollow fibers. A water flux of 0.8–3.3 kg/m^{2}h was achieved with inlet brine temperature in the range of 25–65°C. Similarly, the authors developed a water desalination system using hollow fiber PV membranes [94], as shown in Fig. 10.4.

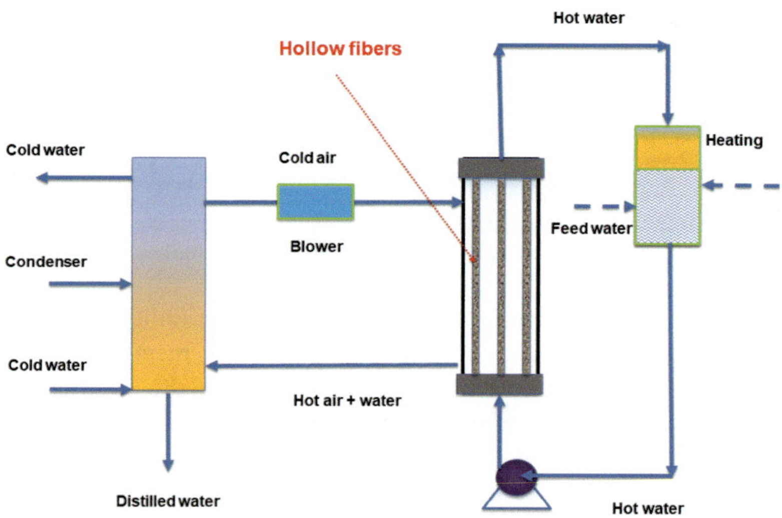

Figure 10.4 Water desalination system installed with hollow fiber PV membranes. Adapted from Ref. [94].

Such an integrated system was able to provide good water flux (1.5–3.0 L/m^{2}h) through the hollow fibers when the water temperature was in the range of 45–65°C. Interestingly, the authors stated that the energy demand for recycling hot water was dependent on water temperature and on hollow fibers diameter and length. The energy requirement for air recycling depended on the air temperature and on the pressure difference of the system. In a further analysis of the feasibility of the desalination PV process, the calculated energy requirement for pumping air and water in a pilot plant of a capacity of 6.3 L/h, installed with anion-exchange hollow fibers of 4 m^2, was found to be about 2 kWh/m^3, when hot water temperature was about 60°C [94].

10.7 Concluding Remarks

This chapter reviews the current available literature on PV applications using hollow fiber membranes. To date, different types of separation applications have been carried out, including water/organic, organic/water, and water desalination, being the dehydration of organics the most explored approach. It is important to point out that there is no report about the use of hollow fiber membranes in any organic–organic separation. Here, new researchers may find a new field of study.

The study of specific aspects of membranes can lead to the preparation of membranes displaying higher separation performance, such as the case of thermal and crosslinked asymmetric Matrimid® hollow fibers [54], crosslinked Torlon/P84 copolyamide–imide blend [56], zeolite LTA deposited on ceramic hollow fibers [64], thin-film composite polyamide fibers [70], sPPSU/PPSU dual-layer fibers [71], carbon-based hollow fiber [72], and NaA zeolite supported on α-Al_2O_3 [74]. Through the study of different works, it turned out that hollow fiber membranes have demonstrated to display great separation performance in PV, also in long-term operations, paving the way for a wider application at the industrial level.

References

1. Wang, Y., Widjojo, N., Sukitpanneenit, P., and Chung, T. S. (2013) Membrane pervaporation, in *Separation and Purification Technologies in Biorefineries* (eds. Ramaswamy, S., Huang, H., and Ramarao, B.), John Wiley & Sons, Ltd, Oxford, pp. 259–299.

2. Kudasheva, A., Sorribas, S., Zornoza, B., Téllez, C., and Coronas, J. (2015) Pervaporation of water/ethanol mixtures through polyimide based mixed matrix membranes containing ZIF-8, ordered mesoporous silica and ZIF-8-silica core-shell spheres. *J. Chem. Technol. Biotechnol.*, **90**(4), 669–677.

3. Castro-Muñoz, R., Buera-Gonzalez, J., de la Iglesia, O., Galiano, F., Fíla, V., Malankowska, M., Rubio, C., Figoli, A., Tellez, C., and Coronas, J. (2019) Towards the dehydration of ethanol using pervaporation cross-linked poly(vinyl alcohol)/graphene oxide membranes. *J. Memb. Sci.*, **582**, 423–434.

4. Han, Y., Wang, K., Lai, J., and Liu, Y. (2014) Hydrophilic chitosan-modified polybenzoimidazole membranes for pervaporation dehydration of isopropanol aqueous solutions. *J. Memb. Sci.*, **463**, 17–23.

5. Mangindaan, D. W., Min Shi, G., and Chung, T. S. (2014) Pervaporation dehydration of acetone using P84 co-polyimide flat sheet membranes modified by vapor phase crosslinking. *J. Memb. Sci.*, **458**, 76–85.

6. Badiger, H., Shukla, S., Kalyani, S., and Sridhar, S. (2014) Thin film composite sodium alginate membranes for dehydration of acetic acid and isobutanol. *J. Appl. Polym. Sci.*, **131**(6), 1–9.

7. Ong, Y. K., Shi, G. M., Le, N. L., Tang, Y. P., Zuo, J., Nunes, S. P., and Chung, T. S. (2016) Recent membrane development for pervaporation processes. *Prog. Polym. Sci.*, **57**, 1–31.

8. Zhang, W., Ying, Y., Ma, J., Guo, X., Huang, H., Liu, D., and Zhong, C. (2017) Mixed matrix membranes incorporated with polydopamine-coated metal-organic framework for dehydration of ethylene glycol by pervaporation. *J. Memb. Sci.*, **527** (December 2016), 8–17.

9. Prasad, N. S., Moulik, S., Bohra, S., Rani, K. Y., and Sridhar, S. (2016) Solvent resistant chitosan/poly(ether-block-amide) composite membranes for pervaporation of n-methyl-2-pyrrolidone/water mixtures. *Carbohydr. Polym.*, **136**, 1170–1181.

10. Han, Y. J., Su, W. C., Lai, J. Y., and Liu, Y. L. (2015) Hydrophilically surface-modified and crosslinked polybenzimidazole membranes for pervaporation dehydration on tetrahydrofuran aqueous solutions. *J. Memb. Sci.*, **475**, 496–503.

11. Shin, C., Chen, X. C., Prausnitz, J. M., and Balsara, N. P. (2017) Effect of block copolymer morphology controlled by casting-solvent quality on pervaporation of butanol/water mixtures. *J. Memb. Sci.*, **523** (September 2016), 588–595.

12. Qin, F., Li, S., Qin, P., Nazmul, M., and Tan, T. (2014) A PDMS membrane with high pervaporation performance for the separation of furfural and its potential in industrial application. *Green Chem.*, **16**, 1262–1273.
13. Singha, N. R., Das, P., and Ray, S. K. (2013) Recovery of pyridine from water by pervaporation using filled and crosslinked EPDM membranes. *J. Ind. Eng. Chem.*, **19**(6), 2034–2045.
14. Manjunath, D., Madhumala, M., Kalyani, S., Prasad, R., and Sridhar, S. (2013) Extraction of tetrahydrofuran and ethylene dichloride solvents from aqueous solutions by pervaporation through thin film composite PDMS membranes extraction. *Sep. Sci. Technol.*, **48**, 706–715.
15. Shen, J.-N., Chu, Y.-X., Ruan, H.-M., Wu, L.-G., Gao, C.-J., and van der Bruggen, B. (2014) Pervaporation of benzene/cyclohexane mixtures through mixed matrix membranes of chitosan and Ag+/carbon nanotubes. *J. Memb. Sci.*, **462**, 160–169.
16. Wang, L., Han, X., Li, J., Zheng, D., and Qin, L. (2013) Modified MCM-41 silica spheres filled polydimethylsiloxane membrane for dimethylcarbonate/methanol separation via pervaporation. *J. Appl. Polym. Sci.*, **127**(6), 4662–4671.
17. Zereshki, S., Figoli, A., Madaeni, S. S., Simone, S., Esmailinezhad, M., and Drioli, E. (2011) Pervaporation separation of MeOH/MTBE mixtures with modified PEEK membrane: Effect of operating conditions. *J. Memb. Sci.*, **371**(1–2), 1–9.
18. Castro-Muñoz, R., Galiano, F., Fíla, V., Drioli, E., and Figoli, A. (2018) Matrimid®5218 dense membrane for the separation of azeotropic MeOH-MTBE mixtures by pervaporation. *Sep. Purif. Technol.*, **199**, 27–36.
19. Castro-Muñoz, R., Galiano, F., de la Iglesia, Ó., Fíla, V., Tellez, C., Coronas, J., and Figoli, A. (2019) Graphene oxide: Filled polyimide membranes in pervaporative separation of azeotropic methanol–MTBE mixtures. *Sep. Purif. Technol.*, **224** (February), 265–272.
20. Cai, D., Hu, S., Miao, Q., Chen, C., Chen, H., Zhang, C., Li, P., Qin, P., and Tan, T. (2017) Two-stage pervaporation process for effective in situ removal acetone-butanol-ethanol from fermentation broth. *Bioresour. Technol.*, **224**, 380–388.
21. Castro-Muñoz, R. (2019) Pervaporation: The emerging technique for extracting aroma compounds from food systems. *J. Food Eng.*, **253** (February), 27–39.
22. Castro-Muñoz, R., Galiano, F., and Figoli, A. (2019) Chemical and biochemical reactions assisted by pervaporation technology. *Crit. Rev. Biotechnol.*, **39**(7), 884–903.

23. Castro-Muñoz, R. (2019) Pervaporation-based membrane processes for the production of non-alcoholic beverages. *J. Food Sci. Technol.*, **56**(5), 2333–2344.

24. Buonomenna, M. G., Golemme, G., Tone, C. M., De Santo, M. P., Ciuchi, F., Perrotta, E., Zappone, B., Galiano, F., and Figoli, A. (2011) Ordering phenomena in nanostructured poly(styrene-b-butadiene-b-styrene) (SBS) membranes for selective ethanol transport. *J. Memb. Sci.*, **385–386**(1), 162–170.

25. Luis, P. and van Der Bruggen, B. (2015) The driving force as key element to evaluate the pervaporation performance of multicomponent mixtures. *Sep. Purif. Technol.*, **148**, 94–102.

26. Castro-Muñoz, R., Galiano, F., Fíla, V., Drioli, E., and Figoli, A. (2019) Mixed matrix membranes (MMMs) for ethanol purification through pervaporation: Current state of the art. *Rev. Chem. Eng.*, **35**(5), 565–590.

27. Purkait, K., Sinha, M., Mondal, P., and Singh, R. (2018) pH-responsive membranes, in *Stimuli Responsive Polymeric Membranes* (eds. Purkait, K., Sinha, M., Mondal, P., and Singh, R.), Elsevier Ltd., Oxford, pp. 2–244.

28. Wijmans, J. G. and Baker, R. W. (1995) The solution-diffusion model: A review. *J. Memb. Sci.*, **107**(1–2), 1–21.

29. Basile, A., Galiano, F., Santoro, S., and Figoli, A. (2016) Pervaporation and membrane contactors, in *Membrane Reactor Engineering: Applications for a Greener Process Industry* (eds. Basile, A., De Falco, M., Centi, G., and Iaquaniello, G.), John Wiley & Sons, West Sussex, United Kingdom, pp. 280–312.

30. Baker, R. W., Wijmans, J. G., and Huang, Y. (2010) Permeability, permeance and selectivity: A preferred way of reporting pervaporation performance data. *J. Memb. Sci.*, **348**(1–2), 346–352.

31. Castro-Muñoz, R., Fíla, V., and Dung, C. T. (2017) Mixed matrix membranes based on PIMs for gas permeation: Principles, synthesis, and current status. *Chem. Eng. Commun.*, **204**(3), 295–309.

32. Galiano, F., Falbo, F., and Figoli, A. (2016) Polymeric pervaporation membranes: Organic–organic separation, in *Nanostructured Polymer Membranes*, (eds. Visakh, O. and Nazarenko, P.), Scrivener Publishing LLC, Massachusetts, United States, pp. 281–304.

33. Castro-Muñoz, R. and González-Valdez, J. (2019) New trends in biopolymer-based membranes for pervaporation. *Molecules*, **24**(19), 3584.

34. Kujawski, W. and Krajewski, S. R. (2004) Sweeping gas pervaporation with hollow-fiber ion-exchange membranes. *Desalination*, **162**(1–3), 129–135.

35. Santoro, S., Galiano, F., Jansen, J. C., and Figoli, A. (2017) Strategy for scale-up of SBS pervaporation membranes for ethanol recovery from diluted aqueous solutions. *Sep. Purif. Technol.*, **176**, 252–261.

36. Ying, L., Wang, Y., Chung, T., Yi, X., and Lai, J. (2009) Polyimides membranes for pervaporation and biofuels separation. *Prog. Polym. Sci.*, **34**, 1135–1160.

37. Maeda, Y., Tsuyumoto, M., Karakane, H., and Tsugaya, H. (1991) Separation of water–ethanol mixture by pervaporation through hydrolyzed polyacrylonitrile hollow fiber membranes. *Polym. J.*, **23**(5), 501–511.

38. Seok, D. R., Kang, S. G., and Hwang, S. T. (1987) Use of pervaporation for separating azeotropic mixtures using two different hollow fiber membranes. *J. Memb. Sci.*, **33**(1), 71–81.

39. Koops, G. H., Nolten, J. A. M., Mulder, M. H. V., and Smolders, C. A. (1994) Integrally skinned polysulfone hollow fiber membranes for pervaporation. *J. Appl. Polym. Sci.*, **54**(3), 385–404.

40. Korin, E., Ladizhensky, I., and Korngold, E. (1996) Hydrophilic hollow fiber membranes for water desalination by the pervaporation method. *Chem. Eng. Process. Process Intensif.*, **35**(6), 451–457.

41. Jian, K. and Pintauro, P. N. (1997) Asymmetric PVDF hollow-fiber membranes for organic/water pervaporation separations. *J. Memb. Sci.*, **135**(1), 41–53.

42. Tsai, H. A., Hong, M. J., Huang, G. S., Wang, Y. C., Li, C. L., Lee, K. R., and Lai, J. Y. (2002) Effect of DGDE additive on the morphology and pervaporation performances of asymmetric PSf hollow fiber membranes. *J. Memb. Sci.*, **208**(1–2), 233–245.

43. Tsai, H. A., Chen, H. C., Chou, W. L., Lee, K. R., Yang, M. C., and Lai, J. Y. (2004) Pervaporation of water/alcohol mixtures through chitosan/cellulose acetate composite hollow-fiber membranes. *J. Appl. Polym. Sci.*, **94**(4), 1562–1568.

44. Tsai, H. A., Ciou, Y. S., Hu, C. C., Lee, K. R., Yu, D. G., and Lai, J. Y. (2005) Heat-treatment effect on the morphology and pervaporation performances of asymmetric PAN hollow fiber membranes. *J. Memb. Sci.*, **255**(1–2), 33–47.

45. Tsai, H. A., Chen, W. H., Kuo, C. Y., Lee, K. R., and Lai, J. Y. (2008) Study on the pervaporation performance and long-term stability of aqueous

iso-propanol solution through chitosan/polyacrylonitrile hollow fiber membrane. *J. Memb. Sci.*, **309**(1–2), 146–155.

46. Tsai, H. A., Ye, Y. L., Lee, K. R., Huang, S. H., Suen, M. C., and Lai, J. Y. (2011) Characterization and pervaporation dehydration of heat-treatment PAN hollow fiber membranes. *J. Memb. Sci.*, **368**(1–2), 254–263.

47. Tsai, H. A., Chen, Y. L., Lee, K. R., and Lai, J. Y. (2012) Preparation of heat-treated PAN hollow fiber membranes for pervaporation of NMP/H_2O mixtures. *Sep. Purif. Technol.*, **100**, 97–105.

48. Tsai, H. A., Wang, T. Y., Huang, S. H., Hu, C. C., Hung, W. S., Lee, K. R., and Lai, J. Y. (2017) The preparation of polyamide/polyacrylonitrile thin film composite hollow fiber membranes for dehydration of ethanol mixtures. *Sep. Purif. Technol.*, **187**, 221–232.

49. Wei, Y. M., Xu, Z. L., Qusay, F. A., and Wu, K. (2005) Polyvinyl alcohol/polysulfone (PVA/PSF) hollow fiber composite membranes for pervaporation separation of ethanol/water solution. *J. Appl. Polym. Sci.*, **98**(1), 247–254.

50. Liu, R., Qiao, X., and Chung, T. S. (2005) The development of high performance P84 co-polyimide hollow fibers for pervaporation dehydration of isopropanol. *Chem. Eng. Sci.*, **60**(23), 6674–6686.

51. Dong, Y. Q., Zhang, L., Shen, J. N., Song, M. Y., and Chen, H. L. (2006) Preparation of poly(vinyl alcohol)-sodium alginate hollow-fiber composite membranes and pervaporation dehydration characterization of aqueous alcohol mixtures. *Desalination*, **193**(1–3), 202–210.

52. Xu, Z. K., Dai, Q. W., Liu, Z. M., Kou, R. Q., and Xu, Y. Y. (2003) Microporous polypropylene hollow fiber membranes: Part II. Pervaporation separation of water/ethanol mixtures by the poly(acrylic acid) grafted membranes. *J. Memb. Sci.*, **214**(1), 71–81.

53. Liu, R. X., Qiao, X. Y., and Chung, T. S. (2007) Dual-layer P84/polyethersulfone hollow fibers for pervaporation dehydration of isopropanol. *J. Memb. Sci.*, **294**(1–2), 103–114.

54. Ying, L., Chung, T., and Rajagopalan, R. (2008) Dehydration of alcohols by pervaporation through polyimide Matrimid – asymmetric hollow fibers with various modifications. *Chem. Eng. Sci.*, **63**, 204–216.

55. Wang, K. Y., Chung, T. S., and Rajagopalan, R. (2007) Dehydration of tetrafluoropropanol (TFP) by pervaporation via novel PBI/BTDA-TDI/MDI co-polyimide (P84) dual-layer hollow fiber membranes. *J. Memb. Sci.*, **287**(1), 60–66.

56. Teoh, M. M., Chung, T. S., Wang, K. Y., and Guiver, M. D. (2008) Exploring Torlon/P84 co-polyamide-imide blended hollow fibers and their chemical cross-linking modifications for pervaporation dehydration of isopropanol. *Sep. Purif. Technol.*, **61**(3), 404–413.

57. Liu, Y. L., Yu, C. H., Ma, L. C., Lin, G. C., Tsai, H. A., and Lai, J. Y. (2008) The effects of surface modifications on preparation and pervaporation dehydration performance of chitosan/polysulfone composite hollow-fiber membranes. *J. Memb. Sci.*, **311**(1–2), 243–250.

58. Zhou, F. and Koros, W. J. (2006) Pervaporation using hollow-fiber membranes for dehydrating acetic acid and water mixtures. *Ind. Eng. Chem. Res.*, **45**(5), 1787–1796.

59. Castro-Muñoz, R., Martin-Gil, V., Ahmad, M. Z., and Fíla, V. (2018) Matrimid® 5218 in preparation of membranes for gas separation: Current state-of-the-art. *Chem. Eng. Commun.*, **205**, 161–196.

60. Galiano, F., Briceño, K., Marino, T., Molino, A., Christensen, K. V., and Figoli, A. (2018) Advances in biopolymer-based membrane preparation and applications. *J. Memb. Sci.*, **564** (May), 562–586.

61. Wang, Y., Gruender, M., and Chung, T. S. (2010) Pervaporation dehydration of ethylene glycol through polybenzimidazole (PBI)-based membranes. 1. Membrane fabrication. *J. Memb. Sci.*, **363**(1–2), 149–159.

62. Qiu, W., Kosuri, M., Zhou, F., and Koros, W. J. (2009) Dehydration of ethanol-water mixtures using asymmetric hollow fiber membranes from commercial polyimides. *J. Memb. Sci.*, **327**(1–2), 96–103.

63. Wang, Y., Goh, S. H., Chung, T. S., and Na, P. (2009) Polyamide-imide/polyetherimide dual-layer hollow fiber membranes for pervaporation dehydration of C1-C4 alcohols. *J. Memb. Sci.*, **326**(1), 222–233.

64. Zhengbao, W., Qinqin, G., Jia, S., and Yushan, Y. (2009) High performance zeolite LTA pervaporation membranes on ceramic hollow fibers by dip coating-wiping seed deposition. *J. Am. Chem. Soc.*, **131**(20), 6910–6911.

65. Guo, J., Zhang, G., Wu, W., Ji, S., Qin, Z., and Liu, Z. (2010) Dynamically formed inner skin hollow fiber polydimethylsiloxane/polysulfone composite membrane for alcohol permselective pervaporation. *Chem. Eng. J.*, **158**(3), 558–565.

66. Widjojo, N. and Chung, T.S. (2009) Pervaporation dehydration of C_2-C_4 alcohols by 6FDA-ODA-NDA/Ultem® dual-layer hollow fiber membranes with enhanced separation performance and swelling resistance. *Chem. Eng. J.*, **155**(3), 736–743.

67. Shan, L., Shao, J., Wang, Z., and Yan, Y. (2011) Preparation of zeolite MFI membranes on alumina hollow fibers with high flux for pervaporation. *J. Memb. Sci.*, **378**(1–2), 319–329.

68. Yuan, H. K., Ren, J., Ma, X. H., and Xu, Z. L. (2011) Dehydration of ethyl acetate aqueous solution by pervaporation using PVA/PAN hollow fiber composite membrane. *Desalination*, **280**(1–3), 252–258.

69. Shi, G. M., Wang, Y., and Chung, T. (2012) Dual-layer PBI/P84 hollow fibers for pervaporation dehydration of acetone. *AICHE J.*, **58**(4), 1133–1145.

70. Zuo, J., Wang, Y., Sun, S. P., and Chung, T. S. (2012) Molecular design of thin film composite (TFC) hollow fiber membranes for isopropanol dehydration via pervaporation. *J. Memb. Sci.*, **405–406**, 123–133.

71. Tang, Y., Widjojo, N., Chung, T., Weber, M., and Maletzko, C. (2013) Nanometric thin skinned dual-layer hollow fiber membranes for dehydration of isopropanol. *AICHE J.*, **59**(8), 2943–2956.

72. Yoshimune, M., Mizoguchi, K., and Haraya, K. (2013) Alcohol dehydration by pervaporation using a carbon hollow fiber membrane derived from sulfonated poly(phenylene oxide). *J. Memb. Sci.*, **425–426**, 149–155.

73. Wang, Y. (2015) Pervaporation dehydration of ethyl acetate via PBI/PEI hollow fiber membranes. *Ind. Eng. Chem. Res.*, **54**(11), 3082–3089.

74. Liu, D., Zhang, Y., Jiang, J., Wang, X., Zhang, C., and Gu, X. (2015) High-performance NaA zeolite membranes supported on four-channel ceramic hollow fibers for ethanol dehydration. *RSC Adv.*, **5**(116), 95866–95871.

75. Dong, Z., Liu, G., Liu, S., Liu, Z., and Jin, W. (2014) High performance ceramic hollow fiber supported PDMS composite pervaporation membrane for bio-butanol recovery. *J. Memb. Sci.*, **450**, 38–47.

76. Liu, D., Liu, G., Meng, L., Dong, Z., Huang, K., and Jin, W. (2015) Hollow fiber modules with ceramic-supported PDMS composite membranes for pervaporation recovery of bio-butanol. *Sep. Purif. Technol.*, **146**, 24–32.

77. Wang, X., Jiang, J., Liu, D., Xue, Y., Zhang, C., and Gu, X. (2017) Evaluation of hollow fiber T-type zeolite membrane modules for ethanol dehydration. *Chinese J. Chem. Eng.*, **25**(5), 581–586.

78. Zhang, X., Wang, M., Ji, C. H., Xu, X.R., Ma, X. H., and Xu, Z. L. (2018) Multilayer assembled CS-PSS/ceramic hollow fiber membranes for pervaporation dehydration. *Sep. Purif. Technol.*, **203** (April), 84–92.

79. Zhao, J., Wang, F., Pan, F., Zhang, M., Yang, X., Li, P., Jiang, Z., Zhang, P., Cao, X., and Wang, B. (2013) Enhanced pervaporation dehydration performance of ultrathin hybrid membrane by incorporating bioinspired multifunctional modifier and $TiCl_4$ into chitosan. *J. Memb. Sci.*, **446**, 395–404.

80. Jia, Z. and Wu, G. (2016) Metal-organic frameworks based mixed matrix membranes for pervaporation. *Microporous Mesoporous Mater.*, **235**, 151–159.

81. Penkova, A., Polotskaya, G., and Toikka, A. (2015) Pervaporation composite membranes for ethyl acetate production. *Chem. Eng. Process. Process Intensif.*, **87**, 81–87.

82. Castro-Munoz, R., Iglesia, Ó. D. La, Fíla, V., Téllez, C., and Coronas, J. (2018) Pervaporation-assisted esterification reactions by means of mixed matrix membranes. *Ind. Eng. Chem. Res.*, **57**(47), 15998–16011.

83. Hu W. W., Zhang, X. H., Zhang, Q. G., Liu, Q. L., Zhu, A. M. (2012) Pervaporation dehydration of water/ethanol/ethyl acetate mixtures using poly(vinyl alcohol)–silica hybrid membranes. *J. Appl. Polym. Sci.*, **126**, 778–787.

84. Yin, J., Kim, E., Yang, J., and Deng, B. (2012) Fabrication of a novel thin-film nanocomposite (TFN) membrane containing MCM-41 silica nanoparticles (NPs) for water purification. *J. Memb. Sci.*, **423–424**, 238–246.

85. Korikov, A. P., Kosaraju, P. B., and Sirkar, K. K. (2006) Interfacially polymerized hydrophilic microporous thin film composite membranes on porous polypropylene hollow fibers and flat films. *J. Memb. Sci.*, **279**(1–2), 588–600.

86. He, X. and Hägg, M. B. (2012) Structural, kinetic and performance characterization of hollow fiber carbon membranes. *J. Memb. Sci.*, **390–391**, 23–31.

87. Kusuki, Y., Shimazaki, H., Tanihara, N., Nakanishi, S., and Yoshinaga, T. (1997) Gas permeation properties and characterization of asymmetric carbon membranes prepared by pyrolyzing asymmetric polyimide hollow fiber membrane. *J. Memb. Sci.*, **134**(2), 245–253.

88. David, L. I. B., and Ismail, A. F. (2003) Influence of the thermastabilization process and soak time during pyrolysis process on the polyacrylonitrile carbon membranes for O_2/N_2 separation. *J. Memb. Sci.*, **213**(1–2), 285–291.

89. Barbosa-Coutinho, E., Salim, V. M. M., and Borges, C. P. (2003) Preparation of carbon hollow fiber membranes by pyrolysis of polyetherimide. *Carbon N. Y.*, **41**(9), 1707–1714.

90. Yoshimune, M., Fujiwara, I., and Haraya, K. (2007) Carbon molecular sieve membranes derived from trimethylsilyl substituted poly(phenylene oxide) for gas separation. *Carbon N. Y.*, **45**(3), 553–560.

91. Tan, X. M. and Rodrigue, D. (2019) A review on porous polymeric membrane preparation. Part II: Production techniques with polyethylene, polydimethylsiloxane, polypropylene, polyimide, and polytetrafluoroethylene. *Polymers (Basel).*, **11**(8).

92. Le-The, H., Tibbe, M., Loessberg-Zahl, J., Palma Do Carmo, M., Van Der Helm, M., Bomer, J., Van Den Berg, A., Leferink, A., Segerink, L., and Eijkel, J. (2018) Large-scale fabrication of free-standing and sub-μm PDMS through-hole membranes. *Nanoscale*, **10**(16), 7711–7718.

93. Sukitpaneenit, P., Chung, T. S., and Jiang, L. Y. (2010) Modified pore-flow model for pervaporation mass transport in PVDF hollow fiber membranes for ethanol-water separation. *J. Memb. Sci.*, **362**(1–2), 393–406.

94. Korngold, E., Korin, E., and Ladizhensky, I. (1996) Water desalination by pervaporation with hollow fiber membranes. *Desalination*, **107**(2), 121–129.

Chapter 11

Hollow Fiber and Nanofiber Membranes in Bioartificial Organs and Tissue Engineering

Sabrina Morelli,[a] Antonella Piscioneri,[a] Simona Salerno,[a] Ricardo Matos,[b] Paula I. P. Soares,[b] Jorge Carvalho Silva,[c] João Paulo Borges,[b] and Loredana De Bartolo[a]

[a]*Institute on Membrane Technology, National Research Council of Italy, (ITM-CNR), via P. Bucci, Cubo 17/C Rende (CS), Italy*

[b]*CENIMAT/i3N, Departamento de Ciência dos Materiais, Faculdade de Ciências e Tecnologia, Universidade Nova de Lisboa, Caparica, Portugal*

[c]*CENIMAT/i3N, Departamento de Física, Faculdade de Ciências e Tecnologia, Universidade Nova de Lisboa, Caparica, Portugal*

l.debartolo@itm.cnr.it

Membrane systems offer interesting approaches for the design of bio/artificial organs, tissue analogues as well as for drug and cell delivery, offering new specific cell therapy with high reliability and efficacy. Membranes in hollow fiber (HF) configuration can be used for the creation of organs/tissue analogous (e.g., liver, kidney, skin,

Hollow Fibers and Nanofibers in Membrane Science: Preparation, Characterization, and Applications
Edited by Alberto Figoli, Mir Saeed Seyed Dorraji, and Francesco Galiano

ISBN 978-981-4968-03-4 (Hardcover), 978-1-003-25663-2 (eBook)
www.jennystanford.com

neuronal system, bone) because of the selective transport properties that allow high control at molecular level of cell microenvironment. Hollow fibers enable the compartmentalization of cells in micro- and nano-structured complexes, the adhesion of cells offering a wide surface area, and a selective mass transfer of molecules to and from cells.

Membranes composed of nanofibers are used as scaffolds in tissue engineering since they are designed to have a large porous structure that facilitates the migration of cells and the transport of nutrients during tissue regeneration. This chapter aims to provide a valuable overview on the application of HFs and nanofibers in the field of bioartificial organs and tissue engineering looking to the emerging applications and future perspectives of membrane science and engineering.

11.1 Introduction

Membrane systems play a pivotal role in the treatment of patients with acute and chronic organ failure diseases. All current extracorporeal blood purification and oxygenation methods are based on the use of membranes, especially in fiber configuration, assembled in a module. In these processes, membranes act as highly selective barriers able to remove endogenous and exogenous toxins from patient's blood (e.g., hemodialysis, hemofiltration, etc.) or to promote gas exchange with blood (blood oxygenation) [1]. Membrane technology offers interesting approaches for the development of bio/artificial organs, tissue analogues as well as for drug and cell delivery providing new specific cell therapy with high reliability and efficacy. Nano- and micro-structured membranes combined with functional cells can be used for the creation of tissue analogous (e.g., liver, kidney, skin, neuronal system, bone) because of the high molecular control of cellular microenvironment. Tailor-made hollow fiber (HF) membranes designed according to precise engineering criteria can provide adequate transport of molecules throughout the cellular compartment, and appropriate biomechanical, chemical, and physical stimuli at the developing tissue. HFs compartmentalize cells in a micro- and nano-structured environment, providing a wide surface area for cell adhesion to volume ratio [2, 3].

Cells can be compartmentalized in the extracapillary space or in the lumen of the fibers, which through its porous structure ensures a continuous and selective transport of nutrients and metabolites to and from cells protecting them from shear stress. Membranes with specific physicochemical, morphological, and transport properties would be able to modulate the adhesion, proliferation, and differentiation of cells, which are important for tissue regeneration. Moreover, membrane systems governing the selective mass transfer of molecules generate the formation of a precisely controlled microenvironment that mimics the specific features of in vivo environment. The mass transfer of molecules occurs in response to existing transmembrane concentration or pressure gradients by diffusion or convection, respectively, or a combination of both mechanisms. Since in bioartificial devices used for the organ replacement therapy, membranes perform immunoprotection functions, the most common suitable HF membranes have a molecular weight cut-off (MWCO) ranging from 70 to 100 kDa to allow the permeation of small molecules excluding proteins with high MW such as immunoglobulins and cells, which are responsible for immune response. In such devices, considering that molecules are simultaneously transported and consumed/produced by cells, the transport of species depends not only by both diffusion/convection but also by kinetic mechanisms. This is further exacerbated by the large range of solutes that must be transported across the membrane: species with different molecular sizes (from small electrolytes to large proteins) and physicochemical properties (hydrophilic, hydrophobic species) [4].

Membrane surface properties can have a direct influence on cell behavior since the topographical, chemical and mechanical inputs can affect cell spreading and migration [5]. Numerous strategies are devoted to modifying membrane surface to offer the specific biochemical and topographical cues for promoting cell adhesion and tissue regeneration. Bioinspired to the ECM structure, various approaches are aimed at developing membranes constituted by electrospun nanofibers that fit the characteristic length scales of in vivo physiologic subcellular structures. Accordingly, nanofiber membranes are widely used as scaffolds to promote the in vitro and in vivo tissue regeneration. They are designed to have a large volume fraction of interconnected pores to favor cell migration and the permeation of nutrients during tissue regeneration. Electrospun fiber membranes can be composed entirely of closed-packed random or oriented fibers, which confer a porous structure that can be controlled by modulating the fiber diameter.

A wide array of polymers are used for the preparation of fibers, including both nondegradable and degradable materials as well as synthetic and natural polymers on the basis of the final application [5]. For the development of scaffolds, either inorganic or organic materials, natural and synthetic, biodegradable polymers and related copolymers, or a combination of both are used. Differently, for bioartificial organs or bioreactors, nondegradable polymers, alone or blended, are preferred. The polymer blending allows to overcome certain limitations of a single material, as poor mechanical properties, inappropriate degradation rate, inadequate hydrophilic/hydrophobic properties, or hampered biocompatibility. In the polymer choice, it is important to take into account the desired physicochemical characteristics of the membrane, for example, polymers with moderate wettability are preferred for the production of HF or nanofibers in contact with cells to promote cell adhesion and an adequate perfusion. On the contrary, hydrophobic membranes (e.g., polypropylene) are used as means for the exchange of oxygen and carbon dioxide to enhance cell/medium oxygenation. Preparation techniques to produce HF and nanofiber membranes include electrospinning or dip coating extrusion or spinning techniques. Wet phase-inversion spinning is widely utilized for the preparation of HFs since the control of parameters involved in the preparation process. Therefore, membrane properties can be tailored by modulating the different working variables in order to achieve a desired morphology, porosity, pore size, wall thickness as well as physicochemical and mechanical properties.

Electrospinning, utilizing a high applied voltage and a grounded target, allows the preparation of membranes with a micro- or nanofibrous structure and high porosities.

To engineer biological tissues in vitro, novel concepts of bioreactor design and membranes are applied in order to enhance the ability to trigger biological signals that promote the morphogenesis of tissue. The use of membranes for cell culture is of key interest in the development of several devices for in vitro and therapeutic applications. In membrane bioreactors, cells are cultured in a more in-vivo-like environment than those in traditional two-dimensional cell culture. Despite the insight achieved in the development of various 3D cell culture systems, mass transfer efficiency and limitations, especially with reference to oxygen supply, still remain controversial topics. Critical process parameters such as shear rate, permeate flux, cell load and distribution, must be considered together with oxygen concentration profile. In this respect, modeling approaches including computational tools are useful for the prediction, allowing the consequent control of parameters that mutually influence cells and organs. These can help the understanding and control of fluid dynamics behavior and mass transfer conditions, increasing reproducibility and robustness of the devices [6]. Several bioreactors with different configurations have been developed by using fibers packed in parallel or cross-assembled, and various modeling studies have been reported to describe the oxygen supply, metabolic rates, and transport phenomena under different operational conditions [7].

Here, we review the HF and electrospun fibers used in bioartificial organs, tissue-engineered constructs, and in vitro bioreactors focusing on liver, neurons, and bone.

11.2 Fibers for Bioartificial Liver and Tissue Engineering

11.2.1 HF Membrane Systems for Bioartificial Liver

HF membranes represent a keystone in the design of bioartificial livers (BALs) and in engineered liver tissues to be used in medical clinical and preclinical practices, and for pioneering therapies and technologies. HF membranes were employed in the most of the BALs approved for clinical trials as extracorporeal supports and temporary bridges, in patients pending for liver transplant or waiting for the regeneration and recovery of their native or post-transplant organ. The use of HF membranes allows the biofabrication of 3D liver tissue under high perfused environment, serving likewise as reliable and predictable tools for testing effects and toxicity of new drugs and therapeutics, alternatively or complementary to animal trials. Additionally, the development of polymeric biodegradable

HF membranes allows the creation of implantable hepatic tissues, which undergo a full integration into the host with a controlled and nontoxic resorption of the membrane scaffold, and without its surgical removal.

Polymeric membranes are notably suitable as supporting material in designing BALs, owing to their intrinsic properties of permeability and selectivity, efficiency, simple operating functions, stability under mild operational conditions, easy scale-up, and robust automation. Moreover, by changing operational parameters, polymeric membranes can be synthetized and easily mass produced with tuned structural and physicochemical properties for targeted applications.

In the first BAL systems, HF membranes were used as selective barriers ensuring solute exchange and selective mass transport of nutrients and metabolites, and in the meantime the compartmentalization of cells, their protection from perfusion shear stress, and their immune isolation by the patient's hosting cells. Successively, considering that hepatocytes are anchorage-dependent cells, membranes were selected in order to perform a further task as supporting scaffold for cell adhesion, mimicking the natural extracellular matrix, which is important for the polarization of hepatocytes and their organization in a 3D architecture. Several papers have demonstrated that membranes offer to hepatocytes specific cues for recapitulating the in vivo milieu of the natural microenvironments, driving and supporting 3D tissue formation as well as ECM protein deposition [5, 8–10].

Most bioartificial livers were realized by parallelly assembling HF membranes for the compartmentalization of hepatocytes in the shell of the fibers. One of the first bioartificial livers, which was involved in phase I and II of the first larger clinical trial, loaded hepatocytes in free suspension in the extracapillary space of a polyvinylchloride (PVC) HF membrane cartridge [11]. The function and viability of these BALs were limited to a few hours, considering that hepatocytes are anchorage-dependent cells. To improve cell viability, other bioartificial livers maintained hepatocytes in adhesion in the extracapillary space of parallel-assembled HFs, encapsulated in agarose microdroplets [12], attached to microcarriers [13], entrapped in collagen gel [14], or in the basement membrane matrix [15].

Few bioreactors compartmentalized liver cells in the lumen of parallel-assembled fibers. In this device, hepatocytes were entrapped in collagen gel in the lumen of PSf HF membranes with MWCO of 100 kDa, and the patient's blood continuously recirculated in the extracapillary space [16].

Several bioartificial livers in clinical trials employed polysulfone (PSf) HF membranes. Hepatocytes encapsulated in agarose microdroplets were maintained in the extracapillary space of PSf HFs with 0.2 μm pore size [12]. *TECA hybrid artificial liver support system*, which was tested in phase I clinical trials, was constituted of PSf HF with 100 kDa cut-off, loaded with hepatocytes spheroids in extra-fiber space [17, 18]. PSf fibers with 0.15 μm pore size were employed in *HepatAssist*™ Circe Biomedical, which was tested on a large scale in a phase III clinical trial and utilized hepatocytes attached to collagen-coated dextran microcarriers into the extracapillary space. The membrane pore size prevented hepatocytes and cell debris from passing through the fiber wall and entering a patient's bloodstream, which flows through the lumens of the fibers [13, 19].

Extensive phase III clinical trials employed the *Extracorporeal Liver Assist Device* (Vital Therapies, Inc.), a BAL consisting of four cellulose acetate (CA) HF membrane cartridges with a cut-off of 70 kDa, which separates 200 g of hepatoblastoma cell line in the extracapillary space from patient's plasma flowing in membrane lumens [20]. This device was successively improved, increasing the CA membrane cut-off (up to 120 kDa) together with the cell mass (up to 400 g for each cartridge), enhancing the oxygenation, and adding filters for preventing cancer cell migration risks [21]. HF cartridges of CA membranes with a nominal 100 kDa diffusive size cut-off constitute the bioreactor of *The Excorp Medical Bioartificial Liver Support System*, which was involved in phase II trials. In this system, 70–120 g of hepatocytes were embedded in a collagen matrix in the extracapillary space. In contrast to other devices that perfuse patient plasma through the bioreactor, the BLSS perfused whole blood through the bioreactor, with significant advantages in affecting the patient outcome [22].

Modified PES multibore fibers, consisting of seven capillaries (960 μm internal diameter) clustered in a single fiber through a foamy porous (0.2 μm pore size) and highly permeable support structure, were employed in a peculiar configuration that conferred high stability and mechanical resistance to the hepatic tissue. Primary human hepatocytes were compartmentalized in the lumen of the different capillaries that were at the same time in communication with each other, thanks to the porous wall of the multibore fibers, and the diffusion of their secreted paracrine factors [23]. A similar setup was achieved by using a multicapillary polyurethane (PU) foam. In this device, hepatocytes were loaded in the microporous foam structure between capillaries with an internal diameter of 1.5 mm and forming a single flow channel for the culture medium [24]. The same authors, realized a liver lobule-like structure regularly arranging close together polyethylene (PE) HFs coated with ethylene vinyl alcohol, among which hepatocytes organoids were formed under a centrifugal force [24].

Taking into account that oxygen represents one of the crucial limiting nutrients in membrane BALs, because of its low solubility, the high uptake rate of hepatocytes, and vascularization concerns, many attempts have been made for oxygenation improvement. In this context, several devices introduced specific HFs devoted to the direct oxygenation of hepatocytes. A mat of porous polypropylene (PP) oxygenating HF membranes (0.2 μm pore size) was crosswise alternated at a constant distance of 200 μm to a mat of polyethylene (PE) heat exchange HF membranes in an oxygenating HF bioreactor. High-cell-density hepatocytes were cultured in direct contact with the perfused medium among the fibers that supplied oxygen requirements and temperature control [25].

A multiple HF network was developed by interweaving four independent capillary membranes of polyamide (PA) for medium/plasma inflow, PSf for medium/plasma outflow, hydrophobic PP for oxygen/carbon dioxide exchanges, and hydrophilic PP for sinusoidal endothelial co-culture [26]. Hepatocytes were cultured on the outer surface and between HFs under a countercurrent perfusion and internal oxygenation, and the device was involved in a phase I clinical trial (*Liver Support System*, Charitè-Humboldt University, Germany). An upgrade of the concept co-cultured the endothelial cells in direct contact with the hepatocytes in the extracapillary space created by two separate bundles of PES HF membranes with MWCO > 400 kDa for the countercurrent perfusion, and multilaminate HF membranes for gas exchanges. The device underwent phase I/II clinical trials [27, 28]. The miniaturized cell bioreactor, utilizing PES HF membranes with MWCO 400–500 kDa, was used as in vitro model to test liver detoxification of different drugs and their specific biotransformation pathway [29, 30].

A multicoaxial HF bioreactor was developed by assembling four concentric tubes with increasing diameter placed one inside the other. Hepatocytes were cultured under radial flow in a sandwich configuration in the space between the two innermost semipermeable tubes, consisting of polysulfone HFs with 1 mm outer diameter/0.65 μm pore size, and 3 mm outer diameter/0.1 μm pore size, respectively. Integral oxygenation occurred in the outermost compartment created by a third coaxial gas permeable silastic tube and the polycarbonate housing [31].

Another configuration based on the use of different fibers was realized by cross-assembling in an alternating manner two bundles of HFs with specific physicochemical, morphological, and transport properties for medium inflow and outflow, respectively, and creating an extracapillary 3D network for cell adhesion and culture, with high mass exchange and oxygenation through the crossflow of the medium.

A crossed HF membrane bioreactor was realized by assembling two HF bundles of polyether ether ketone modified with cardo group (PEEK-WC) with an MWCO of 190 kDa, devoted to the feeding of oxygenated medium, and polyethersulfone (PES) with 0.2 μm of pore size and high hydraulic permeance ($J = 15.2\ \mathrm{L/m^2h{\cdot}mbar}$), devoted to the removal of waste medium (Fig. 11.1). The distance among the fibers was set on the basis of the modeled mass transport of nutrients and cell products. The optimized fluid dynamics ensured a well-mixing microenvironment in which primary human hepatocytes retained their differentiated morpho-functional behavior up to 19 days of culture [32].

A 3D liver system was realized by co-culturing primary human sinusoidal endothelial cells, stellate cells, and hepatocytes in a crossed MBR constituted of two bundles of only PES HF membranes. Cells self-assembled forming tissue-like structures around and between the fibers, with hepatocytes forming aggregates, promoted by stellate cells, and surrounded by sinusoidal tube-like structures [33]. Within the bioreactor, a proper oxygenation and supply of nutrients was provided to the cells, ensuring a physiological amount that supported moreover the formation and long-term maintenance of functional human liver microtissue spheroids, which retained functional features up to 25 days [7, 34].

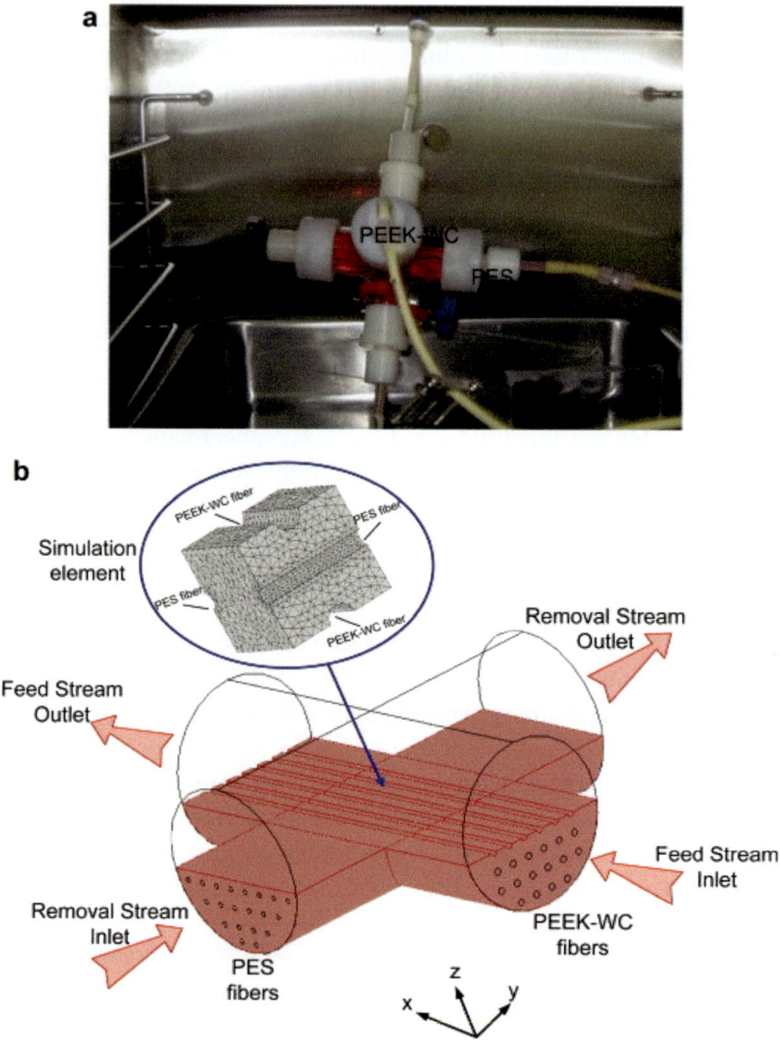

Figure 11.1 Crossed hollow fiber membrane bioreactor: (a) photograph; (b) vertical section scheme. Reprinted from Ref. [32] with permission of Elsevier.

More recently, a human vascularized hepatic tissue was developed by using biodegradable polycaprolactone (PCL) HF membranes. Tailoring operational parameters, fibers were properly synthetized with microporous interconnected pores along the membrane walls, and pores with a mean diameter of 0.4 and 1.3 μm on the external and internal membrane surface, respectively. A 3D compartmentalized co-culture was realized by loading primary human endothelial cells in the lumen of PCL HF membranes, and primary human hepatocytes in adhesion over and between the fibers, in the extralumen space. Endothelial cells in a such configuration colonized the lumen of the PCL HFs forming vascular channels and communicated with hepatocytes through their secreted factors that permeated across the microporous membrane walls. The developed PCL HF membranes together with fluid dynamic conditions created a physiologically relevant microenvironment with a large integration of the cells that retained functional activity and drug biotransformation up to 18 days, establishing a bottom-up tissue engineering approach to create 3D vascularized liver tissues [35].

11.2.2 Hollow Fibers to Induce Hepatic Differentiation

Fiber-membrane-based systems have the advantage to incorporate 3D complexities, thus holding great potential as reliable cell culture models for the development and testing of new liver therapeutic treatments. The validation of in vitro liver preclinical platforms requires the setting up of high-density culture in combination with a wide variety of experiments. These essential prerequisites leave unsolved one of the major issues in the field, which is related to the scarcity of a stable and clinically applicable hepatic cell source. Primary human hepatocytes represent the optimal cell model, but their scarcity sustains the urgent need for an alternative contender to serve as an unlimited source of vital and functional hepatic cells. The use of progenitor/fetal liver cells and/or adult and pluripotent stem cells is an alternative renewable source of human hepatocytes for fiber-based liver tools [36, 37]. Despite the tremendous improvements in terms of differentiation protocols, the achievement of standardized methods is still a huge challenge. Within the years, many techniques have been implemented to optimize the differentiation process and efficacy, in order to achieve a functional hepatic phenotype. The attempts of improving the differentiation potential have the common character to act on cell surrounding facilitating cellular recruitment, interaction, and displacement. The hepatic niche can be mimicked by using fiber membranes with different characteristics, which can also be adjusted to resemble those of the native tissue. For an ideal microenvironment and cell surrounding, which lead to a full cell differentiation, it is necessary to combine the effects of an adequate chemical exposure within an appropriate spatial and temporal frame. To accomplish such a challenging task, several HF membrane bioreactors have been adapted and their use revised to boost the differentiation of progenitor or stem cells toward the hepatic lineage. Monga et al. [38] investigated the use of fetal hepatocytes as an alternative cell source in a membrane bioreactor. The device is constituted by a tight network of HFs with an interwoven arrangement. The tight and repetitive membrane units of three independent sets of hollow fiber disposition give rise to four compartments with different functions. The perfusion of the fresh media is carried out by two sets of hydrophilic and microporous HF membranes, organized according to a countercurrent flow configuration and made up of PES with an MWCO of approximately 500 kDa (mPES; Membrana, Wuppertal, Germany; ID 300 ± 40 μm, wall thickness 100 ± 25 μm, pore size 0.5 ± 0.1 μm). The third bundle is made up of hydrophobic multilaminate HF membrane capillaries (MHF; Mitsubishi, Tokyo, Japan; ID 200 ± 10 μm, wall thickness 42 ± 3 μm) for gas supply in terms of cell oxygenation and CO_2 removal. The extracapillary space hosts the fetal hepatocytes, which are exposed to a proper mass and gas exchange, thanks to membrane arrangement that mimics the hepatic vasculature. Fetal hepatic cells within the bioreactor acquired liver-specific functions and formed a tissue-like structure between the HFs, where the formation of ribbon-like structures was also visible. Within the system, a balance between differentiation and proliferation was even possible to reach, holding the potential of solving hepatic source issues. The same four-compartments membrane bioreactor was used to study its potential to enhance hepatic differentiation in human embryonic stem cells in comparison with traditional culture models. The authors combined a stepwise differentiation protocol with the 3D dynamic perfusion system of the membrane bioreactor; they further established a twin bioreactor setup in which two devices are acting under the same operational setting but differing for the media composition. The twin approach had the aim to highlight the positive effects of the 3D perfusion on hESCs expansion and differentiation, because of the optimal biochemical and topological stimuli offered by the membrane functions. Both bioreactor systems sustained cell proliferation, but a real efficient differentiation was better achieved in cells supplied with specific cytokine under 3D fiber membrane perfusion. Although the only action of the 3D perfusion conditions did not lead to a complete induction of the hepatic maturation without the use of a proper pattern of specific growth factors, the tight capillary network provided an ideal environment with a proper mass exchange. This is corroborated by the enhanced effect of the cytokine pool in terms of hepatic maturation achieved in the bioreactor compared to results obtained in the traditional 2D systems [39]. The improvements in hESC differentiation towards the hepatic lineage by using a 3D HF membrane technology were also confirmed in the study of Sivertsson et al., in which the same bioreactor technology provided encouraging data for the derivation of functional hepatocytes from an undifferentiated cell source [40]. A further exploitation of this system consisted of the simultaneous cultivation of total human fetal liver cells (hepatic progenitor, hematopoietic, and endothelial progenitor). The well-perfused environment enabled an optimal differentiation process for the entire pool of inoculated cells. This multiple cell culture approach suggests that the 3D perfused bioreactor is a suitable tool for performing organotypic culture experiments. This finding opens a new perspective for a more reliable investigational device and an integrated research tool in regenerative medicine [41].

More recently, the same dynamic system was also assessed for its capacity to induce a better hepatic maturation of human induced pluripotent stem cells (hiPSC). The experimental outcomes actually emphasized the potential of the dynamic 3D culture by improving the formation of differentiated tissue in terms of morphological and functional behavior [42, 43].

Another kind of approach was used by Amimoto et al. [44] in which the guiding substrates are six HF membranes usually employed for plasma separation. The study led to the development of an HF/organoid culture method to boost the differentiation of mouse embryonic stem cells toward the hepatic lineage. Undifferentiated cells were inoculated in the lumen of 0.2 μm porous membranes made up of cellulose triacetate (ID: 285 μm; outer diameter: 387 μm). After injecting the cells into the lumen, membranes underwent centrifugation for cell spheroids self-assembling. Within the HF/organoid system, cells grew in the longitudinal direction of the HF. The differentiation pattern was properly achieved and evidenced by the expression of hepatocyte markers together with the ability to perform specific functions.

Matsushita et al. [45] set up a high-cell-density method by cultivating induced pluripotent stem cells (iPS) in the lumen of HF membranes. A set of 6 HFs, made up of polyethylene treated with ethylene vinyl alcohol (ID 330 μm; thickness 50 μm; pore size 0.3 μm), was lumen loaded with the cell suspension. Afterward, the endodermal differentiation, typical of hepatic lineage, was induced and successfully pursued.

Taking advantage of its well-consolidated performance, the multicompartment crossed HF membrane bioreactor developed by De Bartolo [32] was also used as a tool to promote and study the full hepatic maturation of rat embryonic liver cells [46]. Thanks to the optimal cell culture conditions, cells, spread over the entire membrane surface, formed a cord-like structure similar to that of living liver tissue. The acquisition of a functional differentiation was evidenced through the assessment of acquired synthetic and biotransformation functions.

A defined maturation process was further highlighted by the change in terms of specific marker expressions; indeed, cells lost a typical hepatoblast marker as alfa-fetoprotein (AFP) and gained the albumin one. Considering the importance of recruiting cellular hepatic sources to satisfy the high cell number demand, the crossed HF bioreactor [32] was used to induce the differentiation of human hepatic mesenchymal stem cells [47], and the consequent formation of a functional liver construct.

The biomimetic affinity of the PES HF membranes was enhanced by applying a thin collagen layer over the membrane surface. The modification had the advantage to improve the cell-membrane interaction without altering membrane transport properties. This is a fundamental aspect for a proper supply of gases and nutrients at the cellular interface, which determines the maintenance of cell viability and functionality.

The membrane bioreactor performance in terms of oxygen delivery to the cells was evaluated through the cellular oxygen consumption rate (OCR) measurements, highlighting a proper oxygen uptake within the bioreactor chamber avoiding the formation of any hypoxic zone. Stem cells within the membrane system went through morphological changes and were positive for specific intracellular markers of mature hepatocytes. Glycogen storage, synthetic and detoxification functions were positively assessed, confirming the acquisition of a hepatic phenotypic profile. The overall results reinforce the evidence of the useful approach of membrane technology for the realization and implementation of a novel investigational platform.

The huge membrane technology potential resides in its ability to establish a proper microenvironment that promotes cell adhesion and proliferation while directing the precise acquisition of target cell genotype and phenotype. Moreover, the gathered investigational studies confirmed that stem cells have a great potential in the creation of bioartificial liver devices. Improvements in terms of quality control and efficient cell differentiation within a fiber-based liver device require a stable metabolic performance together with a predictable cell behavior. These challenging issues can be faced only through the constant seeking and synergistic integration of innovative auxiliary biomaterials and high degree of differentiation methods.

11.2.3 Nanofibers in Liver Tissue Engineering

An efficient tissue-engineered approach aims to create an interactive environment between cells and scaffolds, as normally occurs when cells are surrounded by the natural extracellular environment. In this sense, the unique features of nanofibers provide enormous potential in the field of tissue engineering; their peculiar intrinsic structure offers specific surface cues at nano- and micro-scale level and a large surface area of interaction. The combination of their intrinsic nanotopography, mechanical and physicochemical properties creates a perfect mimicked extracellular matrix milieu, where cells can restore in vitro their 3D cytoarchitecture. This is because the structure disposition of nanofibers promotes the interaction with integrin-binding proteins, which in turn boost cell attachment and the acquisition of the appropriate morphology. Besides the aforementioned properties, most of the electrospun fibers also exhibit high interconnected porosity, great absorbance and gas permeability, which contribute to enhance their suitability for different tissue-engineered applications [48, 49].

Here are summarized some interesting approaches about the in vitro maintenance of functional liver cells within nanofiber scaffold systems. An example of nanofibrous mats in liver tissue engineering technology is provided by Feng et al. [50] through the development of galactosylated chitosan (GC) nanofibers for rat primary hepatocytes culture. The electrospinning process gave arise to nanofibers with an average diameter of 160 nm, with appropriate structural and mechanical properties. The authors investigated the effect of GC nanofibers on primary hepatocytes interaction and the effect on their adhesion and morphology. The study included an overlook of the integration of hepatic spheroids over the scaffold, together with the maintenance evaluation of liver-specialized functions. The results highlighted the optimal cell response because of the interaction with functionalized galactose nanofiber structures.

The GC nanotopografy not only enhanced a high level of specific functions but also promoted a better cell-material integration when compared with flat substrates prepared from the same functionalized polymer.

In another investigation, an in vitro toxicological model was established by culturing primary rat hepatocytes on nanofibrous poly-(L-lactic acid) (PLLA) [51]. The scaffold was prepared through the phase-inversion technique giving rise to an interconnected macroporous network that mimics the collagen fibrillary disposition within the ECM structure. The 3D scaffold well sustained hepatocytes growth that displayed a cytoskeletal structural integrity, growing along the macroporous interstices and keeping proper hepatic functionalities.

Blended PCL/chitosan nanofibers, with excellent shape and orientation, were used for liver tissue engineering purposes [52]. The experimental approach comprised the preparation of a wide set of different blended nanofibers. The most suitable ones for the culture of epithelial liver mouse cells (Hepa 1-6) exhibit a diameter of 243 ± 32 nm, 79% porosity, and average pore size of 12 ± 5 μm. The matrix showed good appropriateness in terms of biocompatibility and nontoxic response, validating its use for the realization of future tissue-engineered liver constructs.

Hepatocytes in the in vitro culture achieve successful results when defined factors are properly provided to the cells in terms of microenvironmental, physical, chemical, and mechanical stimuli. The simultaneous satisfaction of these vital factors is accomplished by using a 3D scaffold combined with ECM proteins. In this way, the scaffold works as a biomimetic matrix where cells can keep an in-vivo-like morphology. A 3D liver tissue model for the long-term maintenance of functional hepatocytes was realized using a nanofiber scaffold made up of chitosan [53].

Structural analysis investigations evidenced fibers randomly oriented with a diameter comprised between 50 and 300 nm (average value of 156 ± 48 nm). The mean pore size was estimated at 2.2 ± 0.4 μm. Before cell colonization, a FN (fibronectin) coating was applied and the precoated scaffold was co-cultured with fibroblast and hepatocytes.

During the culture time, both fibroblasts and hepatocytes preserved their morphology, showing a well-combined integration between them and within the scaffold mats. Cells migrated along the whole scaffold thickness as an indication of a good substrate interaction. Moreover, hepatocytes showed a growing albumin secretion during the whole culture time and high levels of CYP 450 A1 enzyme activity. The 3D hepatic co-culture system opened a successful perspective towards the development of nanofiber-based system for drug testing and liver tissue engineering.

The development of an innovative tissue engineering strategy implies the continuous seeking of optimized techniques that improve scaffold stability and bioactivity. In this direction, Bual et al. [54] developed a technique aimed to improve the bioactive performance of gelatin/PCL nanofibers blend by adding in the bulk solution liver-specific extracellular matrix (L-ECM) derived from porcine liver. Different scaffold compositions have been realized by varying the presence and the percentage of the three components: PCL, gelatin, and L-ECM. The nanofibers exhibited a smooth surface and high porosity with values ranging between 88.6% and 95.3%. The combination of the two natural components and PCL had the effect to generate a final scaffold with an improved cell interaction besides the preservation of an excellent mechanical stability, which is typical of the PCL. Primary hepatocytes cultured on the different scaffolds acquired their typical morphology and were able to retain specific functions within the entire culture time.

The introduction of tissue-specific signal molecules within the polymer structure provided a balanced microenvironment to the hepatic cells, which also promoted further investigations related to the effects of the different tissue-specific ECM elements for nanofiber organ constructs.

A sponge-like structure, highly porous and with less compact fiber structure, is more appropriate for the hepatocytes culture maintenance. These characteristics promote a better cell migration throughout the scaffold matrices, and it can be achieved through the wet electrospinning technique. Within the liver tissue engineering scenario, this approach was used by Brown et al. [55], which realized a nanofibrous scaffold made of poly(l-lactide-co-glycolide) (PLGA). The resulting PLGA scaffolds are extremely porous with a mean pore size of 27.5 ± 7.5 μm (range of pore size: 9.7–44.5 μm), recapitulating the same pore size of the decellularized liver ECM and thus facilitating hepatocytes distribution along and within the scaffold. While the wet electrospinning methodology strongly affects the pore size in a significant manner, it has a minor effect on the average fiber diameter, which measured 870 ± 300 nm.

The porous nanofibrous matrix was chemically modified with different concentrations of collagen I and fibronectin without affecting the porosity.

To test the ECM-modified nanofiber matrix on liver cell response, a hepatic culture model was established and maintained for 14 days. Coupling the larger pore size of the nanofibers with the collagen incorporation in the mates' structure improved the in vitro performance of the primary hepatocytes. Cellular state indicators of cell viability and above all the hepatic specific functions were higher on cells cultured in the collagen-modified scaffold with promising future applications in the field.

In a similar work by Das et al. [56], by using the same kind of membranes mentioned earlier, different ratios of collagen I to fibronectin were investigated and optimized in order to identify the most useful combination of the two proteins in sustaining hepatic cell functionality.

The ratio of collagen I to fibronection at 3:1 was identified as the most suitable one, in terms of uniform distribution over the electrospun surface and the most efficient in supporting a proper liver cell functionality. Besides pointing out the importance of integrating multiple biomimetic stimuli within the scaffold structure, the study highlighted the importance of the optimal ratio of the incorporated signals to maximize cell interaction and behavior.

In a recent work [57], a challenging approach consisted of the direct integration of the full set ECM proteins with the PLLA polymeric solution for the production of a bioactive electrospun scaffold. The incorporation of the entire ECM pool of proteins, derived from decellularized unused liver, was compared with the integration of single ECM protein as collagen, fibronectin, and lamin-521. The difference in composition had a strong effect on Young's modulus values, thus moving the attention on stiffness properties of the material that need to be modified on the basis of the specific application. The preparation process of the nanofiber scaffold did not influence protein integrity as showed by a positive staining localized through specific antigens. The different nanofibrous environments were validated by using the TH-3 liver cells, by testing their viability and adhesion. Protein scaffolds maintained both cell survival and adhesion with the lowest percentage detected on the polymer-only scaffold. The different cell response indirectly confirmed protein integration within the nanofibrous mats, which positively affected cell behavior. Gene expression assessments, both for liver functional genes and ECM genes, confirmed that the most performing scaffold is the one in which the whole ECM extract was solubilized, confirming that individual ECM elements are not able to fully recapitulate the native liver environment. The common feature of all these systems is their great promise in representing valuable tissue models for future tissue-engineered applications and patient treatment, not only in the field of liver but also to a wide variety of tissue and organs being easily adaptable for different purposes.

11.3 Fiber Membranes for Neuronal Tissue Regeneration

Driven by the urgent need to understand and treat disorders of the nervous system, different kinds of in vitro and in vivo tools have been developed to guide neuronal repair and regeneration. Membrane biohybrid systems represent advanced devices that provide an adequate microenvironment for neuronal growth and differentiation [58–63]. Membrane devices are successfully employed for the repair and regeneration of the nervous system, as well as to realize in vitro brain tissue models for pharmacological screening, and neurodegenerative disease treatments [64–66].

At present, autologous nerve graft is the gold standard method for peripheral nerve repairing. Nevertheless, this approach is restricted by the poor availability of donors. Therefore, scientists in the field of tissue engineering have fabricated synthetic nerve guidance conduits (NGC) for the repair of peripheral nerve injuries. These nerve guides are typically biocompatible, biodegradable, and semipermeable hollow tube where the injured nerves are allocated in the lumen in order to guide and induce the regeneration.

In this context, polymeric membranes in fiber configuration are promising materials for nerve repair and regeneration and are currently used as NGC. HF membranes compartmentalize neuronal processes and guide the elongation and orientation of neuronal growth by creating a 3D environment that enhances the formation of neural network. Such kind of fiber device can give neurotrophic support for damaged nerves, retard the fibrous tissue infiltration, and drive axonal regeneration, and thus could be as effective as autografts [67].

Hollow fiber membranes used as NGC include nondegradable and degradable materials. The first example of HF membrane application in the CNS was made of a nondegradable acrylic copolymer, the Diaflo HF membrane microdialysis tube, which was successfully implanted in the rat hippocampus [68]. Reynolds et al. realized a porous HF made of nondegradable poly(2-hydroxyethyl methacrylate-co-methyl methacrylate (PHEMA) for the nerve regeneration in rat after spinal cord transection surgery [69]. Semipermeable nondegradable PSf HF membranes enabled the regeneration of hamster sciatic nerve [70]. HFs made of another nondegradable PES were fabricated for studying 3D neuronal network in vitro [71], and for the regeneration of the nigrostriatal nerves in an animal model of Parkinson's disease [72].

HF membranes of polyacrylonitrile and polyvinyl chloride copolymer (PAN-PVC) were realized for encapsulating PC-12 cells to investigate the influence of membrane transport properties on neuronal behavior [73]. The study suggests that membrane permeability influences biomass size, cellular proliferation, and drug release.

PEEK-WC and PAN HF membranes were employed for the long-term culture of neuronal cells isolated from the hippocampus of golden hamster for investigating the effect of membrane permeability on neuronal behavior (Fig. 11.2) [60, 61].

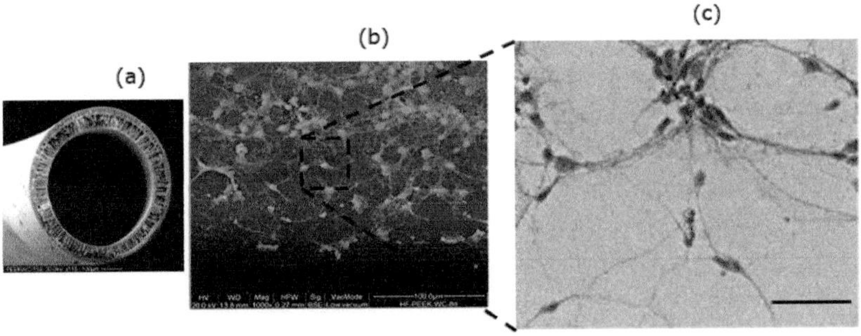

Figure 11.2 Hippocampal neurons cultured on PEEK-WC HF membranes. SEM micrographs of membranes (a) and neurons (b, c) in culture on PEEK-WC fibers. Scale bar: 50 μm. Adapted from Ref. [61] with permission of Wiley; and from Ref. [60] with permission of Springer.

This study demonstrates that membrane transport properties play a key role in controlling the cellular microenvironment by regulating the supply of nutrients and gases. HF membranes, thanks to permeability, could regulate cellular morphology, axonal growth, and specific functions of neurons.

The use of nondegradable biomaterials as nerve guidance conduits requires a secondary surgery to remove it to circumvent chronic tissue response and infection. Recently, nerve guides made with biodegradable polymers that degrade without toxic effect have been fabricated to overcome these drawbacks.

Several types of natural and synthetic biodegradable polymers, including PCL [74], PLGA [75, 76], chitosan [77], and blend [78], have been developed in fiber configuration as nerve guides.

Several papers report different methods to improve the porosity of novel nerve conduits. Porous fibers of PLGA and Pluronic F127 were prepared by Oh et al., by the phase-inversion technique in order to realize HF membranes with interconnected pores [75]. This nerve conduit induced the regeneration of rat sciatic nerve.

Recently, a microporous chitin-based conduit with internal chitosan fibers displayed suitable mechanical strength and porosity to induce the regeneration of a 10 mm defect of sciatic nerve in rats

[77]. To improve and guide the orientation of axonal outgrowth, HF membranes made by blending PLGA with PU (PLGA/PU) were modified by creating aligned grooves on their inner surfaces [78].

During the last decade, various methods, including electrospinning and microfluidic spinning, have been employed for the fabrication of micro- and nanofibers with controlled morphological, structural, and physicochemical characteristics, in order to provide an environment similar to the native ECM.

Several papers report on the development of electrospun fibers with a random orientation that can be employed as neuronal conduits. Randomly oriented electrospun nanofibers of chitosan-graft-poly(ε-caprolactone)/polycaprolactone (CS-PCL/PCL) [79] and silk fibroin (SF) blended with poly(L-lactic acid-co-ε-caprolactone) (SF/PLCL) [80] were used for retinal tissue engineering.

The potentiality of the aligned fibers for the regeneration of injured nerves is due to their structural peculiarities and similarities with the natural peripheral nerve bundles, which are formed by the assembly of multi-aligned neurocytes. Highly oriented fibers guide neuronal growth by offering the cells contact guidance for neurite regeneration and enhancing the diffusion of neurotrophic factors. Aligned micro- and nanofibers have been successfully employed for driving the orientation of different cells, including neural stem cells, neurons, and glial cells.

Electrospun chitosan-PCL fibers guided the orientation of neuronal cells that followed the fiber alignment direction [81]. PLGA-aligned nanofiber membranes enhanced the axonal regeneration in a transected spinal cord rat model [82].

During the last decade, the electrospinning method has been modified to enhance fibers alignment. 3D cylindrical constructs of highly aligned PCL nanofibers have been realized by two-pole air-gap electrospinning and induced the regeneration of 10 mm defect of sciatic nerve in rodent [83].

PLLA-aligned fibers incorporating cells and growth factors were prepared by using the bioelectrospinning method [84]. Recently, a coaxial electrospinning process was employed to realize PLLA highly aligned microtube array (MTA) membranes structured along with a single layer. The electrospun PLLA-MTA membranes were integrated within a dynamic system as the bioreactor to provide a well-controlled microenvironment for neuronal growth in order to realize a novel neuronal conduit device [85]. The membrane properties and the uniform dynamic environment created by PLLA-MTA membrane bioreactor modulated the growth and differentiation of neurons. This advanced membrane device enhanced neuronal phenotype by guiding the orientation of neuronal growth that followed the direction of the microtubes, generating a 3D neuronal tissue-like construct.

Wang et al. encapsulated neuronal growth factor (NGF) within aligned PLGA nanofibers and demonstrated the importance of combining topographical and biochemical cues to enhance peripheral nerve regeneration [86].

Since electric fields enhance the regeneration of injured nerves, a chitosan nanofiber mesh tube was electrically polarized by the immobilization of β-tricalcium phosphate (β-TCP) particles and employed to perform bridge grafting into sciatic nerve defects in rats. This approach created a suitable nerve conduit that induced the regeneration of both motor and sensory nerves after implantation comparably with isografting [87].

The mechanical properties of the synthetic nerve conduits have been improved by increasing their wall porosity. To this purpose, porous tubular conduits were developed in order to enhance the diffusion of nutrients that are essential for nerve regeneration. A novel nerve guide conduit with a helix-flexible structure prepared by electrospinning exhibited mechanical properties that enable peripheral nerve regeneration with less tension during operation [88].

Different approaches have been combined to realize advanced and very complex multi-structural nerve conduits. Highly oriented microfibers were located on the top of a porous flat-sheet membrane, coiled and glued lengthwise, successfully regenerating the injured nerve [89].

Multi-channel design with external porous wall and intraluminal channels was a popular approach to design multi-structural nerve guide conduits [90].

The injection-molding technique was combined with thermally induced phase separation in order to realize multiple-channeled nanofibrous membranes that provided a beneficial luminal microenvironment that enhanced the adhesion of neurons and guided axonal orientation and regeneration [91]. Jeffries and Wang

[92] used electrospinning to fabricate aligned fibrous mats and rolled them up to multi-channel NGC.

Novel bilayer electrospun nanofibrous nerve conduits were realized by Kim et al. [93]. These constructs were characterized by a combination of aligned fibers located on the inner layer and randomly oriented fibers on the outer surface, which enhanced the orientation of neuronal growth and mechanical strength, respectively [93].

In the last years, microfluidic technology has emerged as an advanced method for fabricating fibers at micro and nano scale for peripheral nerve repair. Highly aligned collagen fibers produced by microfluidic method display diameters of only 3 μm, which allowed neuronal cell adhesion [94]. Due to the low diameters and to the high tensile strength and elastic modulus, which exceed those obtained with electrospinning method, these kinds of fibers provided neuronal alignment and growth along the fiber axes showing great potential for peripheral nerve repair [94].

Very aligned collagen-based microfibers were fabricated using the microfluidic technique equipped with a capillary system and a rotary receiving pool [95]. These microfibers were used for co-cultures of neuronal and endothelial cells to realize valid 3D pre-vascularized tissue constructs that guided neuronal growth alignment and promoted neuronal functions.

Microfluidic collagen fibers were placed within a spider silk nonwoven mesh tube, to realize a novel nerve conduit, which allowed the orientation of neuronal processes and ensured adequate nutrient supply [96].

To date, fiber membranes have been employed to realize in vitro models of blood–brain barrier (BBB) to have advanced tools that create an environment similar to that of in vivo brain and microvascular tissue. The most employed BBB model is constituted by a dynamic apparatus consisting of microporous PP HFs that were used for the co-culture of endothelial cells (ECs) from various origins with astrocytes. ECs are seeded in the lumen of HFs, astrocytes in the extraluminal fiber side. The HFs are connected to a perfusion system for oxygen and nutrient supply under the exposure of pulsatile laminar flow. This HF dynamic BBB model has been used to investigate BBB functions and for drug-delivery studies.

The system enables to demonstrate the key role played by the crosstalk between glial and endothelial cells in maintaining a viable

BBB [97], the capacity of a nucleoside to improve specific BBB functions [98], and the brain-penetration ability of antiepileptic drugs [99].

Despite several appealing advantages provided by this dynamic BBB over conventional static systems, some drawbacks have to be addressed to expand the adoption of this BBB model.

Recently, novel advanced fiber-based devices have been realized for creating engineered models of neuronal tissue for Alzheimer's disease (AD) treatment [64, 66]. More specifically, PAN HF membranes were assembled within a bioreactor to establish intraluminal compartments for medium flow and extraluminal ones for the neuronal cultures that communicate through the pores of the membrane walls [64]. The developed PAN HF membrane bioreactor enabled the creation of an in vitro functional neuronal tissue analogue, which was employed for investigating the neuroprotective action of the carotenoid, crocin, against Aβ-induced neurotoxicity linked to AD.

More recently, a neuronal PLLA-MTA membrane bioreactor was used as an investigational device to assess the capacity of isoflavone glycitein, to inhibit the aggregation of Aβ, which plays a pivotal role in the neurodegeneration process of AD [66]. Glycitein inhibited intracellular ROS secretion, the expression of specific protein of apoptotic process, and neuronal cell death and ensured the maintenance of neuronal functions [66].

The studies reported in the present paper highlight the key role of fiber-based devices in guiding nerve repair and regeneration by also providing investigational platforms for neuronal tissue engineering applications.

11.4 Fibers for Bone Tissue Engineering

Bones are a key component of the body, not only by enabling body movements and providing protection of internal organs, but also in the production of blood cells, calcium homeostasis, and acid/base buffering [100]. Due to its high regeneration capacity, most fractures or bone defects heal spontaneously without the need for major medical intervention. This regeneration capacity is even more evident in younger individuals. When a bone fracture occurs, the inflammatory phase begins immediately, producing a hematoma at the injury site. Usually, this phase continues 1 to 4 days and is mediated by various growth factors (i.e., ILs, TNFα, FGFs, BMPs, PDGF, VEGF, etc.) released from platelets, macrophages, and fibroblasts. When the inflammatory phase terminates, an immature callus is generated. Then, the immature callus is replaced by immature woven bone for several weeks. This repair phase occurs in several steps. First, the fibroblasts present in the periosteum undergo massive proliferation, originating the soft callus. The necrotic bone affected by the injury is resorbed by osteoclasts. Then, the recruitment and differentiation of mesenchymal stem cells (MSCs) into chondrocytes lead to the formation of a new cartilaginous tissue, which is gradually replaced by woven bone by endochondral ossification. Last, the woven bone is remodeled into lamellar bone, restoring the mechanical stability of the fractured bone [101, 102]. Nevertheless, large bone defects cannot self-regenerate, requiring surgical intervention to repair them. A critical-size defect (CSD), i.e., a segmental bone deficiency larger than 2–2.5 times the diameter of the affected bone, usually results from injuries, bone tumors, or other bone diseases [103, 104]. The incidence of specific pathologies involving the musculoskeletal system, like osteoarthritis and osteoporosis, is expected to increase due to the higher life expectancy. Bone defects affect more than 20 million people worldwide each year. Graft procedures can be an option for CSDs treatment: autologous bone graft in which the receiver/donor bone comes from the same person; allografts in which the graft comes from an individual of the same species; and xenografts where the grafts come from an individual of another species. However, all three kinds of bone grafts present limitations: the quantity or quality of available autologous bone tissue, and the number of surgical sites can increase the risk of hematoma, infections, bleeding, and chronic pain. Allografts present limited osteoinductive properties, and their availability is limited; xenografts present safety issues [105, 106]. In a parallel pathway, metallic devices are still not long-term successfully implanted in the human body. Therefore, metals present some drawbacks, such as the in vivo corrosion, poor osseointegration, high elastic modulus (leading to stress shielding), and the release of toxic ions. On the other hand, the use of porous metallic scaffolds has been gaining attention in bone tissue engineering (BTE) applications due to its excellent physical properties. In this type of scaffolds, the most widely used materials are titanium (Ti) and its alloys [101]. So far, Ti_6Al_4V is an exceptional metal alloy due to its mechanical properties: Young's modulus is similar to natural bone, thus improving the mechanical shielding to the living tissue [107]. However, the major limitation of these metal composites is their non-biodegradability, thus limiting their application in BTE. Moreover, metals lack bioactive properties, which means that they do not form chemical bonds with living tissues [108]. The BTE approach has emerged over the last decade as a promising alternative therapy to the autologous surgery and metallic devices [104]. Using this approach, smart bone scaffolds have an active role in repairing osseous lesions rather than simply function as space holders to ensure mechanical support [101]. For instance, the natural bone healing conditions can be mimicked by adding growth factors to the scaffold, thus stimulating the replication and differentiation of seed cells to the phenotypically desirable type and helping in host cell recruitment.

Figure 11.3 highlights several key players that scaffolds should retain to make them suitable for BTE. A suitable cell supporting matrix is required, which acts as a bridging material called "scaffold," to provide a 3D substrate for cells to populate on and function appropriately during the formation of bone analogue tissue [109]. Therefore, biomaterial scaffolds should be osteoinductive, osteoconductive, biocompatible, osteogenic, angiogenic, and resorbable/degradable. Considering that bone tissue is highly vascularized, thus receiving about 10 to 20% of the cardiac output, the scaffold should promote vascularization providing an adequate and timely supply of oxygen and nutrients [107]. Furthermore, considering that bone remodels in response to the loadings it experiences, scaffolds should provide a suitable mechanical environment adequate for fracture healing [101, 110]. Thus, the elastic property of the scaffold should not exceed that of natural bone, to maintain a proper mechanical stimulation of the peripheral bone, which in turn depends on the loading conditions [102]. Additionally, the biocompatible requirement is related to different factors behind rejection caused by inflammation, edema, and consequently necrosis [111]. Furthermore, osteoconduction, which has been defined as the capacity of a material to allow bone growth on its surface, is an essential property of smart scaffolds for

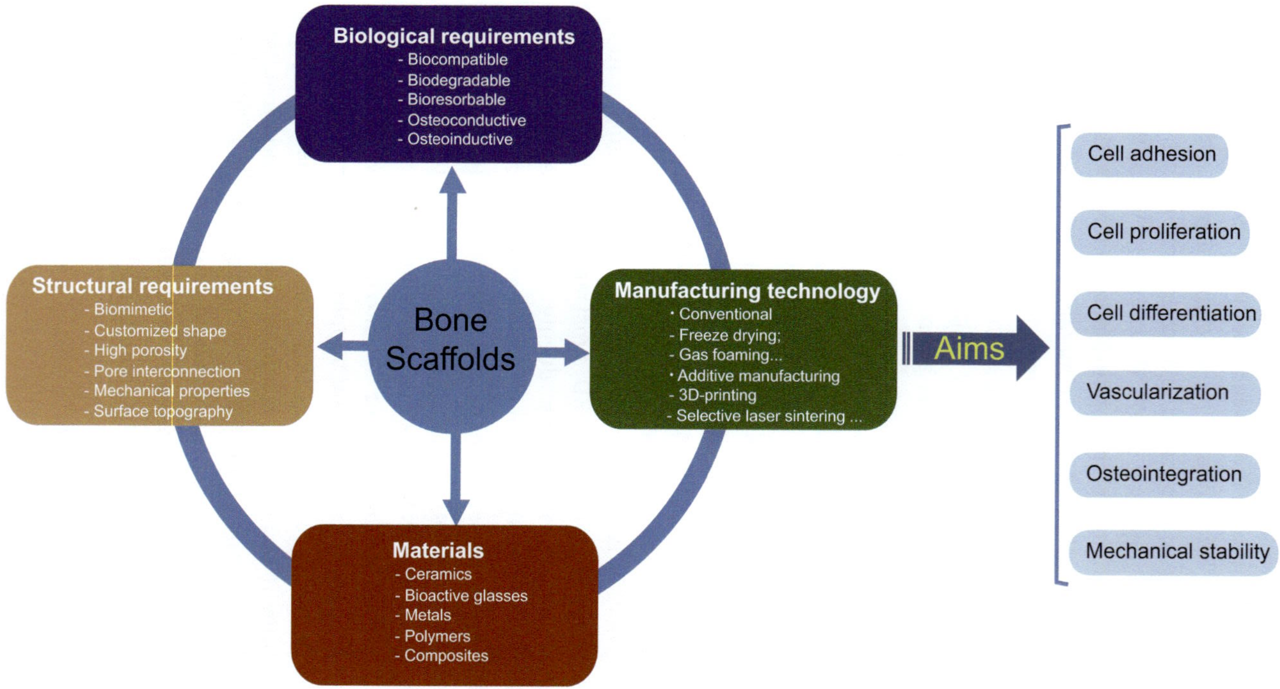

Figure 11.3 Schematic representation of the four main requirements of smart bone scaffolds for BTE applications: biological requirements, structural requirements, manufacturing technology, and materials. The right part of the scheme shows the main biological purpose of bone scaffolds. Reprinted from Ref. [101] with permission of Elsevier.

BTE. The production of physiological bone tissue by the recruitment and the activity of osteoblasts, often defined as osteogenesis, is mandatory for bone regeneration. Thus, an osteoinductive scaffold can stimulate the differentiation and the activity of bone cells, thus promoting osteogenesis. Degradable/resorbable scaffolds must be replaced by bone tissue, and their degradation products must not be toxic [104, 112]. This requisite comes from the fact that bone undergoes constant remodeling. The mismatch between scaffold degradation and new bone tissue formation can lead to a mechanical weakening or even failure of the overall osseous structure. In human beings, the injured bone remodels to achieve a stable biomechanical condition for at least 6 months. Therefore, the degradation kinetics of bone scaffolds should be tailored to last that specific time [101].

To fulfil the requirements, the nanofibrous matrices produced by electrospinning process have recently attracted more attention, mainly due to the structural similarity to the tissue extracellular matrix (ECM), being a simple setup and low-cost technique, as well as the processing availability to a wide range of materials [109, 113]. The bone-mimicking materials should play active roles in helping cells to follow the effective processes leading to bone formation. Thus, a variety of growth factors and key proteins are present in the bone matrix and are involved in bone formation. These materials should also be considered in the design of ECM-mimicking materials, alongside a polymeric matrix and inorganic mineral nanocomponents [109]. Nowadays, the most used materials for scaffolds are natural or synthetic polymers such as polysaccharides, poly(a-hydroxy ester), thermoplastic elastomers, or hydrogels [114–116]. The incorporation of polymers into the scaffolds offers outstanding controllability on their physiochemical properties, such as solubility, porosity, and wettability [101].

Bioactive ceramics such as bioactive glasses (BAGs), calcium phosphates ceramics (CPCs), or glass ceramics are other important categories of materials for BTE. A strategy currently under research is the blending of polymers and ceramics aiming to increase the mechanical stability of the scaffold and to improve tissue interaction. In addition, these scaffolds can be tailored to locally release antibiotics or growth factors that enhance bone ingrowth promoting bone regeneration. Moreover, the blending of polymer–ceramic composites can prevent serious problems associated with the acidic by-products of polymers in vivo, leading to inflammatory reactions and tissue necrosis. It happens due to the basic by-products released from CPCs or BAGs that efficiently act as a buffer for the acidic by-products of polymer degradation [101, 117, 118]. Therefore, to produce a scaffold that fulfils as many requirements as possible, composite systems combining ceramics and polymers are a promising choice for BTE.

11.4.1 Ceramics in BTE

Ceramic materials employed within the body are generally termed bioceramics, which are commonly classified as bioinert, bioresorbable, and bioactive on the basis of the following properties: these materials do not cause a reaction with the host tissue; they undergo dissolution over time and actively take part in a physiological process. The bioceramic materials include zirconia, alumina, titania, calcium phosphates, glass ceramics or silica-based glasses, and pyrolytic carbons. The most common ceramics used in BTE to date are calcium phosphate (CaP) and hydroxyapatite (HaP) since they are very similar to the mineral phase of the bone matrix considering the structure and chemical composition. These materials are well known for their osteoconductive properties and the capability to form an interface with the host bone without forming scar tissues [119, 120]. Both classes of ceramics have been extensively used for filling bone defects (granules, cement) or for the coating of metallic materials reducing the risk of implant rejection. In this context, BAGs are also relevant materials for BTE among the ceramic family [121]. BAG materials have only recently appeared in the field of medicine. The first melt-derived bioactive glass, Bioglass 45S5 (46.1 mol% SiO_2, 24.4 mol% Na_2O, 26.9 mol% CaO, and 2.6 mol% P_2O_5), was discovered by Hench in 1969 [122]. The synthesized degradable bioactive glass was able to bond to living tissue and particularly to bones. The Hench 45S5 BAG latterly trademarked as Bioglass® has been involved in more than 1 million bone repairs [123]. The Hench discovery paved the way for the development of many bioactive materials [105]. The Bioglass® discovery was followed by bioactive gel-derived glasses, also proposed by Hench in 1991 [124]. The gel-derived glasses present some advantages over melt-derived glasses:

lower temperatures are required for their synthesis; improved homogeneity; easy powder technology production; higher purity of the powder; a wider range of BAG compositions (e.g., up to 90 mol% SiO_2), and enhanced bioactivity [112, 125].

Most recently, Salinas and Vallet-Regí [126] developed bioactive glasses with ordered mesoporosity. The mesoporous structure implies a higher specific surface area compared to non-mesoporous BAG (typically several hundred m^2/g versus few m^2/g) [127]. Consequently, the mesoporosity also implies a greater solubility of sol–gel-derived BAG among other advantages such as enhanced protein adsorption. The importance of such characteristic on BTE will be discussed in a sub-topic below. The diversity of shapes that sol–gel process offers is another key advantage: BAGs can be fabricated under the form of fibers, nanoparticles, coatings, monoliths, or particulates. The most outstanding feature of BAG discovery was its ability to exhibit very strong interfacial bonds with surrounding tissues, particularly with the bone. The BAG can form a biologically active hydroxy carbonate apatite (HCA) layer, which is structurally and chemically equivalent to the mineral phase in natural bone. Moreover, BAGs present the highest in vivo bioactivity index ($I_B > 8$) among bioceramics [108, 118]. The high bioactivity inherent to BAG is mainly a consequence of two mechanisms: the rapid development in the HCA layer promotes the attachment and proliferation of osteogenic cells as mentioned earlier; the ionic dissolution products from BAG (i.e., Si, Ca, P) upregulate gene expression in human osteoblast cells by acting as chemical stimuli/trigger to promote osteoblast metabolism, subsequently resulting in faster bone regeneration [128]. Moreover, this upregulation of genes also promotes the secretion of relevant growth factors, including insulin-like growth factors (IGFs) and angiogenic growth factors (e.g., VEGF), which in turn favors neovascularization [129]. The angiogenesis is a crucial step for the successful survival of new tissues, as it allows the supply of oxygen and nutrients to cells. Moreover, BAG has been found to exhibit anti-inflammatory and antibacterial properties: BAG can locally inhibit bacterial proliferation by increasing calcium levels and pH at the implant site [130]. All of these attractive characteristics of BAGs reveal their greater potential to be used in BTE compared to HaP [106, 131].

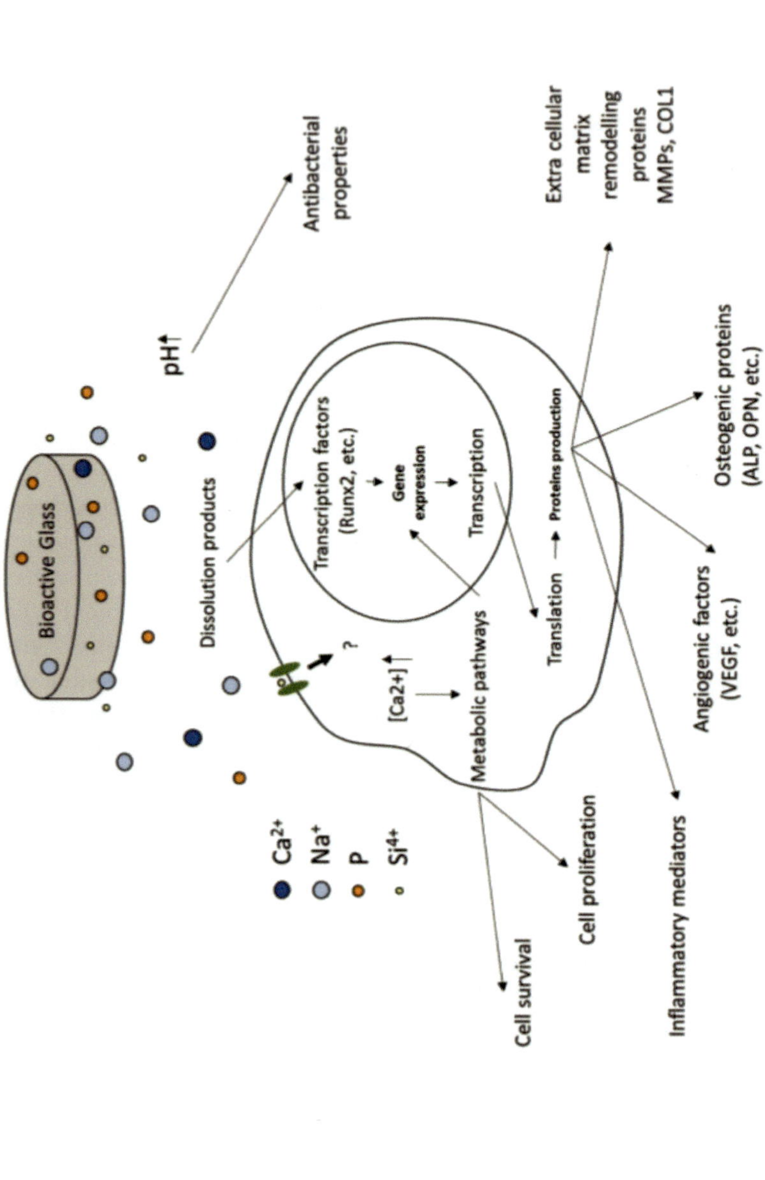

Figure 11.4 Biological properties of a bioactive glass. Reproduced from Ref. [105] with permission of John Wiley and Sons.

The schematic representation of the biological properties of a BAG is shown in Fig. 11.4. However, BAG has limited applications as a scaffold material due to its poor mechanical properties, despite its bone-bonding ability and astonishing biocompatibility [118]. Thus, a major challenge for BTE is now to produce a scaffold presenting both enhanced biological properties and adequate mechanical properties. In this context, organic–inorganic (polymer–ceramic) composites have been developed as reported in the next topic [105].

11.4.2 Polymeric Composite Membranes for BTE

The combination of BAG with biodegradable polymer composite materials has been studied in the last decades as a new set of bioactive materials with enhanced properties for applications ranging from structural implants to TE scaffolds [125]. This strategy is applicable to a wide variety of synthetic and natural polymers: linear aliphatic poly(α-hydroesters) such as PGA, PCL, PLA, and their copolymers (e.g., PLGA) are the most popular synthetic materials used in medicine [105]; natural polymers and proteins most used in biomedical applications include collagen, gelatin, silk, fibrinogen, elastin, and polysaccharides, chondroitin sulphate, alginate, dextran, chitosan, chitin, etc. [112]. These composite materials take advantage of the polymers' flexibility with the strength, stiffness, and high bioactivity of the BAG fillers. The addition of BAG to bioresorbable polymers can also change the acidic polymer degradation kinetics, by allowing the rapid exchange of protons in water for alkali in the BAG, providing a pH buffer effect at the polymer surface, and consequently avoiding an inflammatory response [108, 118, 132, 133]. It is worth noting that the majority of these composite materials uses micron-sized bioactive glass particles as fillers or coatings [134]. Inorganic micron-sized particles generally tend to agglomerate, which hinders the blending/homogenization with polymer solutions. Furthermore, these composites present apatite nucleation mainly at the BAG particles that protrude non-homogeneously from the surface. The irregularities can delay osseointegration in vivo and can also cause structural instability or premature deterioration of the composite [135]. Therefore, the size of the filler particles plays a major role that affects the mechanical/structural properties of composite materials.

These limitations raise the need for a new subclass of composites: nanocomposites composed of an organic matrix with bioactive inorganic nanoparticles or nanostructures (<100 nm) incorporated [105, 109]. In addition, the incorporation of nanoparticles into a polymeric matrix mimics more closely the natural structure of the bone, which consists of nanoscale hydroxyapatite crystallites mixed with the polymeric phase of type I collagen, resulting in the desired mechanical properties of natural bone [136]. The larger specific surface area of the nanoparticles compared to micron-sized particles should also lead to increased interface effects contributing to enhanced bioactivity. Webster et al. compared nanoscale with micron-sized ceramic materials and their respective composites. They observed a significant increase in protein adsorption and osteoblast adhesion on nanoscale materials [137]. In this context, Loher et al. have demonstrated that mechanical properties, degradation rate, and bioactivity of PLGA doped with nanoscale amorphous calcium phosphate are strongly improved when compared to the pure polymer [138]. Hong et al. have studied a family of composites combining PLLA as biodegradable polymer and sol–gel derived BAG nanoparticles. The in vitro results showed that composites containing BAG nanoparticles with lower phosphorous and higher silicon content have better bioactivity compared to that of the BAG with lower silicon and higher phosphorous content [139]. Following a parallel path, instead of introducing bioactive inorganic particles within a polymeric solution, some research focuses on the production of degradable and bioactive nanofibers through the hybridization approach of using inorganic and organic phases in the electrospinning solution. Kim et al. were the pioneer team to develop a nanocomposite of PLA and sol–gel derived BAG by means of electrospinning. It was reported that BAG nanofibers were uniformly dispersed in the PLA matrix. The synthesized nanocomposites exhibited excellent bioactivity, inducing HCA precipitation within 24 h after soaking in SBF [140]. Lu et al. have produced BAG (70S30C, 70 mol% SiO_2, 30 mol% CaO) nanofibers by the electrospinning method with a highly porous microstructure with interconnected macropores and mesopores. It is worth noting that such scaffolds had an elastic modulus close to that of trabecular bone [136, 141]. Another study focused on using electrospinning to produce PLA–nHA fiber scaffolds with interconnected pores and high porosity (80–90%). With the purpose of mimicking the natural bone tissue remodeling process, these scaffolds were used to create an in vitro model system with osteoblast and osteoclast progenitor cells [142]. The composite scaffolds induced the differentiation of hMSCs into osteoblasts and the osteoclastogenesis especially on scaffolds with 20 wt.% of nHA. This investigation demonstrated the importance of the cellular crosstalking on the differentiation and activation processes, thus improving the knowledge over the needed microenvironment for BTE.

In general, although some limitations associated with these scaffolds, such as morphological/compositional control and ensure desirable mechanical properties, the nanofibers–ceramic composites are expected to produce interesting outcomes in the near future.

11.4.3 Scaffold Surface Functionalization and Porous Network

Bone tissue can be divided into two subtypes: compact and spongy. The main differences rely on their density, or in the way the tissue is tightly packed together. The former, also called cortical bone, makes up the outer layer of the bone; it is a dense structure with high stiffness and mechanical strength. The latter, also referred to as cancellous or trabecular bone, is a highly porous and vascularized structure close to that of a honeycomb material along the cross section, ideal for making and storing bone marrow where the production of blood cells takes place. Therefore, it a less dense structure and presents weaker mechanical properties compared to cortical bone [143]. In addition, spongy bone can be converted to compact bone by the action of osteoblasts. For these reasons, an ideal BTE scaffold should have high degree of porosity, with a 3D structure, to ideally mimic the porosity, interconnectivity, and pore size of native spongy bone. The inherent processing nature of electrospinning gives rise to membranes containing pores with sizes of a few micrometers; thus, the introduction of larger sized pores within the nanofibrous network must be considered to improve the role of the 3D scaffolding matrix [109]. Researchers have developed several membranes postprocessing strategies that can be applied to enlarge the pore size: cryogenic electrospinning, salt leaching, sacrificial fibers ultrasound, etc. Macropores play major roles in vitro responses considering cell penetration, or in vivo responses with respect to tissue ingrowth, osteointegration, and vascularization. The availability of free space prevents peripheral cellular growth or fibrosis. On one hand, if pores are too large, the specific surface area is significantly reduced, thus reducing the ligand density available for cells to attach. On the other hand, if pores are too small, it results in the formation of a cellular capsule around the edges of the scaffold and osteochondral ossification, since cell migration is limited. However, it is worth noting that the nanopores and the pore size in the range of a few micrometers tend to modulate cell differentiation and increase the protein adsorption [112]. The research has proved that the adsorption of proteins from body fluid is essential for the enhanced biological performance of the scaffolds. After implantation in the body, complex interactions between materials and host cells take place on the material's surface [144]. Therefore, chemical and topographic characteristics of the scaffold's surface determine which molecules/proteins to be adsorbed from body fluid. The host cells interact with the material's surface through integrins, which initiate various downstream signaling cascades upon phosphorylation (e.g., the mitogen-activated protein kinase pathway and the phosphatidylinositol-3-kinase pathway). Generally, the signaling cascades result in transcription factors that control specific gene expressions [101]. Following the aforementioned discovery, rough nanotopography/mesoporous structures enhance the adsorption of specific amino acids/proteins that benefit cell adhesion and differentiation, like vitronectin, fibronectin, and other osteogenic growth factors, which improve the function of osteoblasts [144–146]. In this context, Webster et al. found that the adhesion, proliferation, and calcium deposition of osteoblasts were promoted in vitro, using surfaces with grain sizes of less than 100 nanometers [137]. In addition, Ma et al. reported that porous PLLA/HaP composite scaffolds exhibit superior osteoconductivity properties and promote osteoblastic cell survival, proliferation, and expression of bone-specific markers (i.e., osteocalcin and bone sialoprotein) when compared to plain PLLA scaffolds during 6 weeks of in vitro cultivation [147].

In this way, scaffold postprocessing techniques should preserve the inherent mesoporosity of ceramic materials, thereby obtaining hierarchical porous structures that combine macroporosity for bone growth and oxygenation and mesoporosity to keep enhanced bioactivity due to higher superficial area and allow drug delivery of appropriate therapeutic biomolecules [142]. From a biological point of view and according to cell size, macropores with a minimum diameter of about 100 μm are required, but diameters over 300 μm are recommended to achieve better vascularization. Moreover, scaffolds should have pore interconnectivity with interconnection windows larger than 100 μm to avoid dead spaces [105, 118, 148]. To ensure an appropriate amount of oxygen for cell survival, a diffusion distance between 150 and 200 μm from the supplying blood vessels should be considered. The vasculature also transports nutrients, soluble factors, and numerous cell types to the bone tissues; therefore, the success of 3D constructed tissues strongly depends on angiogenesis [109, 142]. To sum up, scaffolds with high and hierarchical porosity that maintain mechanical strength and structural integrity are required for BTE applications.

As mentioned before, BAGs can stimulate some biological mechanisms due to their dissolution products. Furthermore, considering that nanofibrous matrices have an extremely large surface area relative to volume, its surface can be tailored to induce the appropriate biological reactions. For instance, it is possible to tailor the nanofiber's surface to possess mesoporosity during electrospinning or load the nanofibers with growth factors. However, the proper dosage of the growth factor has been shown to affect the quality of the neo-vasculature. It has been reported that excess amounts of VEGF cause severe vascular leakage and hypotension [149]. However, growth factor release can be tailored by the amount of growth factor added to the scaffold or by changing the degradation rate of the scaffold. It has been demonstrated that the vasculature is well organized via controlled release of VEGF, compared to the vasculature that arose from an uncontrolled VEGF release [102, 150].

Therefore, the surface functionalization of the membranes, as a post-treatment following the electrospinning process, is crucial to improve the potential of nanofibers as a cell matrix [109]. Although scaffolds containing BAG are capable to individually induce some biological reactions that stimulate bone regeneration, several growth factors, such as transforming growth factor-b (TGF-b), bone morphogenetic proteins (BMPs), vascular endothelial growth factor

(VEGF), platelet-derived growth factor (PDGF), and insulin-like growth factors (IGF), combined with BAG scaffolds, should enhance the bone formation ability of the scaffolds by mimicking the in vivo bone repair conditions [104, 151]. To this purpose, novel electrospun scaffolds of PCL/PVAc core–shell fibers were synthesized and decorated with PLGA particles loaded with BMP2 by simultaneous electrospinning and electrospraying [152]. These scaffolds were able to support and improve cell growth and proliferation, facilitating the expression of osteogenic and osteoconductive markers. BMP2 enhanced the expression of bone formation and maturation markers pointing out its key role in the bone repair and regeneration.

11.5 Conclusion and Future Perspectives

This chapter summarized the great potential of HFs and nanofiber membranes for bioartificial organs and tissue engineering applications. Because of their structural, selective permeability, physicochemical and mechanical properties, they can be used for the development of a bioartificial liver, nerve conduits, and bone substitutes. The research results show the advances concerning the realization of bioengineered organs and tissues opening new challenges regarding the design of membranes and membrane devices to meet the requirements of the specific tissue or organ that must be replaced modulating cell behavior at multiple levels, including adhesion, migration, proliferation, and stem cell fate.

Acknowledgments

This work was funded by National Funds through FCT - Portuguese Foundation for Science and Technology, Reference UID/CTM/50025/2019 and FCT/MCTES, and by project PTDC/CTM-CTM/30623/2017 supported by the Lisbon Regional Operational Program (Lisboa 2020) and Alentejo Regional Operational Program (Alentejo 2020), under the PORTUGAL 2020 Partnership Agreement, through the European Regional Development Fund (ERDF). R. Matos also acknowledges FCT for the PhD grant with reference SFRH/BD/140090/2018.

References

1. De Bartolo, L., Curcio, E., and Drioli E. (2017) *Membrane Systems for Bioartificial Organs and Regenerative Medicine.* De Gruyter. ISBN 978-3-11-026798-3; eBook ISBN 978-3-11-026801-0 pp.1–264
2. De Bartolo, L., Bader, A., and Haverich, A. (2001) Review of a flat membrane bioreactor as bioartificial liver. *Ann. Transplant.*, **6**, 40–46.
3. Drioli, E. and De Bartolo, L. (2006) Membrane bioreactor for cell tissues and organoids. *Artif. Organs*, **30**(10), 793–802.
4. Legallais, C., Kim, D., Mihaila, S. M., Mihajlovic, M., Figliuzzi, M., Bonandrini, B., Salerno, S., Yousef Yengej, F. A., Rookmaaker, M. B., Sanchez Romero, N., Sainz-Arnal, P., Pereira, U., Pasqua, M., Gerritsen, K. G. F., Verhaar, M. C., Remuzzi, A., Baptista, P. M., De Bartolo, L., Masereeuw, R., and Stamatialis, D. (2018) Bioengineering organs for blood detoxification. *Adv. Healthcare Mater.* **7**, 1800430.
5. Salerno, S., Morelli, S., and De Bartolo, L. (2017) Advanced membrane systems for tissue engineering. *Curr. Org. Chem.*, **21**(17), 1760–1774.
6. Curcio, E., Piscioneri, A., Salerno, S., Tasselli, F., Morelli, S., Drioli, E., and De Bartolo, L. (2012) Human lymphocytes cultured in 3-D bioreactors: Influence of configuration on metabolite transport and reactions. *Biomaterials*, **33**, 8296–8303.
7. Khakpour, S., Di Renzo, A., Curcio, E., Paolo Di Maio, F., Giorno, L., and De Bartolo, L. (2017) Oxygen transport in hollow fiber membrane bioreactors for hepatic 3D cell culture: A parametric study. *J. Membr. Sci.*, **544**, 312–322.
8. Salerno, S. and De Bartolo, L. (2017) Biohybrid membrane systems and bioreactors as tools for in vitro drug testing. *Curr. Pharm. Des.*, **23**(2), 319–327.
9. Salerno, S., Campana, C., Morelli, S., Drioli, E., and De Bartolo, L. (2011) Human hepatocytes and endothelial cells in organotypic membrane systems. *Biomaterials*, **32**(34), 8848–8859.
10. Salerno, S., Curcio, E., Bader, A., Giorno, L., Drioli, E., and De Bartolo, L. (2018) Gas permeable membrane bioreactor for the co-culture of human skin derived mesenchymal stem cells with hepatocytes and endothelial cells. *J. Memb. Sci.*, **563**, 694–707.
11. Margulis, M. S., Erukhimo, E. A., Andreiman, L. A., and Viksna, L. M. (1989) Temporary organ substitution by hemoperfusion through suspension of active donor hepatocytes in a total complex of intensive therapy in patients with acute hepatic insufficiency. *Resuscitation*, **18**(1), 85–94.

12. Shiraha, H., Koide, N., Hada, H., Ujike, K., Nakamura, M., Shinji, T., Gotoh, S., and Tsuji, T. (1996) Improvement of serum amino acid profile in hepatic failure with the bioartificial liver using multicellular hepatocyte spheroids. *Biotechnol. Bioeng.*, **50**(4), 416–421.
13. Demetriou, A. A., Rozga, J., Podesta, L., Lepage, E., Morsiani, E., Moscioni, A. D., Hoffman, A., McGrath, M., Kong, L., Rosen, H., Villamil, F., Woolf, G., Vierling, J., and Makowka, L. (1995) Early clinical experience with a hybrid bioartificial liver. *Scand. J. Gastroenterol.*, **30** (sup208), 111–117.
14. Naka, S., Takeshita, K., Yamamoto, T., Tani, T., and Kodama, M. (1999) Bioartificial liver support system using porcine hepatocytes entrapped in a three-dimensional hollow fiber module with collagen gel: An evaluation in the swine acute liver failure model. *Artif. Organs*, **23**(9), 822–828.
15. Nagaki, M., Miki, K., Kim, Y., Ishiyama, H., Hirahara, I., Takahashi, H., Sugiyama, A., Muto, Y., and Moriwaki, H. (2001) Development and characterization of a hybrid bioartificial liver using primary hepatocytes entrapped in a basement membrane matrix. *Dig. Dis. Sci.*, **46**(5), 1046–1056.
16. Nyberg, S. L., Shatford, R. A., Peshwa, M. V, White, J. G., Cerra, F. B., and Hu, W.-S. (1993) Evaluation of a hepatocyte-entrapment hollow fiber bioreactor: A potential bioartificial liver. *Biotechnol. Bioeng.*, **41**(2), 194–203.
17. Ding, Y.-T., Qiu, Y.-D., Chen, Z., Xu, Q.-X., Zhang, H.-Y., Tang, Q., and Yu, D.-C. (2003) The development of a new bioartificial liver and its application in 12 acute liver failure patients. *World J. Gastroenterol.*, **9**(4), 829–832.
18. Gan, J.-H., Zhou, X.-Q., Qin, A.-L., Luo, E.-P., Zhao, W.-F., Yu, H., and Xu, J. (2005) Hybrid artificial liver support system for treatment of severe liver failure. *World J. Gastroenterol.*, **11**(6), 890–894.
19. Demetriou, A. A., Brown, R. S., Busuttil, R. W., Fair, J., McGuire, B. M., Rosenthal, P., Am Esch, J. S. 2nd, Lerut, J., Nyberg, S. L., Salizzoni, M., Fagan, E. A., de Hemptinne, B., Broelsch, C. E., Muraca, M., Salmeron, J. M., Rabkin, J. M., Metselaar, H. J., Pratt, D., De La Mata, M., McChesney, L. P., Everson, G. T., Lavin, P. T., Stevens, A. C., Pitkin, Z., and Solomon, B. A. (2004) Prospective, randomized, multicenter, controlled trial of a bioartificial liver in treating acute liver failure. *Ann. Surg.*, **239**(5), 660–670.
20. Sussman, N. L., Chong, M. G., Koussayer, T., He, D.-E., Shang, T. A., Whisennand, H. H., and Kelly, J. H. (1992) Reversal of fulminant hepatic failure using an extracorporeal liver assist device. *Hepatology*, **16**(1), 60–65.

21. Millis, J., Cronin, D., Johnson, R., Conjeevaram, H., Conlin, C., Trevino, S., and Maguire, P. (2002) Initial experience with the modified extracorporeal liver-assist device for patients with fulminant hepatic failure: system modifications and clinical impact. *Transplantation*, **74**(12), 1735–1746.

22. Patzer, I. J., Lopez, R., Zhu, Y., Wang, Z., Mazariegos, G., and Fung, J. (2002) Bioartificial liver assist devices in support of patients with liver failure. *Hepatobiliary Pancreat Dis Int.*, **1**(1), 18–25.

23. De Bartolo, L., Morelli, S., Rende, M., Campana, C., Salerno, S., Quintiero, N., and Drioli, E. (2007) Human hepatocyte morphology and functions in a multibore fiber bioreactor. *Macromol. Biosci.*, **7**(5), 671–680.

24. Mizumoto, H. and Funatsu, K. (2004) Liver regeneration using a hybrid artificial liver support system. *Artif. Organs*, **28**(1), 53–57.

25. Jasmund, I., Langsch, A., Simmoteit, R., and Bader, A. (2002) Cultivation of primary porcine hepatocytes in an OXY-HFB for use as a bioartificial liver device. *Biotechnol Prog.*, **18**(4), 839–846.

26. Gerlach, J., Encke, J., Hole, O., Müller, C., Ryan, C., and Neuhaus, P. (1994) Bioreactor for a larger scale hepatocyte in vitro perfusion. *Transplantation*, **58**(9), 984–988.

27. Mundt, A., Puhl, G., Muller, A., Sauer, I. M., Muller, C., Richard, R., Fotopoulou, C., Doll, R., Gabelein, G., Hohn, W., Hofbauer, R., Neuhaus, P., and Gerlach, J. C. (2008) A method to assess biochemical activity of liver cells during clinical application of extracorporeal hybrid liver support. *Int. J. Artif. Organs*, **25**(6), 542–548.

28. Sauer, I. M., Zeilinger, K., Pless, G., Kardassis, D., Theruvath, T., Pascher, A., Goetz, M., Neuhaus, P., and Gerlach, J. C. (2003) Extracorporeal liver support based on primary human liver cells and albumin dialysis: Treatment of a patient with primary graft non-function. *J. Hepatol.*, **39**(4), 649–653.

29. Mueller, D., Tascher, G., Müller-Vieira, U., Knobeloch, D., Nuessler, A. K., Zeilinger, K., Heinzle, E., and Noor, F. (2011) In-depth physiological characterization of primary human hepatocytes in a 3D hollow-fiber bioreactor. *J. Tissue Eng. Regen. Med.*, **5**(8), e207–e218.

30. Hoffmann, S. A., Müller-Vieira, U., Biemel, K., Knobeloch, D., Heydel, S., Lübberstedt, M., Nüssler, A. K., Andersson, T. B., Gerlach, J. C., and Zeilinger, K. (2012) Analysis of drug metabolism activities in a miniaturized liver cell bioreactor for use in pharmacological studies. *Biotechnol. Bioeng.*, **109**(12), 3172–3181.

31. Wolfe, S. P., Hsu, E., Reid, L. M., and Macdonald, J. M. (2002) A novel multi-coaxial hollow fiber bioreactor for adherent cell types. Part 1: Hydrodynamic studies. *Biotechnol. Bioeng.*, **77**(1), 83–90.
32. De Bartolo, L., Salerno, S., Curcio, E., Piscioneri, A., Rende, M., Morelli, S., Tasselli, F., Bader, A., and Drioli, E. (2009) Human hepatocyte functions in a crossed hollow fiber membrane bioreactor. *Biomaterials*, **30**(13), 2531–2543.
33. Ahmed, H. M. M., Salerno, S., Morelli, S., Giorno, L., and De Bartolo, L. (2017) 3D liver membrane system by co-culturing human hepatocytes, sinusoidal endothelial and stellate cells. *Biofabrication*, **9**(2), 025022.
34. Ahmed, H. M. M., Salerno, S., Piscioneri, A., Khakpour, S., Giorno, L., and De Bartolo, L. (2017) Human liver microtissue spheroids in hollow fiber membrane bioreactor. *Colloids Surf. B Biointerfaces*, **160**, 272–280.
35. Salerno, S., Tasselli, F., Drioli, E., and De Bartolo, L. (2020) Poly(ε-Caprolactone) hollow fiber membranes for the biofabrication of a vascularized human liver tissue. *Membranes* (Basel)., **10**, 112.
36. Zeilinger, K., Freyer, N., Damm, G., Seehofer, D., and Knöspel, F. (2016). Cell sources for in vitro human liver cell culture models. *Exp. Biol. Med. (Maywood)*, **241**(15), pp 1684–1698.
37. Hosseini, V., Maroufi, N. F., Saghati, S. Asadi, N., Darabi, M., Ahmad, S. N. S., Hosseinkhani, H., and Rahbarghazi, R. (2019). Current progress in hepatic tissue regeneration by tissue engineering. *J. Transl. Med.*, **17**, pp. 383
38. Monga, S. P., Hout, M. S., Baun, M. J., Micsenyi, A., Muller, P., Tummalapalli, L., Ranade, A. R., Luo, J.-H., Strom, S. C, and Gerlach, J. C. (2005). Mouse fetal liver cells in artificial capillary beds in three-dimensional four-compartment bioreactors. *Am. J. Pathol.*, **167**(5), 1279–1292.
39. Miki, T., Ring, A., and Gerlach, J. (2011). Hepatic differentiation of human embryonic stem cells is promoted by three-dimensional dynamic perfusion culture conditions. *Tissue Eng. Part C Methods.*, **17**(5), 557–568.
40. Sivertsson, L., Synnergren, J., Jensen, J., Björquist, P., and Ingelman-Sundberg, M. (2013). Hepatic differentiation and maturation of human embryonic stem cells cultured in a perfused three-dimensional bioreactor. *Stem Cells Dev.*, **22**(4), 581–594.
41. Pekor, C., Gerlach, J. C., Nettleship, I., and Schmelzer, E. (2015). Induction of hepatic and endothelial differentiation by perfusion in a three-dimensional cell culture model of human fetal liver. *Tissue Eng. Part C Methods*, **21**(7), 705–715.

42. Meier, F., Freyer, N., Brzeszczynska, J., Knöspel, F., Armstrong, L., Lako, M., Greuel, S., Damm, G., Ludwig-Schwellinger, E., Deschl, U., Ross, J. A., Beilmann, M., and Zeilinger, K. (2017) Hepatic differentiation of human iPSCs in different 3D models: A comparative study. *Int. J. Mol. Med.*, **40**(6), 1759–1771.

43. Freyer, N., Greuel, S., Knöspel, F., Gerstmann, F., Storch, L., Damm, G., Seehofer, D., Harris, J., Iyer, R., Schubert, F., and Zeilinger, K. (2018). Microscale 3D liver bioreactor for in vitro hepatotoxicity testing under perfusion conditions. *Bioengineering*, **5**(1), 24.

44. Amimoto, N., Mizumoto, H., Nakazawa, K., Ijima, H., Funatsu, K., and Kajiwara, T. (2011). An evaluation of the utility of the hepatic differentiation method using hollow fiber/organoid culture for the development of a hybrid artificial liver device, *Biochem. Eng. J.*, **56**, 69–74.

45. Matsushita, S., Kajiwara, T., and Mizumoto, H. (2019). Expansion and differentiation of human iPS cells in a three-dimensional culture using hollow fibers and separation of the specific population by magnetic-activated cell sorting, *J. Biosci. Bioeng.*, **128**(4), 480–486.

46. Salerno, S., Piscioneri, A., Morelli, S., Al-Fageeh, M. B., Drioli, E., and De Bartolo, L. (2013). Membrane bioreactor for expansion and differentiation of embryonic liver cells. *Ind. Eng. Chem. Res.*, **52**, 10387–10395.

47. Piscioneri, A. Ahmed, H. M. M., Morelli, S., Khakpour, S., and De Bartolo, L. (2018). Membrane bioreactor to guide hepatic differentiation of human mesenchymal stem cells. *J. Membrane Sci.*, **564**, 832–841.

48. Rasouli, R., Barhoum, A., Bechelany, M., and Dufresne, A. (2019). Nanofibers for biomedical and healthcare applications. *Macromol Biosci.*, **19**(2), e1800256.

49. Zhang, J., Zhao, X., Liang, L., Li, J., Demirci, U., and Wang, S. (2018). A decade of progress in liver regenerative medicine. *Biomaterials*, **157**, 161–176.

50. Feng, Z. Q., Chu X., Huang N. P., Wang, T., Wang, Y., Shi, X., Ding, Y., and Gu, Z.-Z. (2009). The effect of nanofibrous galactosylated chitosan scaffolds on the formation of rat primary hepatocyte aggregates and the maintenance of liver function. *Biomaterials*, **30**(14), 2753–2763.

51. Bierwolf, J., Lutgehetmann, M., Feng, K., Erbes, J., Deichmann, S., Toronyi, E., Stieglitz, C., Nashan, B., Ma, P. X., and Pollok, J. M. (2011). Primary rat hepatocyte culture on 3D nanofibrous polymer scaffolds for toxicology and pharmaceutical research. *Biotechnol Bioeng.*, **108**(1), 141–150.

52. Semnani, D., Naghashzargar, E., Hadjianfar, M., Manshadi, F. D., Mohammadi, S., Karbasi, S., and Effaty, F. (2017). Evaluation of PCL/chitosan electrospun nanofibers for liver tissue engineering. *Int. J. Polym. Mater. Polym. Biomater.*, **66**, 149–157.

53. Fe, D., Hussain, A., Yip, D., Parekh, A., Shrirao, A., and Cho, C. H. (2017). Long-term liver-specific functions of hepatocytes in electrospun chitosan nanofiber scaffolds coated with fibronectin. *J. Biomed. Mater. Res. A.*, **105**(8), 2119–2128.

54. Bual, R., Kimura, H., Ikegami, Y., Shirakigawa, N., and Ijima, H. (2018). Fabrication of liver derived extracellular matrix nanofibers and functional evaluation in in vitro culture using primary hepatocytes. *Materialia*, **4**, 518–528.

55. Brown, J. H., Das, P., DiVito, M. D., Ivancic, D., Tan, L. P., and Wertheim, J. A. (2018). Nanofibrous PLGA electrospun scaffolds modified with type I collagen influence hepatocyte function and support viability in vitro. *Acta Biomater.*, **73**, 217–227.

56. Das, P., DiVito, M. D., Wertheim, J. A., and Tan, L. P. (2020). Collagen-I and fibronectin modified three-dimensional electrospun PLGA scaffolds for long-term in vitro maintenance of functional hepatocytes. *Mater. Sci. Eng. C Mater. Biol. Appl.*, **111**, 110723.

57. Grant, R, Hallett, J., Forbes, S., Hay, D., and Callanan, A. (2019). Blended electrospinning with human liver extracellular matrix for engineering new hepatic microenvironments. *Sci. Rep.*, **9**(1), 6293.

58. De Bartolo, L., Rende, M., Morelli, S., Giusi, G., Salerno, S., Piscioneri, A., Gordano, A., Di Vito, A., Canonaco, M., and Drioli, E. (2008) Influence of membrane surface properties on the growth of neuronal cells isolated form hippocampus. *J. Mem. Sci.*, **325**, 139–149.

59. Morelli, S., Salerno, S., Piscioneri, A., Papenburg, B. J., Di Vito, A., Giusi, G., Canonaco, M., Stamatialis, D., Drioli, E., and De Bartolo, L. (2010) Influence of micropatterned PLLA membranes on the outgrowth and orientation of hippocampal neuritis. *Biomaterials*, **31**, 7000–7011.

60. Morelli, S., Piscioneri, A., Salerno, S., Tasselli, F., Di Vito, A., Giusi, G., Canonaco, M., Drioli, E., and De Bartolo, L. (2012) PAN hollow fiber membranes elicit functional hippocampal neuronal network. *J. Mater. Sci. Mater. Med.*, **23**, 149–156.

61. Morelli, S., Piscioneri, A., Salerno, S., Rende, M., Campana, C., Tasselli, F., Di Vito, A., Giusi, G., Canonaco, M., Drioli, E., and De Bartolo, L. (2012) Flat and tubular membrane systems for the reconstruction of hippocampal neuronal network. *J. Tissue Eng. Regen. Med.*, **6**, 299–313.

62. Morelli, S., Piscioneri, A., Messina, A., Salerno, S., Al-Fageeh, M. B., Drioli, E., and De Bartolo, L. (2015) Neuronal growth and differentiation on biodegradable membranes. *J. Tissue Eng. Regen. Med.*, **9**, 106–117.

63. Morelli, S., Piscioneri, A., Drioli, E., and De Bartolo, L. (2017a) Neuronal differentiation modulated by polymeric membrane properties. *Cells Tissues Organs*, **204**, 164–178.

64. Morelli, S., Salerno, S., Piscioneri, A., Tasselli, F., Drioli, E., and De Bartolo, L. (2016) Neuronal membrane bioreactor as a tool for testing crocin neuroprotective effect in Alzheimer's disease. *Chem. Eng. J.*, **305**, 69–78.

65. Piscioneri, A., Morelli, S., Mele, M., Canonaco, M., Bilotta, E., Pantano, P., Drioli, E., and De Bartolo, L. (2015) Neuroprotective effect of human mesenchymal stem cells in a compartmentalized neuronal membrane system. *Acta Biomater.*, **24**, 297–308.

66. Morelli, S., Piscioneri, A., Curcio, E., Salerno, S., Chen, C. C., and De Bartolo, L. (2019) Membrane bioreactor for investigation of neurodegeneration. *Mater. Sci. Eng. C Mater. Biol. Appl.*, **103**, 109793.

67. Yu, X. and Bellamkonda, R. V. (2003) Tissue-engineered scaffolds are effective alternatives to autografts for bridging peripheral nerve gaps. *Tissue Eng.*, **9**, 421–430.

68. Benveniste, H. and Diemer, N. H. (1987) Cellular reactions to implantation of a microdialysis tube in the rat hippocampus. *Acta Neuropathol.*, **74**, 234–238.

69. Reynolds, L. F., Bren, M. C., Wilson, B. C., Gibson, G. D., Shoichet, M. S., and Murphy, R. J. L. (2008) Transplantation of porous tubes following spinal cord transection improves hindlimb function in the rat. *Spinal Cord*, **46**, 58–64.

70. Aebischer, P., Guenard, V., and Brace, S. (1989) Peripheral nerve regeneration through blind-ended semipermeable guidance channels: Effect of the molecular weight cut off. *J. Neurosci.*, **9**, 3590–3595.

71. Brayfield, C. A., Marra, K. G., Leonard, J. P., Cui, X. T., and Gerlach, J. C. (2008) Excimer laser channel creation in polyethersulfone hollow fibers for compartmentalized in vitro neuronal cell culture scaffolds. *Acta Biomater.*, **4**, 244–255.

72. Sajadi, A., Bansadoun, J. C., Schneider, B. L., Lo Bianco, C., and Aebischer, P. (2006) Transient striatal delivery of GDNF via encapsulated cells leads to sustained behavioral improvement in a bilateral model of Parkinson disease. *Neurobiol. Dis.*, **22**, 119–129.

73. Broadhead, K. W., Biran, R., and Tresco, P. A. (2002) Hollow fiber membrane diffusive permeability regulates encapsulated cell line biomass, proliferation, and small molecule release. *Biomaterials*, **23**, 4689–4699.

74. Chiono, V., Vozzi, G., Vozzi, F., Salvadori, C., Dini, F., Carlucci, F., Arispici, M., Burchielli, S., Di Scipio, F., Geuna, S., Fornaro, M., Tos, P., Nicolino, S., Audisio, C., Perroteau, I., Chiaravalloti, A., Domenici, C., Giusti, P., and Ciardelli, G. (2009) Melt-extruded guides for peripheral nerve regeneration. Part I: Poly(epsiloncaprolactone). *Biomed. Microdevices*, **11**, 1037–1050.

75. Oh, S. H., Kim, J. H., Song, K. S., Jeon, B. H., Yoon, J. H., Seo, T. B., Narngung, U., Lee, I. W., and Lee, J. H. (2008) Peripheral nerve regeneration within an asymmetrically porous PLGA/pluronic F127 nerve guide conduit. *Biomaterials*, **29**, 1601–1609.

76. Piscioneri, A., Morelli, S., Drioli, E., De Bartolo, L. (2021) PLGA multiplex membrane platform for disease modelling and testing of therapeutic compounds. *Membranes*, **11**, 112.

77. Jiang, Z., Song, Y., Qiao, J., Yang, Y., Zhang, W., Liu, W., and Hana, B. (2019) Rat sciatic nerve regeneration across a 10-mm defect bridged by a chitin/CM-chitosan artificial nerve graft. *Int. J. Biol. Macromol.*, **129**, 997–1005.

78. Zhang, N., Zhang, C., and Wen, X. J. (2005) Fabrication of semipermeable hollow fiber membranes with highly aligned texture for nerve guidance. *J. Biomed. Mater. Res. A*, **75**, 941–994.

79. Chen, H., Fan, X., Xia, J., Chen, P., Zhou, X., Huang, J., Yu, J., and Gu, P. (2011) Electrospun chitosan-graft-poly (varepsilon-caprolactone)/ poly (varepsilon-caprolactone) nanofibrous scaffolds for retinal tissue engineering. *Int. J. Nanomed.*, **6**, 453–461.

80. Zhang, D., Ni, N., Chen, J., Yao, Q., Shen, B., Zhang, Y., Zhu, M., Wang, Z., Ruan, J., Wang, J., Mo, X., Shi, W., Ji, J., Fan, X., and Gu, P. (2015) Electrospun SF/PLCL nanofibrous membrane: A potential scaffold for retinal progenitor cell proliferation and differentiation. *Sci. Rep.*, **5**, 14326.

81. Cooper, A., Bhattarai, N., and Zhang, M. (2011) Fabrication and cellular compatibility of aligned chitosan-PCL fibers for nerve tissue regeneration. *Carbohydr. Polym.*, **85**, 149–156.

82. Zamani, F., Amani-Tehran, M., Latifi, M., Shokrgozar, M. A., and Zaminy, A. (2014) Promotion of spinal cord axon regeneration by 3D nanofibrous core-sheath scaffolds. *J. Biomed. Mater. Res. A*, **102**, 506–513.

83. Jha, B. S., Colello, R. J., Bowman, J. R., Sell, S. A., Lee, K. D., Bigbee, J. W., Bowlin, G. L., Chow, W. N., Mathern, B. E., and Simpson, D. G. (2011) Two pole air gap electrospinning: Fabrication of highly aligned, three-dimensional scaffolds for nerve reconstruction. *Acta Biomater.*, **7**, 203–215.

84. Shih, Y. H., Yang, J. C., Li, S. H., Yang, W. C. V., and Chen, C. C. (2012) Bio-electrospinning of poly(l-lactic acid) hollow fibrous membrane. *Text. Res. J.*, **82**(6), 602–612.

85. Morelli, S., Piscioneri, A., Salerno, S., Chen, C. C., Chew, C. H., Giorno, L., Drioli, E., and De Bartolo, L. (2017) Microtube array membrane bioreactor promotes neuronal differentiation and orientation. *Biofabrication*, **9**(2), 025018.

86. Wang, C. Y., Liu, J. J., Fan, C. Y., Mo, X. M., Ruan, H. J., and Li, F. F. (2012) The effect of aligned core–shell nanofibres delivering NGF on the promotion of sciatic nerve regeneration. *J. Biomater. Sci. Polym. Ed.*, **23**, 167–184.

87. Wang, W., Itoh, S., Yamamoto, N., Okawa, A., Nagai, A., and Yamashita, K. (2010) Enhancement of nerve regeneration along a chitosan nanofiber mesh tube on which electrically polarized b-tricalcium phosphate particles are immobilized. *Acta Biomater.*, **6**, 4027–4033

88. Quan, Q., Meng, H., Chang, B., Hong, L., Li, R., Liu, G. B., Cheng, X. Q., Tang, H., Liu, P., Sun, Y., Peng, J., Zhao, Q., Wang, Y., and Lu, S. B. (2019) Novel 3-D helix-flexible nerve guide conduits repair nerve defects. *Biomaterials*, **207**, 49–60.

89. Yucel, D., Kose, G. T., and Hasirci, V. (2010) Polyester based nerve guidance conduit design. *Biomaterials*, **31**, 1596–1603.

90. Moore, M. J., Friedman, J. A., Lewellyn, E. B., Mantila, S. M., Krych, A. J., Ameenud-din, S., Knight, A. M., Lu, L., Currier, B. L., and Spinner, R. J. (2006) Multiple-channel scaffolds to promote spinal cord axon regeneration. *Biomaterial*, **27**, 419–429.

91. Sun, C., Jin, X., Holzwarth, J. M., Liu, X., Hu, J., Gupte, M. J., Zhao, Y., and Ma, P. X. (2012) Development of channeled nanofibrous scaffolds for oriented tissue engineering. *Macromol. Biosci.*, **12**, 761–769.

92. Jeffries, E. M. and Wang, Y. (2013) Incorporation of parallel electrospun fibers for improved topographical guidance in 3D nerve guides. *Biofabrication*, **5**(3), 035015.

93. Kim, J. I., Hwang, T. I., Aguilar, L. E., Park, C. H., and Kim, C. S. (2016) A controlled design of aligned and random nanofibers for 3D bi-functionalized nerve conduits fabricated via a novel electrospinning set-up. *Sci. Rep.*, **6**, 2376.

94. Haynl, C., Hofmann, E., Pawar, K., Förster, S., and Scheibel, T. (2016) Microfluidics-produced collagen fibers show extraordinary mechanical properties. *Nano Lett.*, **16**, 5917–5922.
95. Wei, D., Sun, J., Yang, Y., Wu, C., Chen, S., Guo, Z., Fan, H., and Zhang, X. (2018) Cell alignment guided by nano/micro-oriented collagen fibers and the synergistic vascularization for nervous cell functional expression. *Mater. Today Chem.*, **8**, 85–95.
96. Pawar, K., Welzel, G., Haynl, C., Schuster, S., and Scheibel, T. (2019) Recombinant spider silk and collagen-based nerve guidance conduits support neuronal cell differentiation and functionality in vitro. *ACS Appl. Bio Mater.*, **2**, 4872–4880.
97. Stanness, K. A., Westrum, L. E., Fornaciari, E., Mascagni, P., Nelson, J. A., Stenglein, S. G., Myers, T., and Janigro, D. (1997) Morphological and functional characterization of an in vitro blood–brain barrier model. *Brain Res.*, **771**, 329–342.
98. Parkinson, F. E., Friesen, J., Krizanac-Bengez, L., and Janigro, D. (2003) Use of a three-dimensional in vitro model of the rat blood-brain barrier to assay nucleoside efflux from brain. *Brain Res.*, **980**, 233–241.
99. Cucullo, L., Hossain, M., Rapp, E., Manders, T., Marchi, N., and Janigro, D. (2007) Development of a humanized in vitro blood–brain barrier model to screen for brain penetration of antiepileptic drugs. *Epilepsia*, **48**, 505–516.
100. Yuan, H., Barbieri, D., Luo, X., Van Blitterswijk, C. A., and De Bruijn, J. D. (2017) Calcium phosphates and bone induction. *Comprehensive Biomaterials II*, 333–349
101. Huang, Y., Xie, H., and Li, X. (2020) Scaffolds in bone tissue engineering: Research progress and current applications. *Encyclopedia of Bone Biology*, pp. 204–215, Elsevier Inc.
102. Amini, A. R., Laurencin, C. T., and Nukavarapu, S. P. (2012) Bone tissue engineering: Recent advances and challenges. *Crit. Rev. Biomed. Eng.*, **40**, 363–408.
103. Li, Y., Chen, S., Li, L., and Qin, L. (2015) Bone defect animal models for testing efficacy of bone substitute biomaterials. *J. Orthop. Transl.*, **3**, 95–104.
104. El-rashidy, A. A., Roether, J. A., Harhaus, L., Kneser, U., and Boccaccini, A. R. (2017) Regenerating bone with bioactive glass scaffolds: A review of in vivo studies in bone defect models. *Acta Biomater.*, **62**, 1–28.
105. Granel, H., Bossard, C., Nucke, L., Wauquier, F., Rochefort, G. Y., Guicheux, J., Jallot, E., Lao, J., and Wittrantet, Y. (2019) Optimized bioactive glass: The quest for the bony graft. *Adv. Healthcare Mater.*, **8**, 1801542.

106. Bellucci, D., Sola, A., and Cannillo, V. (2016) Hydroxyapatite and tricalcium phosphate composites with bioactive glass as second phase: State of the art and current applications. *J. Biomed. Mater. Res. A*, **104**, 1030–1056.

107. Qu, H., Fu, H., Han, Z., and Sun, Y. (2019) Biomaterials for bone tissue engineering scaffolds: A review. *RSC Adv.*, **9**, 26252–26262.

108. Ibrahim, M. Z., Sarhan, A. A., Yusuf, D. F., and Hamdi, M. (2017) Biomedical materials and techniques to improve the tribological, mechanical and biomedical properties of orthopedic implants: A review article. *J. Alloys Compd.*, **714**, 636–667.

109. Jang, J., Castano, O., and Kim, H. (2009) Electrospun materials as potential platforms for bone tissue engineering. *Adv. Drug Deliv. Rev.*, **61**, 1065–1083.

110. Chen, J.-H., Liu, C., You, L., and Simmons, C. A. (2010) Boning up on Wolff's law: Mechanical regulation of the cells that make and maintain bone. *J. Biomech.*, **43**, 108–118.

111. Sergeeva, N. S., Sviridova, I. K., Frank, G. A., Kirsanova, V. A., Akhmedova, S. A., and Popov, A. A. (2014) Criteria of biocompatibility of materials for bone defect repair. *Bull. Exp. Biol. Med.*, **157**, 689–694.

112. Babaie, E. and Bhaduri, S. B. (2018) Fabrication aspects of porous biomaterials in orthopedic applications: A review. *ACS Biomater. Sci. Eng.*, **4**, 1–39.

113. Hutmacher, D. W. (2000) Scaffolds in tissue engineering bone and cartilage. *Biomaterials*, **21**, 2529–2543

114. Sousa, R. A., Boesel, L. F., Neves, N. M., and Reis, R. L. (2004) Bioinert, biodegradable and injectable polymeric matrix composites for hard tissue replacement: State of the art and recent developments. *Compos. Sci. Technol.*, **64**, 789–817.

115. Ndreu, N., Yang, A., Ylikauppila, Y., Nikkola, H., and Hasirci, L. (2007) Tissue engineering: A new take-off using nanofiber-based scaffolds. *J. Craniofac. Surg.*, **18**, 3–17.

116. Drotleff, S., Lungwitza, U., Breunig, M., Dennis, A., Blunka, T., Tessmarc, J., and Gopferich, A. (2004) Biomimetic polymers in pharmaceutical and biomedical sciences. *Eur. J. Pharm. Biopharm.*, **58**, 385–407.

117. Legeros, R. Z. (2002) Properties of osteoconductive biomaterials: Calcium phosphates. *Clin. Orthop. Relat. Res.*, **395**, 81–98.

118. Rezwan, K., Chen, Q. Z., Blaker, J. J., and Roberto, A. (2006) Biodegradable and bioactive porous polymer/inorganic composite scaffolds for bone tissue engineering. *Biomaterials*, **27**, 3413–3431.

119. Legeros, R. Z., Lin, S., Rohanizadeh, R., Mijares, D., and Legeros, J. P. (2003) Biphasic calcium phosphate bioceramics: Preparation, properties and applications. *J. Mater. Sci. Mater. Med.*, **14**, 201–209.

120. Baino, F., Novajra, G., and Vitale-Brovarone, C. (2015) Bioceramics and scaffolds: A winning combination for tissue engineering. *Front. Bioeng. Biotechnol.*, **17**, 202.

121. Fu, Q., Saiz, E., Rahaman, M. N., and Tomsia, A. P. (2011) Bioactive glass scaffolds for bone tissue engineering: State of the art and future perspectives. *Mater. Sci. Eng. C*, **31**, 1245–1256.

122. Hench, L. L. (2006) The story of bioglass. *J. Mater. Sci. Mater. Med.*, **17**, 967–978.

123. Hench, L. L. (2013) Chronology of bioactive glass development and clinical applications. *New J. Glass Ceramics*, **3**, 67–73.

124. Li, R., Clark, A. E., and Hench, L. L. (1991) An investigation of bioactive glass powders by sol-gel processing. *J. Appl. Biomater.*, **2**, 231–239.

125. Montazerian, M. and Zanotto, E. D. (2016) History and trends of bioactive glass-ceramics. *J. Biomed. Mater. Res. A*, **104**, 1231–1249.

126. Salinas, A. J. and Vallet-Regí, M. (2013) Bioactive ceramics: From bone grafts to tissue engineering. *RSC Adv.*, **3**, 11116–11131.

127. Sepulveda, P., Jones, J. R., and Hench, L. L. (2001) Characterization of melt-derived 45S5 and sol–gel-derived 58S bioactive glasses. *J. Biomed. Mater. Res.*, **58**, 734–740.

128. Xynos, I. D., Edgar, A. J., Buttery, L. D. K., Hench, L. L., and Polak, J. M. (2000) Gene-expression profiling of human osteoblasts following treatment with the ionic products of Bioglass 45S5 dissolution. *J. Biomed. Mater. Res.*, **58**, 734–740.

129. Hoppe, A., Güldal, N. S., and Boccaccini, A. R. (2011) A review of the biological response to ionic dissolution products from bioactive glasses and glass-ceramics. *Biomaterials*, **32**, 2757–2774.

130. Allan, I. (2002) Particulate bioglass reduces the viability of bacterial biofilms formed on its surface in an in vitro model. *Clin. Oral Implants Res.*, **13**, 53–58.

131. Rizwan, M., Hamdi, M., and Basirun, W. J. (2017) Bioglass® 45S5-based composites for bone tissue engineering and functional applications. *J. Biomed. Mater. Res. A*, **105**, 3197–3223.

132. Li, H. and Chang, J. (2005) pH-compensation effect of bioactive inorganic fillers on the degradation of PLGA. *Compos. Sci. Technol.*, **65**, 2226–2232.

133. Boccaccini, A. R., Blaker, J. J., Maquet, V., Jerome, R., Blacher, S., and Roether, J. A. (2005) Biodegradable and bioactive polymer/Bioglass® composite foams for tissue engineering scaffolds. *Mater. Sci. Forum*, **494**, 499–506.

134. Rezwan, K., Chen, Q. Z., Blaker, J. J., and Boccaccini, A. R. (2006) Biodegradable and bioactive porous polymer/inorganic composite scaffolds for bone tissue engineering. *Biomaterials*, **27**, 3413–3431.

135. Blaker, J. J., Bismarck, A., Boccaccini, A. R., Young, A. M., and Nazhat, S. N. (2010) Premature degradation of poly (α-hydroxyesters) during thermal processing of Bioglass®-containing composites. *Acta Biomater.*, **6**, 756–762.

136. Boccaccini, A. R., Erol, M., Stark, W. J., Mohn, D., Hong, Z., and Mano, J. F. (2010) Polymer/bioactive glass nanocomposites for biomedical applications: A review. *Compos. Sci. Technol.*, **70**, 1764–1776.

137. Webster, T. J., Siegel, R. W., and Bizios, R. (1999) Osteoblast adhesion on nanophase ceramics. *Biomaterials*, **20**, 1221–1227.

138. Loher, S., Reboul, V., Brunner, T. J., Simonet, M., Dora, C., Neuenschwander, P., and Stark, W. J. (2006) Improved degradation and bioactivity of amorphous aerosol derived tricalcium phosphate nanoparticles in poly(lactide-co-glycolide). *Nanotechnology*, **17**, 2054.

139. Hong, Z., Reis, R. L., and De Gualtar, C. (2009) Preparation and in vitro characterization of novel bioactive glass ceramic nanoparticles. *J. Biomed. Mater. Res. A*, **88**, 304–313.

140. Kim, B. H., Kim, H., and Knowles, J. C. (2006) Production and potential of bioactive glass nanofibers as a next-generation biomaterial. *Adv. Funct. Mater.*, **16**, 1529–1535.

141. Lu, H., Zhang, T., Wang, X. P., and Fang, Q. F. (2009) Electrospun submicron bioactive glass fibers for bone tissue scaffold. *J. Mater. Sci. Mater. Med.*, **20**, 793–798.

142. Morelli, S., Salerno, S., Holopainen, J., Ritala, M., and De Bartolo, L. (2015) Osteogenic and osteoclastogenic differentiation in polylactic acid-nanohydroxyapatite fiber scaffolds. *J. Biotechnol.*, **204**, 53–62.

143. Baino, F., Fiorilli, S., and Vitale-Brovarone, C. (2016) Bioactive glass-based materials with hierarchical porosity for medical applications: Review of recent advances. *Acta Biomater.*, **42**, 18–32.

144. Li, X., Liu, H., Niu, X., Yu, B., Fan, Y., Feng, Q., and Watari, F. (2012) The use of carbon nanotubes to induce osteogenic differentiation of human adipose-derived MSCs in vitro and ectopic bone formation in vivo. *Biomaterials*, **33**, 4818–4827.

145. Li, X., van Blitterswijk, C. A., Feng, Q., Cui, F., and Watari, F. (2008). The effect of calcium phosphate microstructure on bone-related cells in vitro. *Biomaterials*, **29**, 3306–3316.

146. Li, X., Gao, H., Uo, M., Sato, Y., Akasaka, T., Feng, Q., and Watari, F. (2009). Effect of carbon nanotubes on cellular functionsin vitro. *J. Biomed. Mater. Res. Part A*, **91A**, 132–139.

147. Ma, P. X., Zhang, R., Xiao, G., and Franceschi, R. (2001) Engineering new bone tissue in vitro on highly porous poly(alpha-hydroxyl acids)/hydroxyapatite composite scaffolds. *J. Biomed. Mater. Res.*, **54**, 284–293.

148. Karageorgiou, V. and Kaplan, D. (2005) Porosity of 3D biomaterial scaffolds and osteogenesis. *Biomaterials*, **26**, 5474–5491.

149. Zisch, A. H., Lutolf, M. P., and Hubbell, J. A. (2003) Biopolymeric delivery matrices for angiogenic growth factors. *Cardiovasc. Pathol.*, **12**, 295–310.

150. Ehrbar, M., Djonov, V. G., Schnell, C., Tschanz, S. A., Martiny-Baron, G., Schenk, U., Wood, J., Burri, P. H., Hubbell, J. A., and Zisch, A. H. (2004) Cell demanded liberation of VEGF121 from fibrin implants induces local and controlled blood vessel growth. *Circ. Res.*, **94**, 1124–1132.

151. Roffi, A., Di Matteo, B., Krishnakumar, G. S., Kon, E., and Filardo, G. (2017) Platelet-rich plasma for the treatment of bone defects: from pre-clinical rational to evidence in the clinical practice. A systematic review. *Int. Orthop.*, **41**, 221–237.

152. Aragón, J., Salerno, S., De Bartolo, L., Irusta, S., and Mendoza, G. (2018) Polymeric electrospun scaffolds for bone morphogenetic protein 2 delivery in bone tissue engineering. *J. Colloid Interface Sci.*, **531**, 126–137.

Index

For Product Safety Concerns and Information please contact our EU representative GPSR@taylorandfrancis.com Taylor & Francis Verlag GmbH, Kaufingerstraße 24, 80331 München, Germany